Statistics for Industry and Technology

Progressive Censoring

Theory, Methods, and Applications

N. Balakrishnan
Rita Aggarwala

Birkhäuser
Boston • Basel • Berlin

N. Balakrishnan
Department of Mathematics and Statistics
McMaster University
Hamilton, Ontario L8S 4K1
Canada

Rita Aggarwala
Department of Mathematics and Statistics
University of Calgary
Calgary, Alberta T2N 1N4
Canada

Library of Congress Cataloging-in-Publication Data
Balakrishnan, N., 1956–
Progressive censoring : theory, methods, and applications / N. Balakrishnan, Rita Aggarwala.
p. cm. — (Statistics for industry and technology)
Includes bibliographical references and indexes.
ISBN 0-8176-4001-0 (alk. paper)
1. Nonparametric statistics. 2. Sampling (Statistics) 3. Accelerated life testing—Statistical methods. I. Aggarwala, Rita. II. Title. III. Series.
QA278.8 .B35 2000
519.5—dc21 00-023589
CIP

Printed on acid-free paper.

Birkhäuser

ISBN 0-8176-4001-0
ISBN 3-7643-4001-0 SPIN 10638407

Typeset by Authors in LaTeX.
Cover design by Vernon Press, Boston, MA.
Printed and bound by Edwards Brothers, Inc., Ann Arbor, MI.
Printed in the United States of America.

9 8 7 6 5 4 3 2 1

To my wonderful daughters,
Sarah Malathi and Julia Lakshmi

N.B.

To my constant support team,
Maya, Bhagwan, and Anil

R.A.

Contents

Preface

Censored sampling arises in a life-testing experiment whenever the experimenter does not observe (either intentionally or unintentionally) the failure times of all units placed on a life-test. Inference based on censored sampling has been studied during the past 50 years by numerous authors for a wide range of lifetime distributions such as normal, exponential, gamma, Rayleigh, Weibull, extreme value, log-normal, inverse Gaussian, logistic, Laplace, and Pareto. Naturally, there are many different forms of censoring that have been discussed in the literature.

In this book, we consider a versatile scheme of censoring called *progressive Type-II censoring.* Under this scheme of censoring, from a total of n units placed on a life-test, only m are completely observed until failure. At the time of the first failure, R_1 of the $n-1$ surviving units are randomly withdrawn (or censored) from the life-testing experiment. At the time of the next failure, R_2 of the $n-2-R_1$ surviving units are censored, and so on. Finally, at the time of the m-th failure, all the remaining $R_m = n - m - R_1 - \cdots - R_{m-1}$ surviving units are censored. Note that censoring takes place here progressively in m stages. Clearly, this scheme includes as special cases the complete sample situation (when $m = n$ and $R_1 = \cdots = R_m = 0$) and the conventional Type-II right censoring situation (when $R_1 = \cdots = R_{m-1} = 0$ and $R_m = n - m$).

The ordered failure times arising from such a progressively Type-II right censored sample are called *progressively Type-II right censored order statistics.* These are natural generalizations of the usual order statistics that have been studied quite extensively during the past century. Even though inference under progressively Type-II right censored samples have been discussed in the literature, mathematical properties of these progressively Type-II right censored order statistics have not been studied much until recently. In this book, we present many interesting properties of the progressively Type-II right censored order statistics that generalize the corresponding well-known results for the usual order statistics. These properties are then used to develop exact as well as approximate inferential methods for different parameters of life (such as mean lifetime, quantile, reliability, and tolerance limit) based on progressively

Type-II right censored samples from many lifetime distributions. Using best linear unbiased estimators of scale and location and scale parameters, optimal progressive Type-II censoring schemes can be determined for fixed values of n and m. Such optimal progressive censoring schemes (based on trace-efficiency and determinant-efficiency) are determined for normal, extreme value, extreme value (II), and log-normal distributions. These results, in addition to providing valuable design ideas to reliability practitioners, also provide evidence for the usefulness and efficiency of progressive Type-II right censoring as compared to the conventional Type-II right censoring. Some other important applied inferential problems, such as the prediction of future failures and acceptance sampling plans, are also developed in this book.

The book provides an elaborate discussion on progressive censoring, with a special emphasis on Type-II right censoring. It presents various theoretical results concerning the progressively Type-II censored order statistics and their applications to inferential problems. All the inferential results are illustrated with several numerical examples. Many useful tables giving optimal progressive censoring schemes for four commonly used lifetime distributions are presented. An extensive bibliography on progressive censoring is also included. All these aspects of this book will make it a valuable resource for researchers and graduate students interested in the area of life-testing and reliability and also an important reference guide for reliability practitioners.

We have written this book with the sincere hope that more practitioners will recognize the versatility of progressive censoring and be tempted to employ in their work this type of censoring/sampling scheme and the methodologies based on it. We also hope that the mathematical ideas and results presented in this book will motivate the aspiring researchers (among the statistical and the engineering communities) to explore further into the theoretical aspects of progressive censoring.

The book has been written in a self-contained manner and, therefore, will be quite suitable either as a text for a graduate topic course or as a text for a directed-reading course. Though a one-year mathematical statistics course at the undergraduate level will provide an adequate background to go over this book in its entirety, a basic exposition of order statistics (such as the one based on the book *A First Course in Order Statistics* by B. C. Arnold, N. Balakrishnan and H. N. Nagaraja, John Wiley & Sons, 1992) will make the journey through this book a lot more pleasant !

We express our sincere thanks to Wayne Yuhasz (Executive Editor, Birkhäuser, Boston) for his enthusiasm and keen interest in this project. Our thanks also go to Debbie Iscoe for doing an excellent job in preparing this book in a camera-ready form, and to Ms. Louise Farkas (Senior Production Editor, Birkhäuser, Boston) for helping us with the final production of the volume. Our final appreciation goes to the Natural Sciences and Engineering Research Council of Canada for providing research grants to both of us which certainly facilitated our meeting many times during the course of this project, thus enabling the work to progress smoothly.

Hamilton, Ontario, Canada *N. Balakrishnan*
Calgary, Alberta, Canada *Rita Aggarwala*
August 1999

List of Tables

1

Introduction

1.1 The Big Picture

The importance of product reliability is greater than ever at the present time. As more and more products are introduced to the market, consumers now have the luxury of demanding high quality and long life in the products they purchase. In such a highly demanding and competitive market, one way by which manufacturers (of computers, automobiles, and electronic items, for example) attract consumers to their products is by providing warranties on product lifetimes. In order to design a cost-effective warranty, a manufacturer must have sound knowledge about product failure-time distributions. To gain this knowledge, *life-testing and reliability* experiments are carried out before (and while) products are put on the market. Of course, the information gained through life-testing experiments is also used for other purposes in addition to determining effective warranties; for example, in pharmaceutical applications, the lifetimes of drugs may be studied in order to determine appropriate dosage administration and expiry dates. Furthermore, continuous improvement of products becomes essential and even critical in a competitive market. Life-testing experimentation is one way by which product improvement and product quality can be gauged.

Very roughly, a sample is said to be *censored* if out of n items placed on a life-test, only m ($< n$) of them are actually observed to fail. Of course, there are many different censoring schemes. In this book, a versatile censoring method known as *Progressive Censoring*, which may be employed in life-testing and reliability experimentation, is discussed and evaluated both mathematically and empirically. This method of censoring, incidentally, includes as a special case the "conventional" methods of censoring (to be defined later on in this chapter) that are currently and have been traditionally used by many practitioners. In addition to establishing many interesting mathematical properties of samples obtained from progressively censored experiments and developing pertinent inferential methods, it is shown (see Chapter 10, for example) that the implementation of an efficient progressive censoring scheme can also lead to a significant improvement in the efficiency (precision) of inference procedures as compared to those derived under conventional methods of censoring.

1.2 Genesis

Issues relating to progressive censoring can be dated back more than 40 years. To give a glimpse of its history, here is a query made in 1966 (Query 18, *Technometrics*, August 1966):

> "It is not uncommon in our life-testing for items to fail for reasons quite unrelated to the normal failure mechanism. For example, consider a number of lamps placed simultaneously on life-test. One of the lamps might be accidently broken after the start of the test but before all the lamps had burned out. If all lamps but one had burned out and the last were accidently broken, the population parameters are easily estimated by techniques designed to deal with censoring on the right. Breakage of any lamp but the longest lived one in the sample, however, introduces the problem of how to utilize the information that this lamp burned the observed number of hours before it was destroyed. What procedures can be recommended?"

The response to this enquiry was given by Dr. A. Clifford Cohen of University of Georgia in a subsequent issue of *Technometrics.* Dr. Cohen had written a paper on the subject of progressive censoring a few years earlier [Cohen (1963)]; however, he was not the first to discuss the idea of progressive censoring. Herd (1956), who wrote his Ph.D. thesis entitled *Estimation of the parameters of a population from a multi-censored sample* at Iowa State College (Ames, Iowa), was the first one to have studied this problem. (Note the usage of the phrase *multi-censored sample* instead of *progressive censored sample.*)[1] Yet, the above presented passage, being the first documented practical enquiry about the loss or removal of industrial units from experimentation prior to the termination of the experiment and due to causes other than failure, may be viewed as the genesis of "real-life" problem-based research on the topic of progressive censoring. Now, you are probably wondering, *what exactly is progressive censoring*?

1.3 The Need for Progressive Censoring

There are many scenarios in life-testing and reliability experiments in which units are lost or removed from experimentation before failure. The loss may occur unintentionally, or it may have been designed so in the study. Unintentional loss may occur, for example, in the case of accidental breakage of an experimental unit (as we just saw in the Query with the lamp example), or if an individual under study drops out, or if the experimentation itself must

[1] The term *progressive censoring* has also been used in the past as an alternate term for *sequential testing*, whereby a clinical trial is "statistically" monitored throughout its duration; see Chatterjee and Sen (1973), Majumdar and Sen (1978), and Sinha and Sen (1982, 1983).

cease due to some unforeseen circumstances such as depletion of funds, unavailability of testing facilities, etc. More often, however, the removal of units from experimentation is pre-planned and intentional, and is done so in order to free up testing facilities for other experimentation, to save time and cost, or to exploit the straightforward analysis that often results. In some cases, when there are live units on test, intentional removal of items or termination of the experiment may be due to ethical considerations. The case when the removal of experimental units is pre-planned and intentional will be the focal point of this book. It needs to be mentioned here that practitioners sometimes "pretend" and treat the unintentional loss of experimental units as intentional for the ease of subsequent statistical analysis. Although this presumption can sometimes be statistically justified [see Cohen (1963) and Nelson (1982)], it may often be a "naive" one.

Conventional Type-I and Type-II one-stage right censoring has been studied rather extensively by numerous authors including Harter (1970), Mann, Schafer and Singpurwalla (1974), Bain (1978), Lawless (1982), Nelson (1982), London (1988), Cohen and Whitten (1988), Cohen (1991), Balakrishnan and Cohen (1991), Bain and Engelhardt (1991), and Harter and Balakrishnan (1996). These authors have all considered lifetime studies in industrial as well as actuarial (human lifetime) contexts, in both parametric and non-parametric cases. Consider a sample of n units placed on a life-test at time 0. In conventional Type-I right censoring, a time T, independent of the failure times, is pre-fixed such that beyond this time no failures will be observed, that is, the experimentation terminates at time T. Thus, the number of complete lifetimes (and hence the number of partial lifetimes) observed is a random variable; in fact, it will have a binomial distribution with n as the number of trials and $p = F(T)$ as the probability of success, where $F(\cdot)$ is the cumulative distribution function of the underlying lifetime distribution. On the other hand, in the case of conventional Type-II right censoring, the number of observed failures is fixed, say m $(m \leq n)$, so that at the time of the m-th failure, the experimentation terminates leaving $n - m$ partially observed failure times. Here, the time of termination of the experiment is random; in fact, its distribution will simply be the distribution of the m-th order statistic from a sample of size n drawn from the distribution $F(\cdot)$. Here, "order statistics" refer to the variables arising when n independent and identically distributed variables are arranged in increasing order of magnitude. To be specific, let $X_1, X_2, \cdots, X_n$ be a random sample of size n from a population with cumulative distribution function $F(x)$; then, the ordered sample $X_{1:n} \leq X_{2:n} \leq \cdots \leq X_{n:n}$ yields the n order statistics from a sample of size n from $F(x)$. Interested readers may consult the books of David (1981) and Arnold, Balakrishnan and Nagaraja (1992) for a comprehensive treatment on order statistics. For reasons that will become clear very soon, we will refer to these as *usual order statistics*.

Both Type-I and Type-II one-stage right censoring described above have been generalized to the case of double (left and right) censoring, wherein, in the case of Type-I censoring, observation begins at a pre-fixed time T_L rather than at time 0, and in the case of Type-II censoring, observation of failures begins with the $(r+1)$-th failure where r is fixed, rather than from time 0. These situations arise when, for example, the first few failures occur too quickly to be observed; see Nelson (1982).

If a practitioner desires to remove live experimental units at points other than the final termination point of the experiment, the above described conventional schemes will not be of use to him/her. Neither of the conventional censoring schemes allow for units to be lost or removed from the test at points other than the final termination point. This allowance will be desirable, as in the case of studies of wear, in which the study of the actual aging process requires items to be fully disassembled at various stages in the experiment. Intermediate removal may also be desirable when a compromise between reduced time of experimentation and the observation of at least some extreme lifetimes is sought, or when some of the surviving items in the experiment that are removed early on (particularly when the items under test are very difficult to obtain or very expensive) can be used for some other tests. As we have already discussed, the loss of items at points other than the final termination point may also be unavoidable, as in the case of accidental breakage of experimental units or loss of contact with individuals under study. These reasons and motivations lead reliability practitioners and theoreticians directly into the area of progressive censoring.

1.4 A Relatively Unexplored Idea

Cohen (1963, 1966, 1975, 1976), Ringer and Sprinkle (1972), Wingo (1973), Cohen and Norgaard (1977), Bain (1978), Sherif and Tan (1978), Lawless (1982), Nelson (1982), Gibbons and Vance (1983), Cohen and Whitten (1988), Cohen (1991), Balakrishnan and Cohen (1991), Bain and Engelhardt (1991), Wingo (1993), and Wong (1993) have all discussed the maximum likelihood estimation of the parameters of various lifetime distributions assuming the following general censoring scheme: m censoring times, $T_1, \cdots, T_{m-1}$ are fixed such that at these times, $R_1, \cdots, R_{m-1}$ surviving units are randomly removed (censored) from the test, respectively. Here, the R_i's are fixed, with the provision that there *are* R_i surviving items at time $T_i, i = 1, 2, \cdots, m-1$. Further, the experiment terminates at time T_m with R_m being the number of surviving units at that time. This is a generalization of the Type-I one-stage right censoring described earlier, and is referred to as *Progressive Type-I right censoring* or *Type-I multi-stage right censoring.* Cohen (1963) has mentioned that even if the T_i's are not pre-fixed (as in the case of accidental breakage), the likelihood function is unchanged apart from a function which is independent of the parameters of interest, as long as these T_i's have distribution independent of the lifetimes under study. Sampford (1952) and London (1988) have considered parametric and non-parametric models for this case of random loss wherein patients withdraw from study before its termination for various reasons. Here, R_i's are fixed but T_i's are random. Sen (1986) has mentioned clinical trials with staggered entry times but fixed termination times. This is a situation where the T_i's are random and the R_i's are fixed. His reference to "progressive" censoring is in view of the progressive entry of subjects into the study. Robinson (1983) has considered a case in which the R_i's are random but the T_i's are fixed. These types of schemes (where R_i's and/or T_i's are random) are sometimes referred to as *random censoring* schemes. Gajjar and Khatri (1969) have considered a case of progressive Type-I right censoring in which at each time of removal T_i, the population parameters change (for example, due to adjustment of temperature settings). A summary of many of these developments is presented in Chapter 7.

A generalization of conventional Type-II one-stage right censoring can be put forward in a similar manner. This generalization is naturally referred to as *Progressive Type-II right censoring.* Under this general censoring scheme, n units are placed on test at time zero. Immediately following the first failure, R_1 surviving units are removed from the test at random. Then, immediately following the second observed failure, R_2 surviving units are removed from the test at random. This process continues until, at the time of the m-th observed failure, the remaining $R_m = n - R_1 - R_2 - \cdots - R_{m-1} - m$ units are all removed from the experiment. This censoring scheme may be depicted pictorially as follows:

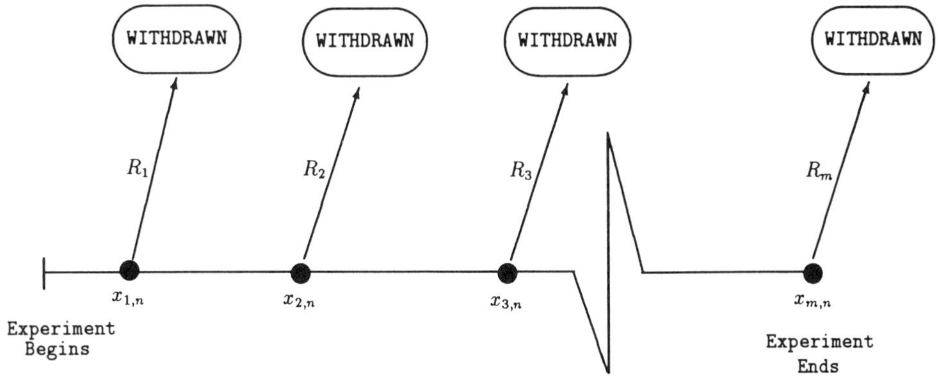

In this censoring scheme, $R_1, R_2, \cdots, R_m$ (and therefore m) are all prefixed. Consequently, here the censoring times (T_i's) are all random, but the numbers of items to fail before each censoring time are all fixed. The resulting m ordered values which are obtained as a consequence of this type of censoring are appropriately referred to as *progressively Type-II right censored order statistics.* Note that if $R_1 = R_2 = \cdots = R_{m-1} = 0$ so that $R_m = n - m$, this scheme reduces to the conventional Type-II one-stage right censoring scheme described earlier, in which case just the first m usual order statistics are observed. Also note that if $R_1 = R_2 = \cdots = R_m = 0$ so that $m = n$, the progressive Type-II right censoring scheme reduces to the case of no censoring (complete sample case), in which case all n usual order statistics are observed. Thus, usual order statistics form a special case of progressively Type-II right censored order statistics, so that any result established for progressively Type-II right censored order statistics become a generalization of the corresponding result for the usual order statistics.

Mann (1969a, 1971), Thomas and Wilson (1972), Cacciari and Montanari (1987), and Viveros and Balakrishnan (1994) have all discussed linear inference under progressive Type-II right censoring when the lifetime distributions are Weibull and exponential. It needs to be mentioned here that maximum likelihood estimation methods proposed for progressive Type-I right censoring (listed above) can be easily modified for progressive Type-II right censoring by replacing the T_i's with the appropriate failure times (this will be discussed further in Chapter 7). However, unconditional linear inference can only be developed for progressively Type-II censored data, since it is only in this case that the number of observed failures and the complete censoring scheme will be completely known to the experimenter in advance; see Arnold, Balakrishnan and Nagaraja (1992). Mann (1971) and Viveros and Balakrishnan (1994), among some others, have used linear inference based on progressive Type-II censoring as an approximation for progressively Type-I censored data with some success. An interesting real application of progressive Type-II right censoring has been illustrated by Montanari and Cacciari (1988). These authors have studied the wear on insulated cables by assuming a Weibull lifetime distribution.

Progressive censoring schemes described above can be generalized in order to accommodate censoring on the left as well. For example, in the case of progressive Type-II censoring, we may assume that the observation of failures begins at the time of the $(r+1)$-th failure (before which r units are known to have failed but their exact failure times are not known) at which time R_{r+1} surviving units are removed from the sample, and so on. Similarly, progressive Type-I censoring can be generalized by starting the observation of failures at a pre-fixed time $T_L > 0$.

As can be seen from the discussion above, most of the work done prior to the 1990's on progressive censoring dealt with the maximum likelihood estimation of parameters which involved solutions of estimating equations by numerical methods. The linear inference discussed by Mann (1969a, 1971) contains cumbersome expressions and requires extesive numerical integration and special tables. Thomas and Wilson (1972), in order to better facilitate Mann's linear inference, have presented an algorithm for computing the moments of progressively Type-II right censored order statistics from arbitrary continuous distributions based on the moments of corresponding usual order statistics. These authors also presented an independence result for the spacings based on progressively Type-II censored order statistics from an exponential distribution. Viveros and Balakrishnan (1994) have derived explicit expressions for the best linear unbiased estimates (BLUE's) of the parameters of both one- and two-parameter exponential distributions based on progressively Type-II right censored samples. In the case of progressive Type-II censoring, several interesting mathematical results as well as efficient inference procedures have been developed recently by the authors; see Balakrishnan and Sandhu (1995, 1996), Aggarwala (1996), and Aggarwala and Balakrishnan (1996, 1998a,b,c). Due to the importance, wide applicability and great scope that such a general progressive Type-II censoring scheme possesses, much of these developments will be presented prominently in this book along with some new examples and applications. We have written this book with the sincere hope that more practioners will recognize the versatility of progressive censoring and be tempted to employ in their work this type of censoring/sampling scheme and the methodologies based on it. It is also our hope that the mathematical ideas and results presented in this book will motivate the aspiring researchers (among the statistical and the engineering communities) to dwell further into the theoretical aspects of progressive censoring.

1.5 Mathematical Notations

We begin with the following assumptions and notations: Suppose n independent units are placed on a life-test with the corresponding failure times $X_1, X_2, \cdots, X_n$ being identically distributed with cumulative distribution function $F(x)$ and probability density function $f(x)$. Suppose further that the prefixed number of failures to be observed is m and that the progressive Type-II right censoring scheme is $(R_1, R_2, \cdots, R_m)$. Then, we shall denote the m completely observed failure times by $X_{i:m:n}^{(R_1,R_2,\cdots,R_m)}$, $i = 1, 2, \cdots, m$. For simplicity in notation, when it is clear as to what the censoring scheme is, we will use the simplified notation $X_{i:m:n}$, $i = 1, 2, \cdots, m$, to denote these failure times bearing in mind that these still depend on the particular choice of $(R_1, R_2, \cdots, R_m)$ used. As mentioned in the last section, we refer to these completely observed failure times as *progressively Type-II right censored order statistics* from $F(x)$ arising from a sample of size n with the censoring scheme $(R_1, R_2, \cdots, R_m)$.

It should be noted that $X_{i:m:n}$ is not the same as $X_{i:n}$, the i-th usual order statistic from a sample of size n from $F(x)$ for $i \geq 2$. This is clearly evident from the fact that there is a possibility that the unit corresponding to the i-th ordered failure time from the original sample of size n may be removed (or censored) before the observation of the i-th progressively Type-II right censored order statistic. However, it is true that $X_{1:m:n} = X_{1:n}$, since no censoring takes place before the first failure.

In the case of usual order statistics from continuous populations, the marginal probability density function of the i-th order statistic takes the simple form [see

David (1981, p. 9) or Arnold, Balakrishnan and Nagaraja (1992, p. 10)]

$$f_{X_{i:n}}(x_i) = \frac{n!}{(i-1)!(n-i)!}\{F(x_i)\}^{i-1}\{1-F(x_i)\}^{n-i}f(x_i). \tag{1.1}$$

Similarly, the joint probability density function of the i-th and j-th order statistics takes the form [see David (1981, p. 10) or Arnold, Balakrishnan and Nagaraja (1992, p. 16)]

$$\begin{aligned} f_{X_{i:n},X_{j:n}}(x_i, x_j) &= \frac{n!}{(i-1)!(j-i-1)!(n-j)!}\{F(x_i)\}^{i-1}\{F(x_j) - \\ &\quad F(x_i)\}^{j-i-1}\{1-F(x_j)\}^{n-j}f(x_i)f(x_j), \ x_i < x_j. \end{aligned} \tag{1.2}$$

It is not immediately evident as to what the corresponding forms for the progressively Type-II right censored order statistics should look like. However, using simple probability arguments, we can write down the joint probability density function of all m progressively Type-II right censored order statistics as

$$\begin{aligned} f_{X_{1:m:n},X_{2:m:n},\cdots,X_{m:m:n}}(x_1, x_2, \cdots, x_m) &= c\prod_{i=1}^{m} f(x_i)\{1-F(x_i)\}^{R_i}, \\ &\quad x_1 < x_2 < \cdots < x_m, \end{aligned} \tag{1.3}$$

where

$$c = n(n-R_1-1)\cdots(n-R_1-R_2-\cdots-R_{m-1}-m+1).$$

Note that immediately preceding the first observed failure, n units are still on test; immediately preceding the second observed failure, $n-R_1-1$ units are still on test, and so on; finally, immediately preceding the m-th observed failure, $n-R_1-R_2-\cdots-R_{m-1}-m+1$ are still on test. Thus, the constant c, in addition to being simply the normalizing constant to make the joint density function integrate to one, is also the number of ways in which the m progressively Type-II right censored order statistics may occur if the observed failure times are $x_1, \cdots, x_m$.

Having already motivated the study of progressively Type-II right censored order statistics from a life-testing point of view, we may now just regard them as a set of special order statistics, and allow the support of the parent distribution $F(x)$ to range over the entire real line. In fact, even in reliability studies, data which take on only positive values are sometimes modelled by normal and extreme value distributions, with location and scale parameters suitably adjusted so that the probability of obtaining a negative value is virtually zero; see Bain and Engelhardt (1991).

It needs to be mentioned here that a progressively Type-II right censored sample is a special case of *generalized order statistics* introduced and studied in detail by Kamps (1995a,b). Specifically, let n be a natural number, $k \geq 1$, $m_1, \cdots, m_{n-1}$ real, and $M_r = \sum_{j=r}^{n-1} m_j$ for $1 \leq r \leq n-1$ be parameters such that $\gamma_r = k + n - r + M_r \geq 1$ for all $r \in \{1, \cdots, n-1\}$; further, let $\tilde{m} = (m_1, \cdots, m_{n-1})$ if $n \geq 2$, $\tilde{m}$ real if $n = 1$. Then, if the random variables $X(r, n, \tilde{m}, k)$, $r = 1, 2, \cdots, n$, possess a joint density of the form

$$
\begin{aligned}
& f_{X(1,n,\tilde{m},k),\cdots,X(n,n,\tilde{m},k)}(x_1, \cdots, x_n) \\
&= k \left(\prod_{j=1}^{n-1} \gamma_j \right) \left\{ \prod_{i=1}^{n-1} \{1 - F(x_i)\}^{m_i} \right\} (1 - F(x_n))^{k-1} \prod_{i=1}^{n} f(x_i), \\
&\qquad -\infty < x_1 < x_2 < \cdots < x_n < \infty,
\end{aligned}
$$

they are called *generalized order statistics*. It is clear that the case of progressive Type-II right censoring is subsumed in the above model by setting $m_i = R_i$ (for $i = 1, 2, \cdots, m-1$), $m_i = 0$ (for $i = m, m+1, \cdots, n-1$) and $k = R_m + 1$. Thus, the general properties established by Kamps (1995a,b) for these generalized order statistics also hold for progressively Type-II right censored order statistics.

As mentioned earlier, any property which holds for progressively Type-II right censored order statistics will also hold for usual order statistics. Incidentally, we can go one step further and say that any property which holds for general progressively Type-II censored order statistics, where both right and left censoring are implemented so that the observation begins at the $(r+1)$-th ordered failure, will also hold for progressively Type-II right censored order statistics, which is the special case when $r = 0$. Therefore, some of the topics we discuss in this book will be for this general progressive Type-II censoring case, which can be described as follows.

Suppose n randomly selected units with independent and identical failure-time distribution $F(x)$ were placed on a life-test; the failure times of the first r units to fail were not observed; at the time of the $(r+1)$-th failure, R_{r+1} number of surviving units are withdrawn (censored) from the test randomly, and so on; at the time of the $(r+i)$-th failure $(1 \leq i \leq m-r)$, R_{r+i} surviving units are randomly withdrawn from the test. We will then denote the lifetimes of the $m - r$ completely observed units to fail by ${}_rX_{r+1:m:n}, {}_rX_{r+2:m:n}, \cdots, {}_rX_{m:m:n}$. We shall call these completely observed lifetimes the *general progressively Type-II censored order statistics* from $F(x)$ arising from a sample of size n with the censoring scheme $(R_{r+1}, \cdots, R_m)$. [As before, when it is unclear as to what the censoring scheme is, the superscript $(R_{r+1}, R_{r+2}, ..., R_m)$ will be added to the notation of these general progressively Type-II censored order statistics.] From the definition, it is apparent that $n = m + R_{r+1} + R_{r+2} + \cdots + R_m$, and that ${}_rX_{r+1:m:n} = X_{r+1:n}$, the $(r+1)$-th usual order statistic from a sample of size n from the distribution $F(x)$, since there has been no random removal of items before this point. If the failure times are from a continuous population with cumulative distribution function $F(x)$ and probability density function $f(x)$, the joint probability density function of ${}_rX_{r+1:m:n}$, ${}_rX_{r+2:m:n}, \cdots$, ${}_rX_{m:m:n}$ is given by

$$f_{{}_rX_{r+1:m:n},{}_rX_{r+2:m:n},\cdots,{}_rX_{m:m:n}}(x_{r+1}, x_{r+2}, \cdots, x_m) = c^* \,\{F(x_{r+1})\}^r$$
$$\times\{1 - F(x_{r+1})\}^{n-r-1} \prod_{i=r+1}^{m} f(x_i)\{1 - F(x_i)\}^{R_i},$$
$$x_{r+1} < x_{r+2} < \cdots < x_m, \qquad (1.4)$$

where

$$c^* = \frac{n!}{r!(n-r-1)!}(n - R_{r+1} - r - 1)\cdots(n - R_{r+1} - \cdots - R_{m-1} - m + 1).$$

It needs to be mentioned here that the above described general progressive Type-II censoring is not usually carried out in practice. However, as pointed out earlier, it is not uncommon to approximate progressively Type-I censored data with progressive Type-II censoring methods; see Mann (1971), Nelson (1982), Viveros and Balakrishnan (1994), and Balakrishnan and Sandhu (1996) for examples. As a result, methods derived for general progressively Type-II censored samples may be used to approximately analyze data obtained from general progressively Type-I censored experiments which are more likely to arise as Nelson (1982) indicated.

1.6 A Friendly Note

Although the material presented in this book may seem daunting at times, we urge you to keep an open mind and go over the details at least at a peripheral level in order to acclimatize yourself with the concept of progressive censoring. The new advances presented in this book will not only provide efficient reliability inferential methods but also can be used to plan time-saving as well as cost-effective life-testing plans.

2

Mathematical Properties of Progressively Type-II Right Censored Order Statistics

2.1 General Continuous Distributions

2.1.1 Introduction

We mentioned earlier in Chapter 1 that marginal distributions of progressively Type-II right censored order statistics (except for the first one, $X_{1:m:n}$), unlike usual order statistics, are quite complicated and cumbersome and are not easily obtained (Try it – you won't like it!). However, certain joint marginal distributions are readily obtained and they do aid in studying mathematical properties of progressively Type-II censored order statistics and also in developing inferential procedures based on progressively censored samples. Furthermore, they aid in providing some efficient algorithms for simulating progressively Type-II censored samples.

A well-known property of usual order statistics from arbitrary continuous distributions is the following Markovian property, which is presented in David (1981, p. 20) and Arnold, Balakrishnan and Nagaraja (1992, pp. 24–25).

Theorem 2.1 *Given $X_{i:n} = x_i$, $X_{j:n}$ $(j > i)$ is independent of the first $i-1$ usual order statistics, $X_{1:n}, \cdots, X_{i-1:n}$.*

It will be of interest to examine whether this Markovian property holds for progressively Type-II right censored order statistics as well. That is, given $X_{i:m:n} = x_i$, is $X_{j:m:n}$ $(j > i)$ independent of the first $i-1$ progressively Type-II right censored order statistics? The answer to this question is affirmative as will be seen in the following section. In proving this result, we will need the joint density of the first i progressively Type-II right censored order statistics, $X_{1:m:n}, \cdots, X_{i:m:n}$.

Another classical result on usual order statistics $X_{1:n}, X_{2:n}, \cdots, X_{n:n}$ from an arbitrary continuous distribution with cumulative distribution function $F(x)$, due to Scheffé and Tukey (1945), is the following.

Theorem 2.2 [Scheffé and Tukey (1945)] *Given $X_{i:n} = x_i$, $X_{j:n}$ $(j > i)$ is distributed as the $(j-i)$-th usual order statistic in a sample of size $n-i$ from $F(x)$ left truncated at x_i, that is, from a population with cumulative distribution function*

$$G(y) = \frac{F(y) - F(x_i)}{1 - F(x_i)}, \quad y > x_i;$$

see also David (1981, p. 20) and Arnold, Balakrishnan and Nagaraja (1992, pp. 25-26). This particular result on the conditional distribution of usual order statistics has been utilized by Horn and Schlipf (1986) in order to develop an efficient algorithm for the generation of conventional Type-II doubly censored samples.

So, a natural question that arises is the following: Does a similar result hold for progressively Type-II right censored order statistics? More precisely, given $X_{i:m:n} = x_i$, are the subsequent progressively Type-II right censored order statistics, $X_{i+1:m:n}, X_{i+2:m:n}, \cdots, X_{m:m:n}$ also distributed as a progressively Type-II right censored sample from the distribution truncated on the left at x_i? The answer to this question is once again affirmative as will be seen in the following section. Naturally, along the lines of Horn and Schlipf (1986), we can utilize this result to develop a method to efficiently generate "trimmed" progressively Type-II right censored samples.

In the following subsection, we establish three useful results for progressively Type-II right censored order statistics $X_{1:m:n}, X_{2:m:n}, \cdots, X_{m:m:n}$ with censoring scheme $(R_1, R_2, \cdots, R_m)$. We shall use the notation presented in Section 1.5 for progressively Type-II right censored samples from arbitrary continuous distributions as there will be no confusion with the censoring scheme involved.

2.1.2 Results

Theorem 2.3
(i) The marginal distribution of $X_{i:m:n}, 1 \le i \le m$, is free of $R_i, R_{i+1}, \cdots, R_m$.
(ii) $X_{1:m:n}, X_{2:m:n}, \cdots, X_{i:m:n}$ form a progressively Type-II right censored sample of size i from n items put on test with the censoring scheme $(R_1, \cdots, R_{i-1}, n - R_1 - \cdots - R_{i-1} - i)$.

Proof In the proof of this result, we will repeatedly use the following formula:

$$\int_a^\infty f(x)\{1 - F(x)\}^k dx = \frac{\{1 - F(a)\}^{k+1}}{k+1}. \tag{2.1}$$

Let us begin with the joint probability density function of $X_{1:m:n}, X_{2:m:n}, \cdots, X_{i:m:n}$ given by

$$\begin{aligned} & f_{X_{1:m:n}, X_{2:m:n}, \cdots, X_{i:m:n}}(x_1, x_2, \cdots, x_i) \\ & = c \prod_{k=1}^{i} f(x_k)\{1 - F(x_k)\}^{R_k} \\ & \times \int_{x_i}^\infty f(x_{i+1})\{1 - F(x_{i+1})\}^{R_{i+1}} \\ & \times \int_{x_{i+1}}^\infty f(x_{i+2})\{1 - F(x_{i+2})\}^{R_{i+2}} \\ & \times \cdots\cdots\cdots \end{aligned}$$

$$\times \int_{x_{m-1}}^{\infty} f(x_m)\{1-F(x_m)\}^{R_m}$$
$$dx_m \cdots dx_{i+2}\, dx_{i+1}, \tag{2.2}$$

where

$$c = n(n-R_1-1)(n-R_1-R_2-\cdots-2)\cdots(n-R_1-R_2-\cdots-R_{m-1}-m+1).$$

Now by repeated use of (2.1) in (2.2), we obtain the joint probability density function of $X_{1:m:n}, X_{2:m:n}, \cdots, X_{i:m:n}$ as

$$\begin{aligned} f_{X_{1:m:n},X_{2:m:n},\cdots,X_{i:m:n}}(x_1,x_2,\cdots,x_i) &= c\left\{\prod_{j=1}^{i-1} f(x_j)\{1-F(x_j)\}^{R_j}\right\} f(x_i) \\ &\quad \times \frac{\{1-F(x_i)\}^{m-i+\sum_{k=i}^{m} R_k}}{\prod_{k=1}^{m-i}\left\{\left(\sum_{\ell=1}^{k} R_{m-\ell+1}\right)+k\right\}}. \end{aligned} \tag{2.3}$$

The factors in the denominator of (2.3) cancel with some of the factors in the constant c due to the fact that $n = m + R_1 + \cdots + R_m$. Furthermore, the summation in the exponent of $\{1-F(x_i)\}$ is simply $n - R_1 - \cdots - R_{i-1} - i$. Using these facts to simplify the expression in (2.3), we now obtain

$$\begin{aligned} & f_{X_{1:m:n},X_{2:m:n},\cdots,X_{i:m:n}}(x_1,x_2,\cdots,x_i) \\ &= n(n-R_1-1)(n-R_1-R_2-2) \\ &\quad \times\cdots(n-R_1-\cdots-R_{i-1}-i+1) \\ &\quad \times \prod_{j=1}^{i-1} f(x_j)\{1-F(x_j)\}^{R_j} f(x_i) \\ &\quad \times\{1-F(x_i)\}^{n-R_1-\cdots-R_{i-1}-i}. \end{aligned} \tag{2.4}$$

Thus, the joint density function of $X_{1:m:n}, X_{2:m:n}, \cdots, X_{i:m:n}$ is completely free of the constants $R_i, \cdots, R_m$. As a matter of fact, Eq. (2.4) reveals that these progressive censored order statistics form a progressively Type-II right censored sample of size i from n items put on test with the censoring scheme $(R_1, \cdots, R_{i-1}, n - R_1 - \cdots - R_{i-1} - i)$. □

Remark 2.1 Theorem 2.3 will enable some simplifications in the numerical as well as algebraic calculations concerning progressively Type-II right censored order statistics; for example, using Theorem 2.3, we have

$$X_{i:m:n}^{(R_1,\cdots,R_m)} \stackrel{d}{=} X_{i:i:n}^{(R_1,\cdots,R_{i-1},n-R_1-\cdots-R_{i-1}-i)}.$$

This result will enable one to reduce the n-fold integration (needed, for example, in the computation of moments of the i-th progressively Type-II right censored order statistic) to an i-fold integration. Theorem 2.3 is also intuitively easy to understand: what happens *after* the observation of the i-th progressively Type-II right censored order statistic should not affect its distribution.

Theorem 2.4 *The progressively Type-II right censored order statistics from an arbitrary continuous distribution form a Markov chain; that is, given* $X_{i:m:n}$, $X_{j:m:n}$ $(j > i)$ *is independent of* $X_{1:m:n}, X_{2:m:n}, \cdots, X_{i-1:m:n}$.

Proof Let us consider the conditional distribution

$$\begin{aligned} &f_{X_{j:m:n}|X_{i:m:n},X_{i-1:m:n},\cdots,X_{1:m:n}}(x_j \mid x_i, x_{i-1}, \cdots, x_1) \\ &= \frac{f_{X_{1:m:n},X_{2:m:n},\cdots,X_{i:m:n},X_{j:m:n}}(x_1, x_2, \cdots, x_i, x_j)}{f_{X_{1:m:n},X_{2:m:n},\cdots,X_{i:m:n}}(x_1, x_2, \cdots, x_i)}. \end{aligned} \tag{2.5}$$

The joint distribution in the denominator of (2.5) is precisely the expression in (2.4). The joint distribution in the numerator of (2.5) is given by

$$\begin{aligned} &f_{X_{1:m:n},X_{2:m:n},\cdots,X_{i:m:n},X_{j:m:n}}(x_1, x_2, \cdots, x_i, x_j) \\ &= c \prod_{\ell=1}^{i} f(x_\ell)\{1 - F(x_\ell)\}^{R_\ell} \\ &\times \int_{x_i}^{x_j} f(x_{j-1})\{1 - F(x_{j-1})\}^{R_{j-1}} \\ &\times \int_{x_i}^{x_{j-1}} f(x_{j-2})\{1 - F(x_{j-2})\}^{R_{j-2}} \\ &\times \cdots\cdots \\ &\times \int_{x_i}^{x_{i+2}} f(x_{i+1})\{1 - F(x_{i+1})\}^{R_{i+1}} dx_{i+1} \cdots dx_{j-1} \\ &\times f(x_j)\{1 - F(x_j)\}^{R_j} \\ &\times \int_{x_j}^{\infty} f(x_{j+1})\{1 - F(x_{j+1})\}^{R_{j+1}} \\ &\times \cdots\cdots \\ &\times \int_{x_{m-1}}^{\infty} f(x_m)\{1 - F(x_m)\}^{R_m}\, dx_m \cdots dx_{j+1}. \end{aligned} \tag{2.6}$$

The second multiple integral in (2.6) can be evaluated by repeated use of the formula in (2.1). The first multiple integral in (2.6) is clearly a function of only x_i and x_j, which we shall denote by $K(x_i, x_j)$. Substitution of this simplified expression in (2.5) gives the conditional density of interest to be

$$f_{X_{j:m:n}|X_{i:m:n},X_{i-1:m:n},\cdots,X_{1:m:n}}(x_j \mid x_i, x_{i-1}, \cdots, x_1)$$

$$= \prod_{\ell=i+1}^{j} \left\{ \left(\sum_{q=\ell}^{m} R_q \right) + m - \ell + 1 \right\} K(x_i, x_j) f(x_j)$$
$$\times \frac{\{1 - F(x_j)\}^{R_j + R_{j+1} + \cdots + R_m + m - j}}{\{1 - F(x_i)\}^{R_{i+1} + R_{i+2} + \cdots + R_m + m - i}}, \quad x_j > x_i. \tag{2.7}$$

This expression is clearly completely independent of $x_1, x_2, \cdots, x_{i-1}$, and it depends only on x_i and x_j. Thus, for continuous distributions, we have

$$f_{X_{j:m:n}|X_{i:m:n}, X_{i-1:m:n}, \cdots, X_{1:m:n}}(x_j \mid x_i, x_{i-1}, \cdots, x_1) = f_{X_{j:m:n}|X_{i:m:n}}(x_j \mid x_i)$$

which establishes that the progressively Type-II right censored order statistics $X_{1:m:n}, \cdots, X_{m:m:n}$ form a Markov chain. □

Remark 2.2 As mentioned earlier, Theorem 2.4 is a generalization of the corresponding well-known result for usual order statistics presented, for example, in David (1981, p. 20) and Arnold, Balakrishnan and Nagaraja (1991, p. 24), and explained briefly in Section 2.1.1.

We shall now use this result to establish the following theorem.

Theorem 2.5 *Given $X_{i:m:n} = x_i, X_{i+1:m:n}, \cdots, X_{m:m:n}$ $(i \geq 1)$ are jointly distributed as a progressively Type-II right censored sample of size $m - i$, with censoring scheme $(R_{i+1}, \cdots, R_m)$, from $n - R_1 - R_2 - \cdots R_i - i$ identically distributed units from the density $f(x)$ left truncated at x_i, namely, with density $f(x)/\{1 - F(x_i)\}$, $x > x_i$.*

Proof Given $X_{i:m:n} = x_i$, the conditional joint density function of $X_{i+1:m:n}$, $\cdots, X_{m:m:n}$ is given by, upon invoking the Markov property established in Theorem 2.4,

$$\begin{aligned} & f_{X_{i+1:m:n}, \cdots, X_{m:m:n}|X_{i:m:n}}(x_{i+1}, \cdots, x_m \mid x_i) \\ &= f_{X_{i+1:m:n}, \cdots, X_{m:m:n}|X_{i:m:n}, \cdots, X_{1:m:n}}(x_{i+1}, \cdots, x_m \mid x_i, \cdots, x_1) \\ &= \frac{f_{X_{1:m:n}, \cdots, X_{m:m:n}}(x_1, \cdots, x_m)}{f_{X_{1:m:n}, \cdots, X_{i:m:n}}(x_1, \cdots, x_i)}. \end{aligned} \tag{2.8}$$

The joint density in the numerator of (2.8) is given in (1.3) while the joint density in the denominator is given in (2.4). Making use of these expressions and simplifying, we obtain the required conditional density function to be

$$\begin{aligned} & f_{X_{i+1:m:n}, X_{i+2:m:n}, \cdots, X_{m:m:n}|X_{i:m:n}}(x_{i+1}, x_{i+2}, \cdots, x_m \mid x_i) \\ &= \left\{ \prod_{j=i+1}^{m} \left(n - \sum_{k=1}^{j-1} R_k - j + 1 \right) \right\} \left[\prod_{k=i+1}^{m} \frac{f(x_k)}{1 - F(x_i)} \left\{ \frac{1 - F(x_k)}{1 - F(x_i)} \right\}^{R_k} \right]. \end{aligned} \tag{2.9}$$

This is simply the joint density function of $m - i$ progressively Type-II right censored order statistics from $n - R_1 - R_2 - \cdots - R_i - i$ identically distributed units with probability density function $f(x)/\{1 - F(x_i)\}$ $(x > x_i)$ placed on a life-test with the progressive censoring scheme $R_{i+1}, \cdots, R_m$. □

Remark 2.3 It should be mentioned here that Theorem 2.5 is a generalization of a result presented originally by Scheffé and Tukey (1945) and explained briefly in Section 2.1.1; see also David (1981, p. 20), Balakrishnan and Cohen (1991, p. 15), and Arnold, Balakrishnan and Nagaraja (1992, p. 23).

Remark 2.4 A dual result to that presented in Theorem 2.5 is known to hold in the case of usual order statistics. Specifically, it is known that the conditional distribution of $X_{i:n}$, given $X_{j:n} = x_j$ for $j > i$, is the same as the distribution of the i-th order statistic in a sample of size $j - 1$ from a population with distribution $F(x)$ truncated on the right at x_j; for example, see Balakrishnan and Cohen (1991, p. 15), and Arnold, Balakrishnan and Nagaraja (1992, pp. 23-24). This result will **not** hold true in general for the case of progressively Type-II censored order statistics. This is evident simply from the fact that the items removed before the observation of the j-th progressively Type-II censored order statistic, which is assumed to be given (say x_j), may have had life times larger than x_j, so they could not possibly be from a distribution right truncated at x_j.

2.2 The Exponential Distribution: Spacings

2.2.1 Introduction

In this section, we will consider the progressive Type-II censoring experiment in which lifetimes of the units are assumed to have come from a (standard) exponential distribution with probability density function

$$f(x) = e^{-x}, \ x > 0, \tag{2.10}$$

and with cumulative distribution function

$$F(x) = 1 - e^{-x}, \ x > 0. \tag{2.11}$$

Due to the well-known "memoryless property" of the above exponential distribution, it is often the lifetime model that statisticians (in the order statistics literature) try to tackle mathematically before others. If tractable results are possible for the exponential distribution, one may be more confident in establishing such results for other distributions such as the uniform, Pareto and power function. So, we too follow the same course here. First, we begin with the following renowned result due to Sukhatme (1937) for usual exponential order statistics which has proven to be very useful in developing several interesting exact inferential results for life-tests based on exponential distributions; see also Rényi (1953) and Epstein and Sobel (1953, 1954). Incidentally, this result has been identified as one of the breakthroughs in the field of Statistics by Kotz and Johnson (1991).

Theorem 2.6 [Sukhatme (1937)] *Let $X_{1:n}, X_{2:n}, \cdots, X_{n:n}$ denote the usual order statistics from the standard exponential distribution in (2.10). Then, the "spacings" defined by*

$$Z_1 = nX_{1:n}, Z_2 = (n-1)(X_{2:n} - X_{1:n}), \cdots, Z_n = X_{n:n} - X_{n-1:n}$$

are independent and identically distributed as standard exponential.

This particular result has been used in a variety of different ways. To name a few, it has been exploited in order to derive explicit expressions for the single and product moments of usual order statistics from the exponential distribution; another important use has been in developing exact inferential methods such as point estimation, interval estimation, tests of hypotheses, prediction intervals, tolerance intervals, and acceptance sampling plans, based on either complete or conventionally Type-II censored samples from exponential distributions. In light of its importance while working with usual order statistics, it would certainly be worthwhile to attempt to find a similar transformation for the progressively Type-II right censored order statistics which will possibly result in independent (exponential?) random variables.

2.2.2 Progressively Type-II Right Censored Spacings

From (1.3), (2.10) and (2.11), we can readily write down the joint density function of $X_{1:m:n}, \cdots, X_{m:m:n}$, a progressively Type-II right censored sample from the standard exponential distribution, with censoring scheme $(R_1, \cdots, R_m)$, as follows:

$$\begin{aligned} & f_{X_{1:m:n}, X_{2:m:n}, \cdots, X_{m:m:n}}(x_1, x_2, \cdots, x_m) \\ &= c\, exp\left\{-\sum_{i=1}^{m}(R_i+1)x_i\right\}, \quad 0 < x_1 < x_2 < \cdots < x_m < \infty, \end{aligned} \tag{2.12}$$

where, as presented earlier in Chapter 1,

$$c = n(n - R_1 - 1)(n - R_1 - R_2 - 2)\cdots(n - R_1 - R_2 - \cdots - R_{m-1} - m + 1).$$

Let us now consider the following transformation:

$$Z_1 = nX_{1:m:n},$$

$$
\begin{aligned}
Z_2 &= (n - R_1 - 1)(X_{2:m:n} - X_{1:m:n}), \\
Z_3 &= (n - R_1 - R_2 - 2)(X_{3:m:n} - X_{2:m:n}), \\
\cdots & \quad \cdots \\
\cdots & \quad \cdots \\
Z_m &= (n - R_1 - \cdots - R_{m-1} - m + 1)(X_{m:m:n} - X_{m-1:m:n}).
\end{aligned} \tag{2.13}
$$

Observe that if we set $R_1 = R_2 = \cdots = R_m = 0$ (so that there is no censoring at all), the above transformation simply reduces to that given in Theorem 2.6.

In order to derive the distribution of $Z_1, Z_2, \cdots, Z_m$ defined in Eq. (2.13), we first consider the inverse transformation given by

$$
\begin{aligned}
X_{1:m:n} &= \frac{Z_1}{n}, \\
X_{2:m:n} &= \frac{Z_1}{n} + \frac{Z_2}{n - R_1 - 1}, \\
X_{3:m:n} &= \frac{Z_1}{n} + \frac{Z_2}{n - R_1 - 1} + \frac{Z_3}{n - R_1 - R_2 - 2}, \\
\cdots & \quad \cdots \\
\cdots & \quad \cdots \\
X_{m:m:n} &= \frac{Z_1}{n} + \frac{Z_2}{n - R_1 - 1} + \cdots + \frac{Z_m}{n - R_1 - \cdots - R_{m-1} - m + 1}.
\end{aligned} \tag{2.14}
$$

The Jacobian of this transformation is, therefore, $\frac{1}{c}$, and consequently, we obtain the joint density function of $Z_1, Z_2, \cdots, Z_m$ as

$$
f_{Z_1, Z_2, \cdots, Z_m}(z_1, z_2, \cdots, z_m) = c\ exp\left(-\sum_{i=1}^{m} z_i\right), \quad z_i > 0, \ i = 1, 2, \cdots, m;
$$

hence, the variables $Z_1, Z_2, \cdots, Z_m$ defined in (2.13) are all independent and identically distributed as standard exponential.

This result was first established by Thomas and Wilson (1972); see also Viveros and Balakrishnan (1994). We shall now state this result formally in the following theorem.

Theorem 2.7 *Let $X_{1:m:n}, \cdots, X_{m:m:n}$ denote a progressively Type-II right censored sample from the standard exponential distribution, with censoring scheme $(R_1, \cdots, R_m)$. The "progressively Type-II right censored spacings" $Z_1, Z_2, \cdots, Z_m$, as defined in Eq. (2.13), are all independent and identically distributed as standard exponential.*

Remark 2.5 From (2.14), it is evident that the progressively Type-II right censored order statistics from an exponential distribution form an additive Markov chain, as noted by Rényi (1953) for the usual order statistics, $X_{1:n}, \cdots, X_{n:n}$.

Remark 2.6 The above result can also be further generalized to the case of general progressively Type-II censored samples [see Balakrishnan and Sandhu (1996)], which will be discussed briefly in Chapter 3.

Theorem 2.7 will prove to be just as useful in facilitating the exact and explicit derivation of single and product moments of progressively Type-II right censored order statistics and also in developing exact inferential methods and efficient simulation algorithms as its counterpart for usual order statistics. We shall explain the derivation of the moments in the following subsection, the simulation algorithms in Chapter 3, and the details on the inferential issues in Chapter 6.

2.2.3 Deriving Moments Using Independent Spacings

The results of Section 2.2.2 can be used to derive exact explicit expressions for the single and the product moments of progressively Type-II right censored order statistics from the standard exponential distribution. We begin with the fact that

$$X_{i:m:n} = \sum_{j=1}^{i} \frac{Z_j}{n - (\sum_{k=0}^{j-1} R_k) - j + 1}, \quad i = 1, \cdots, m,$$

where $R_0 = 0$, and $Z_1, Z_2, \cdots, Z_m$ are all i.i.d. standard exponential random variables. Recognizing now that the progressively Type-II right censored order statistics have all been written as linear combinations of independent standard exponential random variables, we immediately have

$$E(X_{i:m:n}) = \sum_{j=1}^{i} \frac{1}{n - (\sum_{k=0}^{j-1} R_k) - j + 1}, \quad i = 1, \cdots, m, \tag{2.15}$$

and

$$Var\,(X_{i:m:n}) = \sum_{j=1}^{i} \left\{ \frac{1}{n - (\sum_{k=0}^{j-1} R_k) - j + 1} \right\}^2, \quad i = 1, \cdots, m. \tag{2.16}$$

Furthermore, since we can write

$$X_{j:m:n} = X_{i:m:n} + (terms\ which\ are\ independent\ of\ X_{i:m:n})\ (j > i),$$

we also readily have the property that

$$Cov(X_{i:m:n}, X_{j:m:n}) = Var(X_{i:m:n}) \text{ for } j > i.$$

These results, as shall be seen later in Chapter 6, will enable one to develop exact inference for the parameters of both one- and two-parameter exponential distributions based on progressively Type-II right censored samples.

2.3 The Uniform Distribution: Ratios

2.3.1 Introduction

Another distribution that is "friendly" in the usual order statistics literature is the *Uniform(0,1)* distribution with probability density function

$$f(u) = 1, \ 0 < u < 1 \tag{2.17}$$

and with cumulative distribution function

$$F(u) = u, \ 0 < u < 1. \tag{2.18}$$

For convenience in notation, let us denote here the usual order statistics obtained from a random sample of size n from the above uniform distribution by $U_{1:n}, U_{2:n}, \cdots, U_{n:n}$. One of the results that has been used quite extensively is the following result due to Malmquist (1950).

Theorem 2.8 [Malmquist (1950)] *The random variables*

$$\begin{aligned} V_1 &= \frac{1 - U_{n:n}}{1 - U_{n-1:n}}, \\ V_2 &= \frac{1 - U_{n-1:n}}{1 - U_{n-2:n}}, \\ \cdots & \quad \cdots \\ V_{n-1} &= \frac{1 - U_{2:n}}{1 - U_{1:n}}, \\ V_n &= 1 - U_{1:n} \end{aligned} \tag{2.19}$$

are independent Beta random variables, distributed as follows:

$$\begin{aligned} V_1 &\stackrel{d}{=} Beta(1,1), \\ V_2 &\stackrel{d}{=} Beta(2,1), \\ \cdots & \quad \cdots \\ V_{n-1} &\stackrel{d}{=} Beta(n-1,1), \\ V_n &\stackrel{d}{=} Beta(n,1). \end{aligned} \tag{2.20}$$

See also David (1981, p. 21), Balakrishnan and Cohen (1991, p. 33), and Arnold, Balakrishnan and Nagaraja (1992, pp. 79-80). This result has been exploited by Lurie and Hartley (1972) for developing an efficient computer generation algorithm for conventionally Type-II right censored samples from uniform and thus from any arbitrary continuous distribution. We abstain from this discussion here as we will present an extension of Lurie and Hartley's algorithm for the computer generation of progressively Type-II right censored samples from the uniform distribution later in Chapter 3.

A natural question that arises from Theorem 2.8 is whether this elegant result can be generalized to the case of progressively Type-II right censored order statistics. In other words, are the variables

$$\begin{aligned} V_1 &= \frac{1-U_{m:m:n}}{1-U_{m-1:m:n}}, \\ V_2 &= \frac{1-U_{m-1:m:n}}{1-U_{m-2:m:n}}, \\ \cdots & \quad \cdots \\ V_{m-1} &= \frac{1-U_{2:m:n}}{1-U_{1:m:n}}, \\ V_m &= 1-U_{1:m:n} \end{aligned} \tag{2.21}$$

independently (beta?) distributed? The answer to this question is affirmative as shall be seen in the following subsection. Here, $U_{i:m:n}$, $i = 1, 2, \cdots, m$, denote the progressively Type-II right censored order statistics from the *Uniform(0,1)* distribution from a sample of size n with the censoring scheme $(R_1, R_2, \cdots, R_m)$.

2.3.2 Independent Ratios

The generalization of Malmquist's result for the case of progressively Type-II right censored order statistics is presented in the following theorem.

Theorem 2.9 *Let $U_{i:m:n}$, $i = 1, 2, \cdots, m$, denote a progressively Type-II right censored sample from the Uniform(0,1) distribution obtained from a sample of size n with the censoring scheme $(R_1, R_2, \cdots, R_m)$. Let the random variables V_i, $i = 1, 2, \cdots, m$, be as defined before in (2.21). Then, V_i, $i = 1, 2, \cdots, m$ are all statistically independent random variables with*

$$V_i \stackrel{d}{=} Beta\left(i + \sum_{j=m-i+1}^{m} R_j, 1\right), \quad i = 1, 2, \cdots, m.$$

Proof From (1.3), (2.17) and (2.18), the joint density of $U_{i:m:n}$, $i = 1, 2, \cdots, m$, is obtained as

$$\begin{aligned} f_{U_{1:m:n}, U_{2:m:n}, \cdots, U_{m:m:n}}(u_1, u_2, \cdots, u_m) &= c \prod_{i=1}^{m} (1-u_i)^{R_i}, \\ 0 &< u_1 < \cdots < u_m < 1, \end{aligned} \tag{2.22}$$

where, as before,

$$c = n(n - R_1 - 1) \cdots (n - R_1 - \cdots - R_{m-1} - m + 1).$$

Upon inverting the transformation in (2.21), we find that

$$U_{i:m:n} = 1 - \prod_{j=m-i+1}^{m} V_j, \; i = 1, 2, \cdots, m, \tag{2.23}$$

and we find the Jacobian of the transformation to be $\prod_{i=2}^{m} V_i^{i-1}$. From (2.22), we then obtain the joint probability density function of the random variables $V_1, V_2, \cdots, V_m$ as

$$f_{V_1,V_2,\cdots,V_m}(v_1, v_2, \cdots, v_m) = c \prod_{i=1}^{m} v_i^{i-1+\sum_{j=m-i+1}^{m} R_j}, \quad 0 < v_1, \cdots, v_m < 1. \tag{2.24}$$

Using the factorization theorem on (2.24), it is clear that V_i has a *Beta* $\left(i + \sum_{j=m-i+1}^{m} R_j, 1\right)$ distribution, and that the variables $V_1, V_2, \cdots, V_m$ are all statistically independent. □

Remark 2.7 This result can be further generalized to the case of general progressively Type-II censored order statistics [see Aggarwala and Balakrishnan (1998a)], which will be presented briefly in Chapter 3.

Theorem 2.9, like its counterpart for usual order statistics, will prove to be useful in facilitating the exact and explicit derivation of single and product moments of progressively Type-II right censored order statistics and also in developing exact estimation methods as well as efficient simulation algorithms. We shall explain the derivation of the moments in the following subsection, while the details of the simulation algorithms will be presented in Chapter 3, and the corresponding exact estimation will be addressed in Chapter 6.

2.3.3 Deriving Moments Using Independent Ratios

The results of Section 2.3.2 can be used to derive exact and explicit expressions for the single and the product moments of progressively Type-II right censored order statistics from the *Uniform(0,1)* distribution. For this purpose, we first of all adopt the following notations:

$$\begin{aligned}
\textbf{1. } a_i &= i + \sum_{j=m-i+1}^{m} R_j, \; i = 1, \cdots, m \\
\textbf{2. } \alpha_i &= \frac{a_i}{1 + a_i}, \; i = 1, \cdots, m \\
\textbf{3. } \beta_i &= \frac{1}{(a_i + 2)(a_i + 1)}, \; i = 1, \cdots, m \\
\textbf{4. } \gamma_i &= \alpha_i + \beta_i, \; i = 1, \cdots, m.
\end{aligned} \tag{2.25}$$

With these notations, note that

$$E(V_j) = \alpha_j \text{ and } Var(V_j) = \alpha_j\beta_j, \ j = 1, \cdots, m.$$

Now, beginning with

$$E(U_{i:m:n}) = 1 - \prod_{j=m-i+1}^{m} E(V_j), \ i = 1, \cdots, m$$

and

$$Cov(U_{i:m:n}, U_{k:m:n}) = Cov\left(\prod_{j=m-i+1}^{m} V_j, \prod_{j=m-k+1}^{m} V_j\right), \ 1 \le i \le k \le m,$$

we obtain, after some algebraic simplifications, the following expressions:

$$\begin{aligned} E(U_{i:m:n}) &= 1 - \prod_{j=m-i+1}^{m} \alpha_j, \ i = 1, \cdots, m, \\ Var(U_{i:m:n}) &= \left(\prod_{j=m-i+1}^{m} \alpha_j\right)\left(\prod_{j=m-i+1}^{m} \gamma_j - \prod_{j=m-i+1}^{m} \alpha_j\right), \\ Cov(U_{i:m:n}, U_{k:m:n}) &= \left(\prod_{j=m-i+1}^{m} \alpha_j\right)\left(\prod_{j=m-k+1}^{m} \gamma_j - \prod_{j=m-k+1}^{m} \alpha_j\right), \ k < i. \end{aligned} \tag{2.26}$$

These expressions, as shall be seen later in Chapter 6, will enable one to develop exact estimation methods for the parameters of both one- and two-parameter uniform distributions based on progressively Type-II right censored samples.

2.4 The Pareto Distribution: Ratios

2.4.1 Introduction

We will now consider the Pareto distribution and discuss some mathematical properties of progressively Type-II right censored order statistics arising from it. The Pareto distribution has been used commonly to model naturally occurring phenomenon in which the distributions of random variables of interest have long tails; for example, see Arnold (1983) and Johnson, Kotz and Balakrishnan (1994). Although it is not one of the distributions that is frequently used in life-test studies, it is an interesting distribution to examine, particularly from a mathematical viewpoint. Of course, it is also to one's benefit to have a wider choice of distributional results when modelling lifetime data.

Suppose the progressively Type-II right censored order statistics arise from n independent and identical units with the censoring scheme $(R_1, R_2, \cdots, R_m)$

and with individual lifetime distribution as standard *Pareto(ν)* with probability density function

$$f(x) = \nu x^{-\nu-1}, \ x > 1, \ \nu > 0 \tag{2.27}$$

and with cumulative distribution function

$$F(x) = 1 - x^{-\nu}, \ x > 1. \tag{2.28}$$

As usual, let us denote these progressively Type-II right censored order statistics by $X_{1:m:n}, X_{2:m:n}, \cdots, X_{m:m:n}$. In the following subsection, we will establish an independence result for the ratios of these progressively Type-II right censored order statistics from the Pareto distribution similar to that established in the last section for the uniform distribution; see Aggarwala and Balakrishnan (1998c). Once again, this result will enable one to derive exact and explicit expressions for the single and the product moments of progressively Type-II right censored order statistics from the Pareto distribution which in turn will facilitate the linear estimation of the parameters of a Pareto distribution as shall be demonstrated later in Chapter 6.

2.4.2 Independent Ratios

From (1.3), (2.27) and (2.28), we readily have the joint probability density function of $X_{1:m:n}, X_{2:m:n}, \cdots, X_{m:m:n}$ as

$$f_{X_{1:m:n},X_{2:m:n},\cdots,X_{m:m:n}}(x_1, x_2, \cdots, x_m) = c\nu^m \prod_{i=1}^{m} x_i^{-\nu(R_i+1)-1}, \quad 1 < x_1 \le \cdots \le x_m. \tag{2.29}$$

Then, the following independence result holds.

Theorem 2.10 *Let $X_{1:m:n}, X_{2:m:n}, \cdots, X_{m:m:n}$ be a progressively Type-II right censored sample obtained by placing n independent and identical units on a life-test with the censoring scheme $(R_1, R_2, \cdots, R_m)$ and with individual lifetime distribution as standard Pareto(ν) with probability density function in (2.27). Further, let*

$$Y_1 = X_{1:m:n}, \ Y_i = \frac{X_{i:m:n}}{X_{i-1:m:n}}, \quad i = 2, 3, \cdots, m. \tag{2.30}$$

Then, the random variables $Y_1, Y_2, \cdots, Y_m$ are all statistically independent and are distributed as follows:

$$\begin{aligned} Y_1 &\stackrel{d}{=} Pareto(\nu n), \\ Y_i &\stackrel{d}{=} Pareto\left(\nu(n - R_1 - R_2 - \cdots - R_{i-1} - i + 1)\right), i = 2, \cdots, m. \end{aligned} \tag{2.31}$$

Proof From (2.30), we can write the inverse transformation as

$$X_{i:m:n} = \prod_{j=1}^{i} Y_i, \ i = 1, 2, \cdots, m. \tag{2.32}$$

The Jacobian of this one-to-one transformation is $\prod_{i=1}^{m-1} Y_i^{m-i}$. Thus, we obtain the joint probability density function of $Y_1, Y_2, \cdots, Y_m$ as

$$\begin{aligned} g_{Y_1,Y_2,\cdots,Y_m}(y_1, y_2, \cdots, y_m) &= \nu n y_1^{-\nu n-1} \\ &\times \prod_{i=2}^{m} \left\{ \nu(n - R_1 - \cdots - R_{i-1} - i + 1) y_i^{-\nu(n-R_1-\cdots-R_{i-1}-i+1)-1} \right\}, \\ & y_i > 1 \text{ for } i = 1, 2, \cdots, m. \end{aligned} \tag{2.33}$$

By the factorization theorem, it is then clear that the random variables $Y_1, Y_2, \cdots, Y_m$ are all statistically independent, each distributed as *Pareto*, with

$$\begin{aligned} Y_1 &\stackrel{d}{=} Pareto(\nu n), \\ Y_i &\stackrel{d}{=} Pareto\left(\nu(n - R_1 - R_2 - \cdots - R_{i-1} - i + 1)\right), \ i = 2, \cdots, m. \end{aligned} \tag{2.34}$$

□

Remark 2.8 It should be mentioned here that in the special case of conventional Type-II right censoring, that is, $R_1 = R_2 = \cdots = R_{m-1} = 0$, so that $R_m = n - m$, the result presented in Theorem 2.10 reduces to the result

$$Y_i \stackrel{d}{=} Pareto\left(\nu(n - i + 1)\right), \ i = 1, 2, \cdots, m,$$

a known result for usual order statistics from the Pareto distribution; see, for example, Arnold (1983) and Johnson, Kotz and Balakrishnan (1994).

Remark 2.9 The independence result established in this section can also be used to produce an efficient computer-generation algorithm for progressively Type-II right censored samples from the *Pareto(ν)* distribution in much the same way as will be presented in Chapter 3 for uniform and exponential distributions.

2.4.3 Deriving Moments Using Independent Ratios

Using (2.32) and (2.31), we can derive exact and explicit expressions for the single and product moments of progressively Type-II right censored order statistics from the *Pareto(ν)* distribution. For this purpose, let us use the notations

$$\begin{aligned}
\textbf{1. } \alpha_1 &= \nu n \\
\textbf{2. } \alpha_i &= \nu(n - R_1 - R_2 - \cdots - R_{i-1} - i + 1), \ i = 2, 3, \cdots, m \\
\textbf{3. } \beta_i &= \frac{\alpha_i}{\alpha_i - 1}, \ i = 1, 2, \cdots, m, \ \alpha_i > 1 \\
\textbf{4. } \gamma_i &= \frac{\alpha_i}{\alpha_i - 2}, \ i = 1, 2, \cdots, m, \ \alpha_i > 2 \\
\textbf{5. } \delta_i &= \frac{\gamma_i}{\beta_i} = \frac{\alpha_i - 1}{\alpha_i - 2}, \ i = 1, 2, \cdots, m.
\end{aligned} \tag{2.35}$$

We then have the following expressions:

$$\begin{aligned}
E(X_{i:m:n}) &= \prod_{k=1}^{i} \beta_k, \ i = 1, 2, \cdots, m, \\
Cov(X_{i:m:n}, X_{j:m:n}) &= \left\{ \prod_{k=1}^{i} \delta_k - \prod_{k=1}^{i} \beta_k \right\} \prod_{k=1}^{j} \beta_k, \ 1 \le i \le j \le m.
\end{aligned} \tag{2.36}$$

These expressions, as shall be seen later in Chapter 6, will enable one to derive explicit expressions for efficient linear estimators for the parameters of a Pareto distribution.

2.5 Bounds for Means and Variances

In this section, we present the universal bounds for means and variances of progressively Type-II right censored order statistics established by Balakrishnan, Cramer and Kamps (1999). These generalize the corresponding well-known results for the usual order statistics presented by David (1981) and Arnold and Balakrishnan (1989).

First, let us denote

$$\begin{aligned}
\gamma_r &= n - r + 1 + \sum_{i=r}^{m} R_i, \qquad 1 \le r \le m, \\
c_{r-1} &= \prod_{i=1}^{r} \gamma_i, \qquad 1 \le r \le m,
\end{aligned}$$

and

$$a_{i,r} = \prod_{\substack{j=1 \\ j \ne i}}^{r} \frac{1}{\gamma_j - \gamma_i}, \qquad 1 \le i \le r \le m.$$

Then, the cumulative distribution function of $X_{r:m:n}^{(R_1,\cdots,R_m)}$ can be written as

$$F_{X_{r:m:n}}(x) = 1 - c_{r-1} \sum_{i=1}^{r} \frac{a_{i,r}}{\gamma_i} \{1 - F(x)\}^{\gamma_i}, \quad -\infty < x < \infty, \; 1 \le r \le m; \tag{2.37}$$

see, for example, Kamps and Cramer (1999).

For the *Uniform(0,1)* distribution, denoting the progressively Type-II right censored order statistics by $U_{r:m:n}^{(R_1,\cdots,R_m)}$, we have from (2.37)

$$F_{U_{r:m:n}}(u) = 1 - c_{r-1} \sum_{i=1}^{r} \frac{a_{i,r}}{\gamma_i} (1-u)^{\gamma_i}, \; 0 < u < 1, \; 1 \le r \le m. \tag{2.38}$$

From (2.38), we readily obtain the probability density function of $U_{r:m:n}^{(R_1,\cdots,R_m)}$ as

$$f_{U_{r:m:n}}(u) = c_{r-1} \sum_{i=1}^{r} a_{i,r} (1-u)^{\gamma_i - 1}, \; 0 < u < 1, \; 1 \le r \le m. \tag{2.39}$$

Using the expression in (2.39), we have

$$\begin{aligned} \int_0^1 \{f_{U_{r:m:n}}(u)\}^2 du &= c_{r-1}^2 \sum_{i=1}^{r} \sum_{j=1}^{r} a_{i,r} a_{j,r} \int_0^1 (1-u)^{\gamma_i + \gamma_j - 2} du \\ &= c_{r-1}^2 \sum_{i=1}^{r} \sum_{j=1}^{r} a_{i,r} a_{j,r} / (\gamma_i + \gamma_j - 1) \\ &= c_{r-1}^2 \sum_{i=1}^{r} \left\{ a_{i,r} \Big/ \prod_{j=1}^{r} (\gamma_i + \gamma_j - 1) \right\}. \end{aligned} \tag{2.40}$$

Then, Balakrishnan, Cramer and Kamps (1999) have established the following theorem.

Theorem 2.11 *Let $X_{r:m:n}^{(R_1,\cdots,R_m)}$ denote a progressively Type-II right censored order statistic from an arbitrary continuous distribution $F(\cdot)$ with mean 0 and variance 1. Then, for $1 \le r \le m$,*

$$\left| E\left(X_{r:m:n}^{(R_1,\cdots,R_m)} \right) \right| \le \left\{ c_{r-1}^2 \sum_{i=1}^{r} \frac{a_{i,r}}{\prod_{j=1}^{r} (\gamma_i + \gamma_j - 1)} - 1 \right\}^{1/2}. \tag{2.41}$$

Proof For $1 \le r \le m$, consider

$$E\left(X_{r:m:n}^{(R_1,\cdots,R_m)} \right) = \int_0^1 F^{-1}(u) \{ f_{U_{r:m:n}}(u) - \lambda \} du, \tag{2.42}$$

where λ needs to be chosen optimally. Applying Cauchy-Schwarz inequality to (2.42), we obtain

$$\begin{aligned}\left|E\left(X_{r:m:n}^{(R_1,\cdots,R_m)}\right)\right| &\le \left[\int_0^1 \{f_{U_{r:m:n}}(u)-\lambda\}^2\,du\right]^{1/2}\\ &= \left[\int_0^1 \{f_{U_{r:m:n}}(u)\}^2\,du - 2\lambda+\lambda^2\right]^{1/2}. \qquad (2.43)\end{aligned}$$

Upon noting that the right hand side of (2.43) attains its minimum when $\lambda = 1$ and using the expression in (2.40), we get

$$\left|E\left(X_{r:m:n}^{(R_1,\cdots,R_m)}\right)\right| \le \left\{c_{r-1}^2 \sum_{i=1}^{r} \frac{a_{i,r}}{\prod_{j=1}^{r}(\gamma_i+\gamma_j-1)} - 1\right\}^{1/2}.$$

□

Remark 2.10 For $m = n$ (i.e., $R_1 = \cdots = R_m = 0$), we have the usual order statistics and in this case when $r = n$, the upper bound in (2.41) reduces to $\frac{n-1}{\sqrt{2n-1}}$, the well-known Hartley-David-Gumbel bound, which is attained if and only if

$$F(x) = \left\{\frac{1}{n}\left(\frac{n-1}{\sqrt{2n-1}}\,x+1\right)\right\}^{1/(n-1)}, \quad -\frac{\sqrt{2n-1}}{n-1} \le x \le \sqrt{2n-1};$$

see, for example, Arnold, Balakrishnan and Nagaraja (1992, p. 121).

Remark 2.11 In the special case when $R_1 = \cdots = R_{r-1} = R$, we can show that the bound in (2.41) simplifies to

$$\frac{c_{r-1}}{(r-1)!}\left\{\frac{(2r-2)!}{\prod_{i=0}^{2r-2}(2\gamma_r - 1 + i(R+1))} - 1\right\}^{1/2},$$

which was derived by Kamps (1995a, p. 184).

Remark 2.12 Along the lines of Balakrishnan (1990), who presented an improvement to the Hartley-David-Gumbel bound for the mean of largest usual order statistic by using the means of largest usual order statistics in two sample sizes, improvements to the upper bound in (2.41) are possible; see, for example, Balakrishnan, Cramer and Kamps (1999).

Remark 2.13 In general, the upper bound in (2.41) is not sharp. Therefore, the method of the greatest convex minorant due to Moriguti (1953) [see also Balakrishnan (1993)] can be applied in this context in order to derive the sharp bound

$$\left|E\left(X_{r:m:n}^{(R_1,\cdots,R_m)}\right)\right| \le \left\{\int_0^1 g_{r:m:n}^2(u)du\right\}^{1/2},$$

where $g_{r:m:n}(u)$ is the right hand derivative of the greatest convex minorant $G_{r:m:n}(u)$ of $F_{U_{r:m:n}}(u) - u$ on the interval (0,1)

Proceeding in a similar vein and utilizing an upper bound for the variance of a function of a random variable established by Cacoullos and Papathanasiou (1985) [see also Papathanasiou (1990)], Balakrishnan, Cramer and Kamps (1999) have derived an upper bound for the variance of a progressively Type-II right censored order statistic which is presented in the following theorem.

Theorem 2.12 *Let $X_{r:m:n}^{(R_1,\cdots,R_m)}$ be a progressively Type-II right censored order statistic from an absolutely continuous distribution function $F(\cdot)$ and density function $f(\cdot)$. Also, let $F^{-1}(\cdot)$ be a differentiable function. Then, for $1 \le r \le m$,*

$$\begin{aligned} & Var\left(X_{r:m:n}^{(R_1,\cdots,R_m)}\right) \\ \le & \frac{c_{r-1}^2}{\prod_{i=1}^r(\gamma_i+1)} \sum_{i=1}^r \frac{a_{i,r}}{\gamma_i(\gamma_i+1)} \\ & \times \int_0^1 \left\{(F^{-1})'(u)\right\}^2 (1-u)^{\gamma_i}(\gamma_i u - R_i - 1)du. \end{aligned} \tag{2.44}$$

Moreover, equality is attained in (2.44) iff $F(\cdot)$ is a uniform distribution, i.e., $F(u) = d(u-c)$ for some constants $c \in \boldsymbol{R}$ and $d > 0$.

We conclude this chapter with the hope that by now the reader is comfortable with the concept of progressive Type-II censoring and the mathematical notations that go with it.

3

Simulational Algorithms

3.1 Introduction

Many statistical problems are far too complex to be solved explicitly using pure paper-and-pencil mathematics. For example, many industrial systems consist of hundreds of inter-related variables with possibly different probability distributions. In modelling such a complex system mathematically, without the use of heavy computation, many assumptions will be made and many variables assumed constant or negligible in their contribution in order to arrive at a tractable model. However, with high speed computers (getting higher in speed every few months, it seems) readily available to virtually all practitioners and researchers, more and more complex systems may be modelled via statistical simulation in which many variables can be made to vary simultaneously, and the process may be studied with surprisingly high accuracy; we can finally "fit a model to the data" rather than fitting the data to a "friendly" model. As one becomes more familiar with problems in statistics and mathematics, and we see that even innocent-looking problems are sometimes impossible to solve explicitly, the appreciation for simulation and computing power grows rapidly. Many of us have already mused, "It is a wonder that any problems at all were solved without the use of computers."

Someone who is interested in lifetime studies may wish to generate usual order statistics from some continuous population with cumulative distribution function $F(x)$. A direct way of simulating usual order statistics is to generate a pseudo-random sample from the distribution $F(x)$, and then to sort the sample to produce the required order statistics. This direct method will be time-consuming and inefficient due to the sorting involved, and will also produce some unwanted order statistics (if only a partial set of order statistics is required for the problem, like in a conventionally censored sample). For this reason, many authors, including Schucany (1972), Lurie and Hartley (1972), Lurie and Mason (1973), Ramberg and Tadikamalla (1978), and Horn and Schlipf (1986) have discussed various efficient algorithms for generating either a complete or a partial set of usual order statistics without requiring any sorting routines. A brief review of all these developments may be found in Arnold, Balakrishnan and Nagaraja (1992, pp. 95–97) and Tadikamalla and Balakrishnan (1998).

A reliability practitioner who is employing a progressive censoring scheme to study component or system lifetimes may similarly be interested in the simulation of progressively Type-II right censored order statistics from some continuous distribution. Again, he or she may directly simulate a sample of size n from the lifetime distribution of interest, sort this sample, have the computer perform the random censoring at the various points of item removal, and then arrive at the progressively Type-II right censored sample of interest. Obviously, this will be inefficient and time-consuming, especially when sample sizes are moderately large. Sections 3.2 and 3.3 will be mainly concerned with the efficient computer generation of progressively Type-II right censored order statistics from an arbitrary continuous distribution, while Section 3.4 will present results for general progressively Type-II censored order statistics.

3.2 Simulation Using the Uniform Distribution

As noted earlier in Chapter 2, Theorem 2.9 will enable us to efficiently generate progressively Type-II right censored order statistics from the standard *Uniform*(0,1) distribution. In it, by denoting the i-th progressively Type-II right censored order statistic from a random sample of size n and censoring scheme $(R_1, \cdots, R_m)$ from the *Uniform*(0,1) distribution by $U_{i:m:n}$ $(i = 1, 2, \cdots, m)$, we established that the random variables

$$\begin{aligned} V_1 &= \frac{1 - U_{m:m:n}}{1 - U_{m-1:m:n}}, \\ V_2 &= \frac{1 - U_{m-1:m:n}}{1 - U_{m-2:m:n}}, \\ \cdots & \quad \cdots \\ \cdots & \quad \cdots \\ V_{m-1} &= \frac{1 - U_{2:m:n}}{1 - U_{1:m:n}}, \\ V_m &= 1 - U_{1:m:n} \end{aligned}$$

are independent Beta$\left(i + \sum_{j=m-i+1}^{m} R_j, 1\right)$ random variables. Therefore, the random variables

$$W_i = V_i^{i+\sum_{j=m-i+1}^{m} R_j}$$

are independent *Uniform*$(0, 1)$ random variables. We can now, assuming that we are able to generate pseudo-random *Uniform*$(0, 1)$ variables, efficiently generate a progressively Type-II right censored sample from any continuous distribution using the following simple algorithm:

1. Generate m independent *Uniform*$(0, 1)$ observations $W_1, W_2, \cdots, W_m$.
2. Set $V_i = W_i^{1/\left(i+\sum_{j=m-i+1}^{m} R_j\right)}$ for $i = 1, 2, \cdots, m$.
3. Set $U_{i:m:n} = 1 - V_m V_{m-1} \cdots V_{m-i+1}$ for $i = 1, 2, \cdots, m$. Then $U_{1:m:n}$, $U_{2:m:n}, \cdots$, $U_{m:m:n}$ is the required progressively Type-II right censored sample from the *Uniform*$(0, 1)$ distribution.

4. Finally, we set $X_{i:m:n} = F^{-1}(U_{i:m:n})$, for $i = 1, 2, \cdots, m$, where $F^{-1}(\cdot)$ is the inverse cumulative distribution function of the lifetime distribution under consideration. Then $X_{1:m:n}, X_{2:m:n}, \cdots, X_{m:m:n}$ is the required progressively Type-II right censored sample from the distribution $F(\cdot)$.

The above simulational algorithm requires exactly m pseudo-random uniform observations, and does not require any sorting.

Remark 3.1 In the special case $R_1 = R_2 = \cdots = R_{m-1} = 0$, the above algorithm reduces to that given by Lurie and Hartley (1972) for the efficient generation of conventional Type-II right censored samples from an arbitrary continuous distribution $F(\cdot)$.

Example 3.1 The following progressively Type-II right censored sample from the Laplace (or double exponential) distribution was simulated using MapleV and this algorithm with $m = 10$, $n = 20$, location parameter $\mu = 25$, scale parameter $\sigma = 5$ and censoring scheme $(2, 0, 0, 2, 0, 0, 0, 2, 0, 4)$:

19.2117, 21.9736, 23.4178, 23.6625, 23.8022, 24.2302, 25.6207, 25.8699, 26.4800, 27.5534.

Here, the location parameter μ and the scale parameter σ are introduced at the final stage after having generated the required progressively Type-II right censored sample from the standard Laplace distribution by using the algorithm presented above.

This sample will be used later in Chapter 6 when we discuss inference under progressive censoring.

3.3 Simulation Using the Exponential Distribution

Using Theorem 2.7 similarly, we will be able to efficiently generate progressively Type-II right censored order statistics from the exponential distribution. In that theorem, we established the independence of the spacings

$$\begin{aligned}
Z_1 &= nX_{1:m:n}, \\
Z_2 &= (n - R_1 - 1)(X_{2:m:n} - X_{1:m:n}), \\
Z_3 &= (n - R_1 - R_2 - 2)(X_{3:m:n} - X_{2:m:n}), \\
&\vdots \\
Z_m &= (n - R_1 - \cdots - R_{m-1} - m + 1)(X_{m:m:n} - X_{m-1:m:n})
\end{aligned}$$

and we also showed that these spacings each follow a standard exponential distribution. Using this information, we can generate progressively Type-II right censored order statistics from arbitrary distributions using the following simulational algorithm:

1. Simulate m independent exponential random variables $Z_1, Z_2, \cdots, Z_m$. This may be done using the inverse transformation $Z_i = -\ln(1 - U_i)$ where U_i are independent *Uniform*$(0, 1)$ random variables, or by using some other efficient algorithm.

2. Set
$$X_{i:m:n} = \frac{Z_1}{n} + \frac{Z_2}{n - R_1 - 1} + \frac{Z_3}{n - R_1 - R_2 - 2} + \cdots + \frac{Z_i}{n - R_1 - R_2 - \cdots - R_{i-1} - i + 1}.$$
for $i = 1, 2, \cdots, m$. This is the required progressively Type-II right censored sample from the standard exponential distribution.

3. Finally, we set $Y_{i:m:n} = F^{-1}[1 - \exp(-X_{i:m:n})]$, for $i = 1, 2, \cdots, m$, where $F^{-1}(\cdot)$ is the inverse cumulative distribution function of the lifetime distribution under consideration. Then $Y_{1:m:n}, Y_{2:m:n}, \cdots, Y_{m:m:n}$ is the required progressively Type-II right censored sample from the distribution $F(\cdot)$.

The above simulational algorithm requires exactly m pseudo-random observations, and does not require any sorting. In comparing the two methods of simulation, we find that both methods require the same number of pseudo-random observations, and both algorithms require exponentiation. The algorithm using the uniform distribution requires multiplication and division, and the algorithm using the exponential distribution requires division and the initial generation of exponential rather than uniform random variables. Depending upon the form of the inverse distribution function of the population under consideration, one algorithm will be more efficient than the other.

Remark 3.2 In Chapter 2, we also established the independence of the ratios (2.30) for the Pareto distribution. In a similar manner to which we have developed simulation algorithms for the uniform and exponential distributions, using the probability integral transformation in order to generate the independent random variables representing the ratios (in this case, independent Pareto random variables) and the inverse tranformation (2.32) of the ratios, we can develop a simulational algorithm for generating progressively Type-II right censored samples from the Pareto distribution. This short exercise is left to the reader.

3.4 General Progressively Type-II Censored Samples

We have presented to this point a number of results for progressively Type-II right censored order statistics. As mentioned earlier in Chapter 1, many of these results can be generalized to the case of general progressively Type-II censored order statistics. We will present a few of these generalizations now, without proofs, as the proofs are very similar to those already given.

3.4.1 Arbitrary Continuous Distributions

Theorem 3.1 *(Generalization of Theorem 2.3) Consider a general progressively Type-II censored sample* ${}_rX_{r+1:m:n}$, ${}_rX_{r+2:m:n}, \cdots$, ${}_rX_{m:m:n}$ *with censoring scheme* $(R_{r+1}, R_{r+2}, \cdots, R_m)$ *from an arbitrary continuous distribution.*

(i) The distribution of ${}_rX_{r+i:m:n}, 1 \le i \le m$, *is free of* $R_{r+i}, R_{r+i+1}, \cdots, R_m$.

(ii) ${}_rX_{r+1:m:n}$, ${}_rX_{r+2:m:n}, \cdots$, ${}_rX_{r+i:m:n}$ *form a general progressively Type-II censored sample of size i from n items put on test with censoring scheme* $(R_{r+1}, \cdots,\ R_{r+i-1},\ n - R_{r+1} - \cdots - R_{r+i-1} - r - i)$.

Remark 3.3 Notice that for the special case $r = 0$, this result reduces to that given in Theorem 2.3.

Theorem 3.2 *(Generalization of Theorem 2.4) The general progressively Type-II censored order statistics from an arbitrary continuous distribution form a Markov chain; that is, given* ${}_rX_{r+i:m:n}$, ${}_rX_{r+j:m:n}$ $(j > i)$ *is independent of* ${}_rX_{r+1:m:n}$, ${}_rX_{r+2:m:n}, \cdots$, ${}_rX_{r+i-1:m:n}$.

Remark 3.4 Again, upon setting $r = 0$, we obtain the result of Theorem 2.4.

Theorem 3.3 *(Generalization of Theorem 2.5) For general progressively Type-II censored order statistics from arbitrary continuous distributions, given* ${}_rX_{r+i:m:n} = x_{r+i}$, ${}_rX_{r+i+1:m:n}$, $\cdots$, ${}_rX_{m:m:n}$ $(i \ge 1)$ *are jointly distributed as a progressively Type-II right censored sample of size* $m - r - i$ *from* $n - R_{r+1} - R_{r+2} - \cdots - R_{r+i} - r - i$ *identically distributed random variables from the density* $f(x)$ *left-truncated at* x_{r+i}, *that is, with density* $f(x) / [1 - F(x_{r+i})]$ $(x > x_{r+i})$, *and with progressive censoring scheme* $(R_{r+i+1}, \cdots, R_m)$.

3.4.2 The Exponential Distribution

We will now consider general progressively Type-II censored order statistics from the exponential distribution.

Theorem 3.4 *(Generalization of Theorem 2.7) Let* ${}_rX_{r+1:m:n}$, $\cdots$, ${}_rX_{m:m:n}$ *denote a general progressively Type-II censored sample from the standard exponential distribution, with censoring scheme* $(R_{r+1}, \cdots, R_m)$. *The "generalized spacings"*

$$\begin{aligned} Z_{r+1} &= (n-r)\ {}_rX_{r+1:m:n}, \\ Z_{r+2} &= (n-r-R_{r+1}-1)\left({}_rX_{r+2:m:n} - {}_rX_{r+1:m:n}\right), \end{aligned}$$

$$
\begin{aligned}
Z_{r+3} &= (n-r-R_{r+1}-R_{r+2}-2)\left({}_rX_{r+3:m:n} - {}_rX_{r+2:m:n}\right),\\
\cdots & \quad \cdots \\
Z_m &= (n-r-R_{r+1}-\cdots-R_{m-1}-m+r+1)\left({}_rX_{m:m:n} - {}_rX_{m-1:m:n}\right)
\end{aligned} \tag{3.1}
$$

are independent random variables, with $Z_{r+2}, \cdots, Z_m$ being standard exponential random variables and $Z_{r+1}/(n-r) = {}_rX_{r+1:m:n}$ being distributed as the $(r+1)$-th usual order statistic, $X_{r+1:n}$, from a sample of size n from a standard exponential distribution.

Remark 3.5 Notice that, in accordance with Sukhatme (1937), or upon taking $n = r+1$ and $R_1 = R_2 = \cdots = R_{r+1} = 0$ in Theorem 2.7 we are able to write ${}_rX_{r+1:m:n} = \sum_{i=1}^{r+1} \frac{W_i}{n-i+1}$ where W_i $(i = 1, 2, \cdots, r+1)$ are independent standard exponential random variables. Furthermore, ${}_rX_{j:m:n} = {}_rX_{r+1:m:n}+$ *(a linear combination of independent standard exponential random variables)*, $r+2 \le j \le m$. This is used in the following discussions.

Remark 3.6 It is again evident that, if one considers the inverse transformation of (3.1), the general progressively Type-II censored order statistics from an exponential distribution form an additive Markov chain.

Theorem 3.4 will enable us to derive the moments of general progressively Type-II censored order statistics from the exponential distribution in much the same way as we derived the moments of progressively Type-II right censored order statistics from the exponential distribution in Chapter 2. The means, variances and covariances in this case are given by

$$
\begin{aligned}
E\left({}_rX_{r+1:m:n}\right) &= \sum_{i=1}^{r+1} \frac{1}{n-i+1} = \alpha_{r+1:n},\\
E\left({}_rX_{j:m:n}\right) &= \alpha_{r+1:n} + \sum_{i=r+2}^{j} \frac{1}{n-\left(\sum_{k=r+1}^{i-1} R_k\right) - i + 1},\\
& \qquad j = r+2, \cdots, m,\\
Var\left({}_rX_{r+1:m:n}\right) &= \sum_{i=1}^{r+1} \left(\frac{1}{n-i+1}\right)^2 = \beta_{r+1:n},\\
Var\left({}_rX_{j:m:n}\right) &= \beta_{r+1:n} + \sum_{i=r+2}^{j} \left(\frac{1}{n-\left(\sum_{k=r+1}^{i-1} R_k\right) - i + 1}\right)^2,\\
& \qquad j = r+2, \cdots, m,\\
Cov\left({}_rX_{i:m:n}, {}_rX_{j:m:n}\right) &= Var\left({}_rX_{i:m:n}\right), \quad r+1 \le i < j \le m.
\end{aligned}
$$

We will use these moments later on to develop exact inference for one- and two-parameter exponential distributions when general progressively Type-II censored samples are obtained. This will be discusssed in Chapter 6.

Using Theorem 3.4, we can also generalize the simulational algorithm for progressively Type-II right censored samples from arbitrary continuous distributions given in Section 3.3 to the case of general progressively Type-II censored samples. The algorithm is as follows:

1. Simulate $r+1$ independent exponential random variables, $W_1, W_2, \cdots, W_{r+1}$.

2. Set ${}_rX_{r+1:m:n} = \sum_{i=1}^{r+1} \frac{W_i}{n-i+1}$.

3. Simulate $m-r-1$ independent exponential random variables Z_{r+2}, Z_{r+3}, $\cdots, Z_m$.

4. Set

$$\begin{aligned} {}_rX_{r+i:m:n} &= {}_rX_{r+1:m:n} + \frac{Z_{r+2}}{n-r-R_{r+1}-1} \\ &+ \frac{Z_{r+3}}{n-r-R_{r+1}-R_{r+2}-2} + \cdots \\ &+ \frac{Z_{r+i}}{n-r-R_{r+1}-R_{r+2}-\cdots-R_{r+i-1}-i+1} \end{aligned}$$

   for $i = 2, 3, \cdots, m-r$. This is the required general progressively Type-II censored sample from the standard exponential distribution.

5. Finally, we set ${}_rY_{r+i:m:n} = F^{-1}\left[1 - \exp\left(-{}_rX_{r+i:m:n}\right)\right]$ for $i = r+1, r+2, \cdots, m$, where $F^{-1}(\cdot)$ is the inverse cumulative distribution function of the distribution under consideration. Then ${}_rY_{r+1:m:n}$, ${}_rY_{r+2:m:n}$, $\cdots$, ${}_rY_{m:m:n}$ is the required progressively Type-II right censored sample from the distribution $F(\cdot)$.

3.4.3 The Uniform Distribution

In this section, we generalize the results of Section 2.3 to the case of general progressive Type-II censoring. The generalized independence result presented in this section will enable one to efficiently generate general progressively Type-II censored samples from the *Uniform*(0,1) distribution, and consequently, via the probability integral transformation, from any continuous distribution.

Theorem 3.5 *(Generalization of Theorem 2.9) Let ${}_rU_{r+i:m:n}$, $i = 1, 2, \cdots, m-r$, denote a general progressively Type-II censored sample from the Uniform$(0,1)$ distribution. Let*

$$\begin{aligned} V_{r+i} &= \frac{1 - {}_rU_{m-i+1:m:n}}{1 - {}_rU_{m-i:m:n}}, \quad i = 1, 2, \cdots, m-r-1, \\ V_m &= 1 - {}_rU_{r+1:m:n}. \end{aligned}$$

Then,

$$V_{r+i} \stackrel{d}{=} Beta\left(i + \sum_{j=m-i+1}^{m} R_j, 1\right), \; i = 1, 2, \cdots, m-r-1,$$

$$V_m \stackrel{d}{=} Beta\,(n-r, r+1)\,,$$

and $V_j, j = r+1, \cdots, m$, *are all independent random variables.*

Theorem 3.5, like its counterpart for progressively Type-II right censored order statistics, can be used to obtain the single and product moments of general progressively Type-II censored order statistics from the *Uniform*(0,1) distribution. We will adopt the following notation in doing so:

$$\begin{aligned}
a_{r+i} &= i + \sum_{j=m-i+1}^{m} R_j, \; i = 1, \cdots, m-r, \\
\alpha_i &= \frac{a_i}{1+a_i}, \; i = r+1, \cdots, m-1, \\
\alpha_m &= \frac{a_m}{1+r+a_m}, \\
\beta_i &= \frac{1}{(a_i+2)(a_i+1)}, \; i = r+1, \cdots, m-1, \\
\beta_m &= \frac{r+1}{(a_m+r+2)(a_m+r+1)}, \\
\gamma_i &= \alpha_i + \beta_i, \; i = r+1, \cdots, m.
\end{aligned}$$

We then obtain, after some algebra, the following expressions:

$$\begin{aligned}
E\left({}_rU_{r+i:m:n}\right) &= 1 - \prod_{j=m-i+1}^{m} \alpha_j, \; i = 1, \cdots, m-r, \\
Var\left({}_rU_{r+i:m:n}\right) &= \left[\prod_{j=m-i+1}^{m} \alpha_j\right]\left[\prod_{j=m-i+1}^{m} \gamma_j - \prod_{j=m-i+1}^{m} \alpha_j\right], \\
&\quad i = 1, \cdots, m-r,
\end{aligned}$$

and

$$\begin{aligned}
Cov\left({}_rU_{r+i:m:n}, {}_rU_{r+k:m:n}\right) &= \left[\prod_{j=m-i+1}^{m} \alpha_j\right]\left[\prod_{j=m-k+1}^{m} \gamma_j - \prod_{j=m-k+1}^{m} \alpha_j\right], \\
&\quad k < i, \; i = 2, \cdots, m-r.
\end{aligned}$$

Notice that as a result of Theorem 3.5, if we let $W_{r+i} = V_{r+i}^{a_{r+i}}$, $i = 1, 2, \cdots, m-r-1$, then W_{r+i} are independent and identically distributed *Uniform*$(0,1)$ random variables (also independent of V_m). This fact is used in developing the following simulational algorithm for generating general progressively Type-II censored order statistics from an arbitrary continuous distribution $F(\cdot)$ using the *Uniform*$(0,1)$ distribution:

1. Generate a $Beta\,(n-r, r+1)$ random variable V_m. This may be done using any efficient Beta-generating algorithm; see, for example, Johnson, Kotz and Balakrishnan (1995).

2. Independently generate $m-r-1$ independent $Uniform(0,1)$ observations $W_{r+1}, \cdots, W_{m-1}$.

3. Set $V_{r+i} = W_{r+i}^{1/a_{r+i}}$, $i = 1, 2, \cdots, m-r-1$.

4. Set ${}_rU_{r+i:m:n} = 1 - V_{m-i+1}V_{m-i+2}\cdots V_m$, $i = 1, 2, \cdots, m-r$. Then ${}_rU_{r+i:m:n}$, $i = 1, 2, \cdots, m-r$, is the required general progressively Type-II censored sample from the $Uniform(0,1)$ distribution.

5. Finally, set ${}_rX_{r+i:m:n} = F^{-1}\left({}_rU_{r+i:m:n}\right)$ for $i = 1, 2, \cdots, m-r$, where $F^{-1}(\cdot)$ is the inverse distribuiton function of the population from which we desire to generate the sample. Then ${}_rX_{r+1:m:n}$, ${}_rX_{r+2:m:n}$, $\cdots$, ${}_rX_{m:m:n}$ is the required general progressively Type-II censored sample from the distribution $F(\cdot)$.

Remark 3.7 For the special case $r = 0$, this algorithm will reduce to that given in Section 3.2, since in that case, a Beta-generating algorithm for V_m, where V_m has Beta$(n, 1)$ distribution, is to take the n-th power of a $Uniform(0,1)$ random variable.

Another Approach: Conditioning Method of Simulation

Perhaps a more intuitive approach to the problem of efficiently simulating general progressively Type-II censored order statistics from the uniform distribution is to use Theorem 3.3, which tells us that given ${}_rU_{r+1:m:n} = u_{r+1}$, ${}_rU_{r+2:m:n}$, $\cdots$, ${}_rU_{m:m:n}$ are jointly distributed as a progressively Type-II right censored sample of size $m-r-1$ from $n-R_1-r-1$ $Uniform(u_{r+1}, 1)$ variates with progressive censoring scheme $(R_{r+2}, \cdots, R_m)$.

We can, therefore, generate a general progressively Type-II censored sample from an arbitrary continuous distribution with cumulative distribution function $F(x)$ and progressive censoring scheme $(R_{r+1}, R_{r+2}, \cdots, R_m)$ using the following algorithm:

1. Generate ${}_rU_{r+1:m:n}$ as a $Beta\,(r+1, n-r)$ random variate.

2. Generate a progressively Type-II right censored sample from the $Uniform$(0,1) distribution of size $m-r-1$, which we will denote by $U_{(r+2)}, \cdots, U_{(m)}$, with censoring scheme $(R_{r+2}, \cdots, R_m)$ using the algorithm given in Section 3.2.

3. Set ${}_rU_{i:m:n} = U_{(i)}\left(1 - {}_rU_{r+1:m:n}\right) + {}_rU_{r+1:m:n}$, $i = r+2, \cdots, m$, where $U_{(i)}$ $(i = r+2, \cdots, m)$ is as defined in step 2 of this algorithm.

4. Set ${}_rX_{i:m:n} = F^{-1}({}_rU_{i:m:n})$, $i = r+1, \cdots, m$. This is the desired general progressively Type-II censored sample.

Remark 3.8 The simulational algorithm developed in this subsection generalizes the one given by Horn and Schlipf (1986) for generating conventional doubly censored samples.

Remark 3.9 The simulational algorithm given in this subsection, although arrived at in a completely different manner, is mathematically equivalent to that given just earlier. This may be seen simply by noting that if X has a *Beta* distribution with parameters a and b, then $1 - X$ has a *Beta* distribution with parameters b and a.

Example 3.2 The following general progressively Type-II censored sample from the exponential distribution with location parameter $\mu = 25$ and scale parameter $\sigma = 10$ was generated on MapleV using this algorithm with $n = 50$, $r = 5$, $R_6 = 2$, $R_7 = 1$, $R_8 = 0$, $R_9 = 2$, $R_{10} = 1$, $R_{11} = 3$, $R_{12} = 2$, $R_{13} = 3$, $R_{14} = 4$, $R_{15} = 3$, $R_{16} = 0$, $R_{17} = 2$, $R_{18} = 1$, $R_{19} = 2$, $R_{20} = 4$:

$$25.99609, 26.17323, 26.55884, 26.65558, 27.32842, 27.52826, 28.58114, 28.58350, 28.68850, 29.09515, 29.17521, 29.47387, 29.61337, 33.44267, 35.74206.$$

This sample will be used later on in Chapter 6, when we discuss methods of inference based on general progressively Type-II censored samples.

4

Recursive Computation and Algorithms

4.1 Introduction

In the next two chapters, considerable emphasis will be placed on obtaining moments of progressively Type-II censored order statistics. One reason for this will become clear in Chapters 6 and 10 when the focus is on developing efficient inferential procedures based on progressive Type-II censoring. We have seen in the last chapter that moments of progressively censored order statistics from some distributions can be obtained explicitly. However, very often, it is not possible to obtain explicit expressions for these moments. The idea of obtaining moments of usual order statistics in a recursive manner has been explored for a number of distributions; for example, see Balakrishnan, Malik and Ahmed (1988), and Balakrishnan and Sultan (1998). In this chapter, we establish several recurrence relations satisfied by the single and product moments of progressively Type-II right censored order statistics from the exponential, Pareto and power function distributions as well as their truncated forms. It is important when establishing such recurrence relations that the relations be *complete*, in the sense that they may be used in a simple recursive manner in order to compute all the single and product moments of all progressively Type-II right censored order statistics from the distributions of interest for all sample sizes n and all censoring schemes $(R_1, R_2, \cdots, R_m)$, $m \leq n$. This, in fact, is the case for all the distributions mentioned above.

4.2 The Exponential Distribution

In the following two subsections, we will present complete recurrence relations for the single and product moments of progressively Type-II right censored order statistics from the standard exponential distribution. These results, due to Aggarwala and Balakrishnan (1996), will generalize the results of Joshi (1978, 1982) for usual order statistics. Although we have already derived explicit expressions for the first and second moments in this case, the method of obtaining the recursive relationships and the recursive algorithm presented will serve as useful illustrations. In addition, they will suggest ways of proceeding to develop similar results for some other distributions as shall be seen in subsequent sections.

4.2.1 Recurrence Relations for Single Moments

Let $X_1, X_2, \cdots, X_n$ denote a random sample from the standard exponential distribution, that is, with probability density function $f(x) = e^{-x}$ $(x \geq 0)$ and cumulative distribution function $F(x) = 1 - e^{-x}$. Then, as before, the corresponding progressively Type-II right censored order statistics with censoring scheme $(R_1, R_2, \cdots, R_m)$ will be denoted by $X_{1:m:n}^{(R_1,R_2,\cdots,R_m)}, \cdots, X_{m:m:n}^{(R_1,R_2,\cdots,R_m)}$. It will be evident shortly why the superscript mentioned briefly in Chapter 1 is now essential in our notation.

Notice that for the standard exponential distribution, we have the relation $f(x) = 1 - F(x)$ $(x \geq 0)$. This relationship between $f(x)$ and $F(x)$ uniquely defines this distribution, and is therefore the *characterizing differential equation* for the exponential distribution. This relationship will be exploited in this subsection to derive recurrence relations for the single moments of exponential progressively Type-II right censored order statistics.

For any continuous distribution, we have from (1.3)

$$\begin{aligned}\mu_{i:m:n}^{(R_1,\cdots,R_m)^{(k)}} &= E\left\{\left[X_{i:m:n}^{(R_1,\cdots,R_m)}\right]^k\right\} \\ &= A(n, m-1) \int\int_{0<x_1<\cdots<x_m<\infty}\cdots\int x_i^k f(x_1)\left[1 - F(x_1)\right]^{R_1} \\ &\quad \times f(x_2)\left[1 - F(x_2)\right]^{R_2}\cdots f(x_m)\left[1 - F(x_m)\right]^{R_m} dx_1\cdots dx_m, \end{aligned} \tag{4.1}$$

where

$$A(p, q) = p(p - R_1 - 1)(p - R_1 - R_2 - 2)\cdots(p - R_1 - R_2 - \cdots - R_q - q) \tag{4.2}$$

for $q = 0, 1, \cdots, p-1$. When $k = 1$, the superscript in the notation of the mean of the progressively Type-II right censored order statistics may be omitted without any confusion.

Theorem 4.1 *For* $2 \leq m \leq n$ *and* $k > -1$,

$$\begin{aligned}\mu_{1:m:n}^{(R_1,\cdots,R_m)^{(k+1)}} &= \frac{1}{R_1+1}\left[(k+1)\,\mu_{1:m:n}^{(R_1,\cdots,R_m)^{(k)}}\right. \\ &\quad \left. - (n - R_1 - 1)\,\mu_{1:m-1:n}^{(R_1+R_2+1,R_3,\cdots,R_m)^{(k+1)}}\right], \end{aligned} \tag{4.3}$$

and for $m = 1$, $n = 1, 2, \cdots$ *and* $k > -1$,

$$\mu_{1:1:n}^{(n-1)^{(k+1)}} = \frac{k+1}{n}\mu_{1:1:n}^{(n-1)^{(k)}}. \tag{4.4}$$

Proof Let us consider, for $i = 1$ in (4.1), upon employing the characterizing differential equation $f(x_1) = 1 - F(x_1)$,

$$\begin{aligned} \mu_{1:m:n}^{(R_1,\cdots,R_m)^{(k)}} &= A(n, m-1) \iint\cdots\int_{0<x_2<\cdots<x_m<\infty} \left\{ \int_0^{x_2} x_1^k [1 - F(x_1)]^{R_1+1} dx_1 \right\} \\ &\times f(x_2)[1 - F(x_2)]^{R_2} \cdots f(x_m)[1 - F(x_m)]^{R_m} dx_2 \cdots dx_m. \end{aligned} \tag{4.5}$$

Now, integrating the innermost integral by parts, we get

$$\begin{aligned} \int_0^{x_2} x_1^k [1 - F(x_1)]^{R_1+1} dx_1 &= \frac{1}{k+1} \left\{ x_2^{k+1} [1 - F(x_2)]^{R_1+1} \right. \\ &+ \left. (R_1 + 1) \int_0^{x_2} x_1^{k+1} [1 - F(x_1)]^{R_1} f(x_1) dx_1 \right\}. \end{aligned}$$

Substituting the above expression into (4.5), and simplifying the resulting equation, we arrive at

$$\mu_{1:m:n}^{(R_1,\cdots,R_m)^{(k)}} = \frac{n - R_1 - 1}{k+1} \mu_{1:m-1:n}^{(R_1+R_2+1,R_3,\cdots,R_m)^{(k+1)}} + \frac{R_1 + 1}{k+1} \mu_{1:m:n}^{(R_1,\cdots,R_m)^{(k+1)}},$$

which, upon rearrangement, gives the relation in (4.3). Eq. (4.4) is obtained similarly, noting that the integral in (4.5) will be a single integral, ranging over the entire positive real line, and R_1 must be simply $n - 1$, since the equation $R_1 + R_2 + \cdots + R_m + m = n$ must be satisfied. □

Remark 4.1 As mentioned in Chapter 2, it is true that the first progressively Type-II right censored order statistic is the same as the first usual order statistic from a sample of size n, regardless of the censoring scheme employed. We could, therefore, use Joshi's (1978) result on moments of usual order statistics for obtaining the moments of the first progressively censored order statistic. However, Theorem 4.1 has been included here in order to establish completeness of the recurrence relations even without this knowledge.

Theorem 4.2 *For* $2 \le i \le m - 1$, $m \le n$ *and* $k > -1$,

$$\begin{aligned} \mu_{i:m:n}^{(R_1,\cdots,R_m)^{(k+1)}} &= \frac{1}{R_i + 1} \left[(k+1) \mu_{i:m:n}^{(R_1,\cdots,R_m)^{(k)}} \right. \\ &- (n - R_1 - \cdots - R_i - i) \mu_{i:m-1:n}^{(R_1,\cdots,R_{i-1},R_i+R_{i+1}+1,R_{i+2},\cdots,R_m)^{(k+1)}} \\ &+ \left. (n - R_1 - \cdots - R_{i-1} - i + 1) \mu_{i-1:m-1:n}^{(R_1,\cdots,R_{i-2},R_{i-1}+R_i+1,R_{i+1},\cdots,R_m)^{(k+1)}} \right]. \end{aligned} \tag{4.6}$$

Proof Let us consider, substituting $f(x_i) = 1 - F(x_i)$ into (4.1),

$$\begin{aligned}
\mu_{i:m:n}^{(R_1,\cdots,R_m)^{(k)}} &= A(n, m-1) \underset{0<x_1<\cdots<x_{i-1}<x_{i+1}<\cdots<x_m<\infty}{\int \cdots \int \cdots \int} \\
& \left\{ \int_{x_{i-1}}^{x_{i+1}} x_i^k \left[1 - F(x_i)\right]^{R_i+1} dx_i \right\} f(x_1) \left[1 - F(x_1)\right]^{R_1} \\
& \times \cdots f(x_{i-1}) \left[1 - F(x_{i-1})\right]^{R_{i-1}} f(x_{i+1}) \left[1 - F(x_{i+1})\right]^{R_{i+1}} \\
& \times \cdots f(x_m) \left[1 - F(x_m)\right]^{R_m} dx_1 \cdots dx_{i-1} dx_{i+1} \cdots dx_m.
\end{aligned} \tag{4.7}$$

Integrating the innermost integral by parts, we obtain upon simplification,

$$\begin{aligned}
\int_{x_{i-1}}^{x_{i+1}} x_i^k \left[1 - F(x_i)\right]^{R_i+1} dx_i &= \frac{1}{k+1} \left\{ x_{i+1}^{k+1} \left[1 - F(x_{i+1})\right]^{R_i+1} \right. \\
&- x_{i-1}^{k+1} \left[1 - F(x_{i-1})\right]^{R_i+1} \\
&+ \left. (R_i + 1) \int_{x_{i-1}}^{x_{i+1}} x_i^{k+1} \left[1 - F(x_i)\right]^{R_i} f(x_i)\, dx_i \right\}.
\end{aligned}$$

Substituting this expression into (4.7), and simplifying the resulting equation, we obtain

$$\begin{aligned}
& \mu_{i:m:n}^{(R_1,\cdots,R_m)^{(k)}} \\
&= \frac{n - R_1 - \cdots - R_i - i}{k+1} \mu_{i:m-1:n}^{(R_1,\cdots,R_{i-1},R_i+R_{i+1}+1,R_{i+2},\cdots,R_m)^{(k+1)}} \\
&- \frac{n - R_1 - \cdots - R_{i-1} - i + 1}{k+1} \mu_{i-1:m-1:n}^{(R_1,\cdots,R_{i-2},R_{i-1}+R_i+1,R_{i+1},\cdots,R_m)^{(k+1)}} \\
&+ \frac{R_i + 1}{k+1} \mu_{i:m:n}^{(R_1,\cdots,R_m)^{(k+1)}}.
\end{aligned}$$

Rewriting the above equation, we obtain the relation in (4.6). □

Theorem 4.3 *For $2 \leq m \leq n$ and $k > -1$,*

$$\mu_{m:m:n}^{(R_1,\cdots,R_m)^{(k+1)}} = \frac{k+1}{R_m + 1} \mu_{m:m:n}^{(R_1,\cdots,R_m)^{(k)}} + \mu_{m-1:m-1:n}^{(R_1,\cdots,R_{m-2},R_{m-1}+R_m+1)^{(k+1)}}. \tag{4.8}$$

Proof Let us write, upon using the characterizing differential equation $f(x_m) = 1 - F(x_m)$ in (4.1),

$$\begin{aligned}
\mu_{m:m:n}^{(R_1,\cdots,R_m)^{(k)}} &= A(n, m-1) \underset{0<x_1<\cdots<x_{m-1}<\infty}{\int\int \cdots \int} \\
& \times \left\{ \int_{x_{m-1}}^{\infty} x_m^k \left[1 - F(x_m)\right]^{R_m+1} dx_m \right\} f(x_1) \left[1 - F(x_1)\right]^{R_1} \\
& \times \cdots f(x_{m-1}) \left[1 - F(x_{m-1})\right]^{R_{m-1}} dx_1 \cdots dx_{m-1}.
\end{aligned} \tag{4.9}$$

Integrating the innermost integral by parts, we obtain

$$\begin{aligned}
&\int_{x_{m-1}}^{\infty} x_m^k \ [1-F(x_m)]^{R_m+1}\, dx_m \\
&= \frac{1}{k+1}\Big\{-x_{m-1}^{k+1}\,[1-F(x_{m-1})]^{R_m+1} \\
&+ (R_m+1)\int_{x_{m-1}}^{\infty} x_m^{k+1}\,[1-F(x_m)]^{R_m} f(x_m)\, dx_m\Big\}.
\end{aligned}$$

Upon substituting this expression into (4.9), and simplifying the resulting equation, we get

$$\begin{aligned}
\mu_{m:m:n}^{(R_1,\cdots,R_m)^{(k)}} &= -\frac{n-R_1-\cdots-R_{m-1}-m+1}{k+1} \\
&\times \ \mu_{m-1:m-1:n}^{(R_1,\cdots,R_{m-2},R_{m-1}+R_m+1)^{(k+1)}} + \frac{R_m+1}{k+1}\mu_{m:m:n}^{(R_1,\cdots,R_m)^{(k+1)}}.
\end{aligned}$$

After rewriting this equation, we obtain the relation in (4.8). □

Remark 4.2 Using these recurrence relations, we can obtain all the single moments of all progressively Type-II right censored order statistics for all sample sizes and all censoring schemes $(R_1, \cdots, R_m)$ in a simple recursive manner. The recursive algorithm will be described in detail after the next subsection.

Remark 4.3 Simpler recurrence relations have been presented recently by Balakrishnan, Cramer, Kamps and Schenk (1999) as follows:

$$\mu_{1:m:n}^{(R_1,\cdots,R_m)^{(k+1)}} = \frac{k+1}{n}\mu_{1:1:n}^{(n-1)^{(k)}} = \frac{(k+1)!}{n^{k+1}}$$

and, for $2 \le i \le m$

$$\mu_{i:m:n}^{(R_1,\cdots,R_m)^{(k+1)}} = \frac{k+1}{R_i^+}\mu_{i:i:n}^{\left(R_1,\cdots,R_{i-1},R_i^+-1\right)^{(k)}} + \mu_{i-1:i-1:n}^{\left(R_1,\cdots,R_{i-2},R_{i-1}^+-1\right)^{(k+1)}},$$

where $R_i^+ = \sum_{j=i}^{m}(R_j+1)$.

4.2.2 Recurrence Relations for Product Moments

For any continuous distribution, we can write the (i,j)-th product moment of progressively Type-II right censored order statistics from (1.3) as

$$\begin{aligned}
\mu_{i,j:m:n}^{(R_1,\cdots,R_m)} &= E\left[X_{i:m:n}^{(R_1,\cdots,R_m)} X_{j:m:n}^{(R_1,\cdots,R_m)}\right] \\
&= A(n,m-1) \iint\cdots\int_{0<x_1<\cdots<x_m<\infty} x_i x_j f(x_1)\,[1-F(x_1)]^{R_1} \\
&\times \ f(x_2)\,[1-F(x_2)]^{R_2}\cdots f(x_m)\,[1-F(x_m)]^{R_m}\, dx_1\cdots dx_m.
\end{aligned} \tag{4.10}$$

In this subsection, for the case of the standard exponential distribution, we will again exploit the characterizing differential equation $f(x) = 1 - F(x)$ in order to derive recurrence relations for these product moments which will enable us to compute all the product moments of progressively Type-II right censored order statistics for all sample sizes and all possible censoring schemes.

Theorem 4.4 *For $1 \le i < j \le m-1$ and $m \le n$,*

$$\begin{aligned} &\mu_{i,j:m:n}^{(R_1,\cdots,R_m)} \\ &= \frac{1}{R_j+1}\Big[\mu_{i:m:n}^{(R_1,\cdots,R_m)} \\ &- (n - R_1 - \cdots - R_j - j)\,\mu_{i,j:m-1:n}^{(R_1,\cdots,R_{j-1},R_j+R_{j+1}+1,R_{j+2},\cdots,R_m)} \\ &+ (n - R_1 - \cdots - R_{j-1} - j + 1)\,\mu_{i,j-1:m-1:n}^{(R_1,\cdots,R_{j-2},R_{j-1}+R_j+1,R_{j+1},\cdots,R_m)}\Big]. \end{aligned} \tag{4.11}$$

Proof For $1 \le i < j \le m-1$, using the fact that $f(x_j) = 1 - F(x_j)$, we have from (4.1)

$$\begin{aligned} \mu_{i:m:n}^{(R_1,\cdots,R_m)} &= E\left[X_{i:m:n}^{(R_1,\cdots,R_m)}\left\{X_{j:m:n}^{(R_1,\cdots,R_m)}\right\}^0\right] \\ &= A(n, m-1) \underset{0<x_1<\cdots<x_{j-1}<x_{j+1}<\cdots<x_m<\infty}{\int\cdots\int\cdots\int} x_i \\ &\times \left\{\int_{x_{j-1}}^{x_{j+1}} x_j^0\,[1-F(x_j)]^{R_j+1}\,dx_j\right\} f(x_1)\,[1-F(x_1)]^{R_1} \\ &\times \cdots f(x_{j-1})\,[1-F(x_{j-1})]^{R_{j-1}}\, f(x_{j+1})\,[1-F(x_{j+1})]^{R_{j+1}} \\ &\times \cdots f(x_m)\,[1-F(x_m)]^{R_m}\,dx_1\cdots dx_{j-1}dx_{j+1}\cdots dx_m. \end{aligned} \tag{4.12}$$

Integrating the innermost integral by parts, we obtain

$$\begin{aligned} \int_{x_{j-1}}^{x_{j+1}} [1-F(x_j)]^{R_j+1}\,dx_j &= x_{j+1}\,[1-F(x_{j+1})]^{R_j+1} \\ &- x_{j-1}\,[1-F(x_{j-1})]^{R_j+1} \\ &+ (R_j+1)\int_{x_{j-1}}^{x_{j+1}} x_j\,[1-F(x_j)]^{R_j}\, f(x_j)\,dx_j, \end{aligned}$$

which, when substituted into (4.12) and simplified, results in the following equation:

$$\begin{aligned} \mu_{i:m:n}^{(R_1,\cdots,R_m)} &= (n - R_1 - \cdots - R_j - j)\,\mu_{i,j:m-1:n}^{(R_1,\cdots,R_{j-1},R_j+R_{j+1}+1,R_{j+2},\cdots,R_m)} \\ &- (n - R_1 - \cdots - R_{j-1} - j + 1)\,\mu_{i,j-1:m-1:n}^{(R_1,\cdots,R_{j-2},R_{j-1}+R_j+1,R_{j+1},\cdots,R_m)} \\ &+ (R_j+1)\,\mu_{i,j:m:n}^{(R_1,\cdots,R_m)}. \end{aligned}$$

Upon rearrangement of this equation, we obtain the relation in (4.11). □

Remark 4.4 Notice that Theorem 4.4 holds even for $j = i+1$, without altering the proof, provided we realize that $\mu_{i,i:m:n}^{(R_1,\cdots,R_m)} = \mu_{i:m:n}^{(R_1,\cdots,R_m)^{(2)}}$.

Theorem 4.5 *For $1 \leq i \leq m-1$ and $m \leq n$,*

$$\begin{aligned}\mu_{i,m:m:n}^{(R_1,\cdots,R_m)} &= \frac{1}{R_m+1}\Big[\mu_{i:m:n}^{(R_1,\cdots,R_m)} \\ &+ (n-R_1-\cdots-R_{m-1}-m+1)\,\mu_{i,m-1:m-1:n}^{(R_1,\cdots,R_{m-2},R_{m-1}+R_m+1)}\Big].\end{aligned} \tag{4.13}$$

Proof From (4.1), using the fact that $f(x_m) = 1 - F(x_m)$, we have

$$\begin{aligned}\mu_{i:m:n}^{(R_1,\cdots,R_m)} &= E\left[X_{i:m:n}^{(R_1,\cdots,R_m)}\left\{X_{m:m:n}^{(R_1,\cdots,R_m)}\right\}^0\right] \\ &= A(n,m-1)\int\!\!\cdots\!\!\int_{0<x_1<\cdots<x_{m-1}<\infty} x_i \\ &\times \left\{\int_{x_{m-1}}^{\infty} x_m^0\,[1-F(x_m)]^{R_m+1}\,dx_m\right\} f(x_1)\,[1-F(x_1)]^{R_1} \\ &\times \cdots f(x_{m-1})\,[1-F(x_{m-1})]^{R_{m-1}}\,dx_1\cdots dx_{m-1}.\end{aligned} \tag{4.14}$$

Integrating the innermost integral by parts, we obtain

$$\begin{aligned}\int_{x_{m-1}}^{\infty}[1-F(x_m)]^{R_m+1}\,dx_m &= -x_{m-1}\,[1-F(x_{m-1})]^{R_m+1} \\ &+ (R_m+1)\int_{x_{m-1}}^{\infty} x_m\,[1-F(x_m)]^{R_m}\,f(x_m)\,dx_m\end{aligned}$$

which, when substituted into (4.14) and simplified, yields

$$\begin{aligned}\mu_{i:m:n}^{(R_1,\cdots,R_m)} &= -(n-R_1-\cdots-R_{m-1}-m+1)\,\mu_{i,m-1:m-1:n}^{(R_1,\cdots,R_{m-2},R_{m-1}+R_m+1)} \\ &+ (R_m+1)\,\mu_{i,m:m:n}^{(R_1,\cdots,R_m)}.\end{aligned}$$

Upon rewriting this equation, we immediately obtain the relation in (4.13). □

Remark 4.5 Using these recurrence relations, we can obtain all the product moments of progressively Type-II right censored order statistics for all sample sizes and all censoring schemes $(R_1,\cdots,R_m)$.

Remark 4.6 A simpler recurrence relation has been presented recently by Balakrishnan, Cramer, Kamps and Schenk (1999) as

$$\mu_{i,j:m:n}^{(R_1,\cdots,R_m)} = \frac{1}{R_j^+}\mu_{i:i:n}^{(R_1,\cdots,R_{i-1},R_i^+-1)} + \mu_{i,j-1:j-1:n}^{(R_1,\cdots,R_{j-2},R_{j-1}^+-1)}$$

for $1 \leq i < j \leq m$, where $R_i^+ = \sum_{j=i}^{m}(R_j+1)$.

Remark 4.7 For the special case $R_1 = \cdots = R_m = 0$ so that $m = n$ in which case the progressively censored order statistics become the usual order statistics $X_{1:n}, X_{2:n}, \cdots, X_{n:n}$, whose single moments are denoted by $\mu_{i:n}^{(k)}$ for $1 \leq i \leq n$ and product moments are denoted by $\mu_{i,j:n}$ for $1 \leq i < j \leq n$, the relations established in Subsections 4.2.1 and 4.2.2 reduce to the following:

From (4.3): For $k > -1$,

$$\mu_{1:n}^{(k+1)} = (k+1)\,\mu_{1:n}^{(k)} - (n-1)\,\mu_{1:n-1:n}^{(1,0,\cdots,0)^{(k+1)}}.$$

From (4.6): For $2 \leq i \leq n-1$ and $k > -1$,

$$\mu_{i:n}^{(k+1)} = (k+1)\,\mu_{i:n}^{(k)} - (n-i)\,\mu_{i:n-1:n}^{(0,\cdots,0,1,0,\cdots,0)^{(k+1)}} + (n-i+1)\,\mu_{i-1:n-1:n}^{(0,\cdots,0,1,0,\cdots,0)^{(k+1)}},$$

where, in the superscript of the second term on the right hand side, the 1 is in the i-th position, and in the superscript of the third term on the right hand side, the 1 is in the $(i-1)$-th position.

From (4.8): For $k > -1$,

$$\mu_{n:n}^{(k+1)} = (k+1)\,\mu_{n:n}^{(k)} + \mu_{n-1:n-1:n}^{(0,\cdots,0,1)^{(k+1)}}.$$

Now, if $R_1 = R_2 = \cdots = R_{j-1} = 0$ so that there is no censoring before the time of the j-th failure, then the first j progressively Type-II right censored order statistics are simply the first j usual order statistics. Thus, the relations above reduce to:

From (4.3): For $k > -1$,

$$\mu_{1:n}^{(k+1)} = (k+1)\,\mu_{1:n}^{(k)} - (n-1)\,\mu_{1:n}^{(k+1)}$$

which simplifies to

$$n\mu_{1:n}^{(k+1)} = (k+1)\,\mu_{1:n}^{(k)}.$$

From (4.6): For $2 \leq i \leq n-1$ and $k > -1$,

$$\mu_{i:n}^{(k+1)} = (k+1)\,\mu_{i:n}^{(k)} - (n-i)\,\mu_{i:n}^{(k+1)} + (n-i+1)\,\mu_{i-1:n}^{(k+1)},$$

which simplifies to

$$(n-i+1)\,\mu_{i:n}^{(k+1)} = (k+1)\,\mu_{i:n}^{(k)} + (n-i+1)\,\mu_{i-1:n}^{(k+1)}.$$

From (4.8): For $k > -1$,

$$\mu_{n:n}^{(k+1)} = (k+1)\,\mu_{n:n}^{(k)} + \mu_{n-1:n}^{(k+1)}.$$

These recurrence relations are equivalent to those established by Joshi (1978).

Following similar arguments, substituting $R_1 = R_2 = \cdots = R_m = 0$ into (4.11) and (4.13), we obtain for $1 \le i < j \le n$

$$\mu_{i,j:n} = \frac{1}{n-j+1}\mu_{i:n} + \mu_{i,j-1:n},$$

which is the relation established by Joshi (1982).

4.2.3 Recursive Algorithm

Using the recurrence relations established in Sections 4.2.1 and 4.2.2, the means, variances and covariances of all progressively Type-II right censored order statistics from the standard exponential distribution can be readily computed as follows.

Setting $k = 0$, (4.4) will give us the values $\mu_{1:1:n}^{(n-1)} = 1/n, n = 1, 2, \cdots$ which in turn, again using (4.4) with $k = 1$, will give us the values $\mu_{1:1:n}^{(n-1)^{(2)}} = 2/n^2, n = 1, 2, \cdots$. Thus, all first and second moments with $m = 1$ for all sample sizes n will be obtained. Next, using (4.3), we can determine all moments of the form $\mu_{1:2:n}^{(R_1,R_2)}$ for all $R_1, R_2, n = 2, 3, \cdots$, which can in turn be used, with (4.3), to determine all moments of the form $\mu_{1:2:n}^{(R_1,R_2)^{(2)}}, n = 2, 3, \cdots$. Eq. (4.8) can then be used to obtain $\mu_{2:2:n}^{(R_1,R_2)}$ for all R_1, R_2, and $n \ge 2$, and these values can be used to obtain all moments of the form $\mu_{2:2:n}^{(R_1,R_2)^{(2)}}$ for $n \ge 2$ using (4.8) again. Eq. (4.3) can now be used again to obtain $\mu_{1:3:n}^{(R_1,R_2,R_3)}$ and $\mu_{1:3:n}^{(R_1,R_2,R_3)^{(2)}}$ for all R_1, R_2 and R_3, and $n \ge 3$, and (4.6) can be used next to obtain all moments of the form $\mu_{2:3:n}^{(R_1,R_2,R_3)}$ and $\mu_{2:3:n}^{(R_1,R_2,R_3)^{(2)}}$. Finally, (4.8) can be used to obtain all moments of the form $\mu_{3:3:n}^{(R_1,R_2,R_3)}$ and $\mu_{3:3:n}^{(R_1,R_2,R_3)^{(2)}}$. This process can be continued until all desired first and second order moments (and therefore all variances) are obtained for all sample sizes and all censoring schemes.

From (4.13), all moments of the form $\mu_{m-1,m:m:n}^{(R_1,\cdots,R_m)}$, $m = 2, 3, \cdots, n$, can be determined, since only single moments, which have already been obtained, are needed to calculate them. Then, using (4.11), all moments of the form $\mu_{i-1,i:m:n}^{(R_1,\cdots,R_m)}, 2 \le i < m$, can be obtained. From this point, using (4.13), we can obtain all moments of the form $\mu_{m-2,m:m:n}^{(R_1,\cdots,R_m)}$, $m = 3, 4, \cdots, n$, and subsequently, using (4.11), all moments of the form $\mu_{i-2,i:m:n}^{(R_1,\cdots,R_m)}$, $3 \le i < m$, can be determined. Continuing this process, all the desired product moments (and therefore all covariances) can be obtained for all sample sizes and all censoring schemes.

Remark 4.8 As mentioned earlier, the results presented here are for the sake of completeness. However, all of the results are not necessary to actually compute the required moments. By making use of Theorem 2.3, we know that

$$X_{i:m:n}^{(R_1,\cdots,R_m)} \stackrel{d}{=} X_{i:i:n}^{(R_1,\cdots,R_{i-1},n-R_1-\cdots-R_{i-1}-i)},$$

and therefore,

$$\mu_{i:m:n}^{(R_1,\cdots,R_m)^{(k)}} = \mu_{i:i:n}^{(R_1,\cdots,R_{i-1},n-R_1-\cdots-R_{i-1}-i)^{(k)}}.$$

Thus, we need only find moments of the form $\mu_{m:m:n}^{(R_1,\cdots,R_m)}$ and $\mu_{i,m:m:n}^{(R_1,\cdots,R_m)}$, that is, we need to only use relations (4.4), (4.8) and (4.13) to obtain all the single and product moments of progressively Type-II right censored order statistics from the exponential distribution. A similar comment holds for the recurrence relations for moments of progressively Type-II right censored order statistics from other continuous distributions, some of which are discussed in the rest of this chapter.

We will now proceed to derive similar recurrence relations for moments of progressively Type-II right censored order statistics from the truncated exponential distribution, the Pareto distribution and its truncated forms, and the power function distribution and its truncated forms. The proofs of these relations are very similar to the proofs presented in this section for the exponential distribution, and thus will be excluded here. It should be noted that in a number of these cases, moments of progressively Type-II right censored order statistics can not be obtained in any simple explicit form, and hence the method of recursion will be the method of choice for obtaining moments.

4.3 The Doubly Truncated Exponential Distribution

To further generalize the results for the exponential distribution presented in the last section, we will consider the doubly truncated exponential distribution. We will present in this section recurrence relations for the single and product moments of progressively Type-II right censored order statistics from the doubly truncated exponential distribution which are once again complete. Recurrence relations established in this case also generalize the results of Joshi (1978, 1979, 1982) and Balakrishnan and Joshi (1984) for the usual order statistics from the doubly truncated exponential distribution.

4.3.1 Recurrence Relations for Single Moments

The probability density function of the doubly truncated exponential distribution is

$$f(x) = \frac{e^{-x}}{P-Q}, \quad Q_1 < x < P_1.$$

Here, $1-P$ is the proportion of right truncation on the standard exponential distribution and Q is the proportion of left truncation. Thus, $P = 1-e^{-P_1}$ and $Q = 1-e^{-Q_1}$. The cumulative distribution function of the doubly truncated exponential distribution is then given by

$$F(x) = \frac{1}{P-Q}\left(e^{-Q_1} - e^{-x}\right), \; Q_1 < x < P_1.$$

Thus, the characterizing differential equation for this distribution is

$$f(x) = \frac{1-Q}{P-Q} - F(x) = \left(\frac{1-Q}{P-Q} - 1\right) + [1 - F(x)].$$

This relation will now be exploited in order to derive complete recurrence relations for the single and product moments of progressively Type-II right censored order statistics from the doubly truncated exponential distribution.

Theorem 4.6 *For $k > -1$,*

$$\mu_{1:1:1}^{(0)^{(k+1)}} = (k+1)\,\mu_{1:1:1}^{(0)^{(k)}} - \left(\frac{1-P}{P-Q}\right) P_1^{k+1} + \left(\frac{1-Q}{P-Q}\right) Q_1^{k+1}. \tag{4.15}$$

Theorem 4.7 *For $n \geq 2$ and $k > -1$,*

$$\mu_{1:1:n}^{(n-1)^{(k+1)}} = \frac{k+1}{n}\mu_{1:1:n}^{(n-1)^{(k)}} - \left(\frac{1-P}{P-Q}\right)\mu_{1:1:n-1}^{(n-2)^{(k+1)}} + \left(\frac{1-Q}{P-Q}\right) Q_1^{k+1}. \tag{4.16}$$

Theorem 4.8 *For $2 \leq m \leq n-1$, $R_1 \geq 1$ and $k > -1$,*

$$\begin{aligned}
\mu_{1:m:n}^{(R_1,\cdots,R_m)^{(k+1)}} &= \frac{1}{R_1+1}\Bigg\{(k+1)\,\mu_{1:m:n}^{(R_1,\cdots,R_m)^{(k)}} \\
&- \left(\frac{1-P}{P-Q}\right)\left[\frac{n(n-R_1-1)}{n-1}\mu_{1:m-1:n-1}^{(R_1+R_2,R_3,\cdots,R_m)^{(k+1)}}\right. \\
&+ \left.\frac{n}{n-1}R_1\mu_{1:m:n-1}^{(R_1-1,R_2,\cdots,R_m)^{(k+1)}}\right] \\
&- (n-R_1-1)\,\mu_{1:m-1:n}^{(R_1+R_2+1,R_3,\cdots,R_m)^{(k+1)}} + n\left(\frac{1-Q}{P-Q}\right) Q_1^{k+1}\Bigg\}.
\end{aligned} \tag{4.17}$$

Theorem 4.9 *For $2 \leq m \leq n$, $R_1 = 0$ and $k > -1$,*

$$\begin{aligned}
\mu_{1:m:n}^{(0,R_2,\cdots,R_m)^{(k+1)}} &= (k+1)\,\mu_{1:m:n}^{(0,R_2,\cdots,R_m)^{(k)}} - n\left(\frac{1-P}{P-Q}\right)\mu_{1:m-1:n-1}^{(R_2,R_3,\cdots,R_m)^{(k+1)}} \\
&- (n-1)\,\mu_{1:m-1:n}^{(R_2+1,R_3,\cdots,R_m)^{(k+1)}} + n\left(\frac{1-Q}{P-Q}\right) Q_1^{k+1}.
\end{aligned} \tag{4.18}$$

Theorem 4.10 *For $2 \le i \le m-1$, $m \le n-1$, $R_i \ge 1$ and $k > -1$,*

$$\begin{aligned}
&\mu_{i:m:n}^{(R_1,\cdots,R_m)^{(k+1)}} \\
&= \frac{1}{R_i+1}\left\{(k+1)\,\mu_{i:m:n}^{(R_1,\cdots,R_m)^{(k)}}\right. \\
&- \left(\frac{1-P}{P-Q}\right)\left[\frac{A(n,i)}{A(n-1,i-1)}\mu_{i:m-1:n-1}^{(R_1,\cdots,R_{i-1},R_i+R_{i+1},R_{i+2},\cdots,R_m)^{(k+1)}}\right. \\
&- \frac{A(n,i-1)}{A(n-1,i-2)}\mu_{i-1:m-1:n-1}^{(R_1,\cdots,R_{i-2},R_{i-1}+R_i,R_{i+1},\cdots,R_m)^{(k+1)}} \\
&+ \left.\frac{A(n,i-1)}{A(n-1,i-1)}R_i\mu_{i:m:n-1}^{(R_1,\cdots R_{i-1},R_i-1,R_{i+1},\cdots,R_m)^{(k+1)}}\right] \\
&- (n-R_1-\cdots-R_i-i) \\
&\times \mu_{i:m-1:n}^{(R_1,\cdots,R_{i-1},R_i+R_{i+1}+1,R_{i+2},\cdots,R_m)^{(k+1)}} \\
&+ (n-R_1-\cdots-R_{i-1}-i+1) \\
&\times \left.\mu_{i-1:m-1:n}^{(R_1,\cdots,R_{i-2},R_{i-1}+R_i+1,R_{i+1},\cdots,R_m)^{(k+1)}}\right\},
\end{aligned} \tag{4.19}$$

where $A(p,q)$ is as defined in (4.2).

Theorem 4.11 *For $2 \le i \le m-1$, $m \le n$, $R_i = 0$ and $k > -1$,*

$$\begin{aligned}
\mu_{i:m:n}^{(R_1,\cdots R_{i-1},0,R_{i+1},\cdots,R_m)^{(k+1)}} &= (k+1)\,\mu_{i:m:n}^{(R_1,\cdots,R_{i-1},0,R_{i+1},\cdots R_m)^{(k)}} \\
&- \left(\frac{1-P}{P-Q}\right)\frac{A(n,i-1)}{A(n-1,i-2)} \\
&\times \left[\mu_{i:m-1:n-1}^{(R_1,\cdots,R_{i-1},R_{i+1},R_{i+2},\cdots,R_m)^{(k+1)}}\right. \\
&- \left.\mu_{i-1:m-1:n-1}^{(R_1,\cdots,R_{i-2},R_{i-1},R_{i+1},\cdots,R_m)^{(k+1)}}\right] \\
&- (n-R_1-\cdots-R_{i-1}-i) \\
&\times \mu_{i:m-1:n}^{(R_1,\cdots,R_{i-1},R_{i+1}+1,R_{i+2},\cdots,R_m)^{(k+1)}} \\
&+ (n-R_1-\cdots-R_{i-1}-i+1) \\
&\times \mu_{i-1:m-1:n}^{(R_1,\cdots,R_{i-2},R_{i-1}+1,R_{i+1},\cdots,R_m)^{(k+1)}}.
\end{aligned} \tag{4.20}$$

Theorem 4.12 *For $2 \le m \le n-1$, $R_m \ge 1$ and $k > -1$,*

$$\begin{aligned}
\mu_{m:m:n}^{(R_1,\cdots,R_m)^{(k+1)}} &= \frac{1}{R_m+1}\left\{(k+1)\,\mu_{m:m:n}^{(R_1,\cdots,R_m)^{(k)}}\right. \\
&+ \left(\frac{1-P}{P-Q}\right)\left[\frac{A(n,m-1)}{A(n-1,m-2)}\mu_{m-1:m-1:n-1}^{(R_1,\cdots,R_{m-2},R_{m-1}+R_m)^{(k+1)}}\right.
\end{aligned}$$

$$
\begin{aligned}
& - \frac{A(n, m-1)}{A(n-1, m-1)} R_m \mu_{m:m:n-1}^{(R_1, \cdots R_{m-1}, R_m - 1)^{(k+1)}} \Bigg] \\
& + (n - R_1 - \cdots - R_{m-1} - m + 1) \\
& \times \mu_{m-1:m-1:n}^{(R_1, \cdots, R_{m-2}, R_{m-1} + R_m + 1)^{(k+1)}} \Bigg\}.
\end{aligned} \tag{4.21}
$$

Theorem 4.13 *For* $2 \leq m \leq n$, $R_m = 0$ *and* $k > -1$,

$$
\begin{aligned}
\mu_{m:m:n}^{(R_1, \cdots, R_{m-1}, 0)^{(k+1)}} &= (k+1) \mu_{m:m:n}^{(R_1, \cdots, R_{m-1}, 0)^{(k)}} \\
&+ (n - R_1 - R_2 - \cdots - R_{m-1} - m + 1) \\
&\times \mu_{m-1:m-1:n}^{(R_1, \cdots, R_{m-2}, R_{m-1} + 1)^{(k+1)}} \\
&- \left(\frac{1-P}{P-Q} \right) \frac{A(n, m-1)}{A(n-1, m-2)} \\
&\times \left[P_1^{k+1} - \mu_{m-1:m-1:n-1}^{(R_1, \cdots, R_{m-2}, R_{m-1})^{(k+1)}} \right].
\end{aligned} \tag{4.22}
$$

Remark 4.9 The recurrence relations presented in this subsection are complete in the sense that they will enable one to compute all the single moments of all progressively Type-II right censored order statistics from doubly, right ($Q \to 0$), and left ($P \to 1$) truncated exponential distributions for all sample sizes and all censoring schemes. The recursive algorithm will be detailed in Section 4.3.3.

Remark 4.10 It is easily seen that by letting $P \to 1$ and $Q \to 0$, these results reduce to the recurrence relations given in Section 4.2.1 for the standard exponential distribution.

4.3.2 Recurrence Relations for Product Moments

Following steps similar to those in Section 4.2.2 for the product moments of progressively Type-II right censored order statistics from the exponential distribution, we may obtain the following recurrence relations for product moments of progressively Type-II right censored order statistics from the doubly truncated exponential distribution.

Theorem 4.14 *For* $1 \leq i < j \leq m-1$, $m \leq n-1$ *and* $R_j \geq 1$,

$$\mu_{i,j:m:n}^{(R_1, \cdots, R_m)}$$

$$
\begin{aligned}
= & \ \frac{1}{R_j+1}\Bigg\{\mu_{i:m:n}^{(R_1,\cdots,R_m)} \\
- & \left(\frac{1-P}{P-Q}\right)\left[\frac{A(n,j)}{A(n-1,j-1)}\mu_{i,j:m-1:n-1}^{(R_1,\cdots,R_{j-1},R_j+R_{j+1},R_{j+2},\cdots,R_m)}\right. \\
- & \ \frac{A(n,j-1)}{A(n-1,j-2)}\mu_{i,j-1:m-1:n-1}^{(R_1,\cdots,R_{j-2},R_{j-1}+R_j,R_{j+1},\cdots,R_m)} \\
+ & \left.\frac{A(n,j-1)}{A(n-1,j-1)}R_j\mu_{i,j:m:n-1}^{(R_1,\cdots,R_{j-1},R_j-1,R_{j+1},\cdots,R_m)}\right] \\
- & \ (n-R_1-\cdots-R_j-j)\,\mu_{i,j:m-1:n}^{(R_1,\cdots,R_{j-1},R_j+R_{j+1}+1,R_{j+2},\cdots,R_m)} \\
+ & \ (n-R_1-\cdots-R_{j-1}-j+1)\,\mu_{i,j-1:m-1:n}^{(R_1,\cdots,R_{j-2},R_{j-1}+R_j+1,R_{j+1},\cdots,R_m)}\Bigg\}.
\end{aligned}
\tag{4.23}
$$

Theorem 4.15 *For $1 \le i < j \le m-1,\ m \le n$ and $R_j = 0$,*

$$
\begin{aligned}
& \mu_{i,j:m:n}^{(R_1,\cdots R_{j-1},0,R_{j+1},\cdots,R_m)} \\
= & \ \mu_{i:m:n}^{(R_1,\cdots R_{j-1},0,R_{j+1},\cdots,R_m)} - \left(\frac{1-P}{P-Q}\right)\frac{A(n,j-1)}{A(n-1,j-2)} \\
\times & \left[\mu_{i,j:m-1:n-1}^{(R_1,\cdots,R_{j-1},R_{j+1},R_{j+2},\cdots,R_m)} - \mu_{i,j-1:m-1:n-1}^{(R_1,\cdots,R_{j-1},R_{j+1},\cdots,R_m)}\right] \\
- & \ (n-R_1-\cdots-R_{j-1}-j)\,\mu_{i,j:m-1:n}^{(R_1,\cdots,R_{j-1},R_{j+1}+1,R_{j+2},\cdots,R_m)} \\
+ & \ (n-R_1-\cdots-R_{j-1}-j+1)\,\mu_{i,j-1:m-1:n}^{(R_1,\cdots,R_{j-2},R_{j-1}+1,R_{j+1},\cdots,R_m)}.
\end{aligned}
\tag{4.24}
$$

Theorem 4.16 *For $1 \le i \le m-1,\ m \le n-1$ and $R_m \ge 1$,*

$$
\begin{aligned}
\mu_{i,m:m:n}^{(R_1,\cdots,R_m)} = & \ \frac{1}{R_m+1}\Bigg\{\mu_{i:m:n}^{(R_1,\cdots,R_m)} \\
- & \left(\frac{1-P}{P-Q}\right)\left[-\frac{A(n,m-1)}{A(n-1,m-2)}\mu_{i,m-1:m-1:n-1}^{(R_1,\cdots,R_{m-2},R_{m-1}+R_m)}\right. \\
+ & \left.\frac{A(n,m-1)}{A(n-1,m-1)}R_m\mu_{i,m:m:n-1}^{(R_1,\cdots R_{m-1},R_m-1)}\right] \\
+ & \ (n-R_1-\cdots-R_{m-1}-m+1)\,\mu_{i,m-1:m-1:n}^{(R_1,\cdots,R_{m-2},R_{m-1}+R_m+1)}\Bigg\}.
\end{aligned}
\tag{4.25}
$$

Theorem 4.17 *For $1 \le i \le m-1$, $m \le n$ and $R_m = 0$,*

$$\begin{aligned}
\mu_{i,m:m:n}^{(R_1,\cdots,R_{m-1},0)} &= \mu_{i:m:n}^{(R_1,\cdots,R_{m-1},0)} \\
&- \left(\frac{1-P}{P-Q}\right)\frac{A(n,m-1)}{A(n-1,m-2)}\left[P_1\mu_{i:m-1:n-1}^{(R_1,\cdots,R_{m-1})}\right. \\
&- \left.\mu_{i,m-1:m-1:n-1}^{(R_1,\cdots,R_{m-2},R_{m-1})}\right] \\
&+ (n-R_1-\cdots-R_{m-1}-m+1)\,\mu_{i,m-1:m-1:n}^{(R_1,\cdots,R_{m-2},R_{m-1}+1)}.
\end{aligned} \tag{4.26}$$

Remark 4.11 Using these recurrence relations, we can obtain all the product moments of progressively Type-II right censored order statistics from the doubly truncated exponential distribution for all sample sizes and all censoring schemes $(R_1, \cdots, R_m)$. The recursive algorithm is described in Section 4.3.3.

Remark 4.12 Note that for the case of the doubly truncated exponential distribution, explicit expressions for the single and product moments of progressively Type-II right censored order statistics are not readily obtained in closed form. Therefore, the recurrence relations presented in this section will be useful to efficiently calculate the required moments in a simple recursive manner.

Remark 4.13 Letting $P \to 1$ and $Q \to 0$, we readily obtain the relations derived in Section 4.2.2.

Remark 4.14 Following arguments similar to those used in Remark 4.5, we may show that for the special case $R_1 = \cdots = R_m = 0$, the recurrence relations in Sections 4.3.1 and 4.3.2 reduce to the following recurrence relations which are equivalent to those given by Joshi (1979) and Balakrishnan and Joshi (1984) for the usual order statistics from the doubly truncated exponential distribution:

From Eq. (4.15), we obtain for $k > -1$,

$$\mu_{1:1}^{(k+1)} = (k+1)\,\mu_{1:1}^{(k)} - \left(\frac{1-P}{P-Q}\right)P_1^{k+1} + n\left(\frac{1-Q}{P-Q}\right)Q_1^{k+1}.$$

From (4.16) and (4.18), we have for $n \ge 2$ and $k > -1$,

$$\mu_{1:n}^{(k+1)} = \frac{k+1}{n}\mu_{1:n}^{(k)} - \left(\frac{1-P}{P-Q}\right)\mu_{1:n-1}^{(k+1)} + \left(\frac{1-Q}{P-Q}\right)Q_1^{k+1}.$$

From (4.20), we have for $2 \le i \le n-1$ and $k > -1$,

$$\mu_{i:n}^{(k+1)} = \mu_{i-1:n}^{(k+1)} + \frac{1}{n-i+1}\left\{(k+1)\,\mu_{i:n}^{(k)} - n\left(\frac{1-P}{P-Q}\right)\left[\mu_{i:n-1}^{(k+1)} - \mu_{i-1:n-1}^{(k+1)}\right]\right\}.$$

From (4.22), we have for $n \geq 2$ and $k > -1$,

$$\mu_{n:n}^{(k)} = (k+1)\,\mu_{n:n}^{(k)} + \mu_{n-1:n}^{(k+1)} - n\left(\frac{1-P}{P-Q}\right)\left(P_1^{(k+1)} - \mu_{n-1:n-1}^{(k+1)}\right).$$

From (4.24), we obtain for $1 \leq i < j \leq n-1$,

$$\mu_{i,j:n} = \mu_{i,j-1:n} + \frac{1}{n-j+1}\left[\mu_{i:n} - n\left(\frac{1-P}{P-Q}\right)(\mu_{i,j:n-1} - \mu_{i,j-1:n-1})\right].$$

From (4.26), we obtain for $1 \leq i \leq n-1$,

$$\mu_{i,n:n} = \mu_{i,n-1:n} + \mu_{i:n} - n\left(\frac{1-P}{P-Q}\right)(P_1\mu_{i:n-1} - \mu_{i,n-1:n-1}).$$

Remark 4.15 If we let $Q \to 0$ in the above relations, we arrive at recurrence relations equivalent to those given in Joshi (1978, 1982) for the usual order statistics from the right truncated exponential distribution. Alternatively, if we let $P \to 1$, we obtain recurrence relations for the usual order statistics from the left truncated exponential distribution.

4.3.3 Recursive Algorithm

Using the recurrence relations presented in Sections 4.3.1 and 4.3.2, the means, variances and covariances of all progressively Type-II right censored order statistics can be readily computed as follows:

Setting $k = 0$, (4.15) will give us the value $\mu_{1:1:1}^{(0)}$, which in turn, again using (4.15) with $k = 1$, will give us the value $\mu_{1:1:1}^{(0)^{(2)}}$. From these values, we can recursively compute $\mu_{1:1:n}^{(n-1)}$ and $\mu_{1:1:n}^{(n-1)^{(2)}}$ for $n = 2, 3, \cdots$ using (4.16). Thus, all the first and second moments with $m = 1$ for all sample sizes n can be obtained. Next, using (4.18), we can determine all moments of the form $\mu_{1:2:n}^{(0,n-2)}$, $n = 2, 3, \cdots$ which can in turn be used, with (4.18), to determine all moments of the form $\mu_{1:2:n}^{(0,n-2)^{(2)}}$, $n = 2, 3, \cdots$. Eq. (4.17) can then be used to obtain $\mu_{1:2:n}^{(R_1,R_2)}$ for $R_1 = 1, 2, \cdots$ and $n \geq 3$, and these values can be used to obtain all moments of the form $\mu_{1:2:n}^{(R_1,R_2)^{(2)}}$ using (4.17) again. Now, (4.22) can be used again to obtain $\mu_{2:2:n}^{(n-2,0)}$ and $\mu_{2:2:n}^{(n-2,0)^{(2)}}$ for all n and (4.21) can be used next to obtain, for $R_1, R_2 = 1, 2, \cdots$ and $n \geq 3$, all moments of the form $\mu_{2:2:n}^{(R_1,R_2)}$ and $\mu_{2:2:n}^{(R_1,R_2)^{(2)}}$. This process can be continued until all the desired first and second order moments (and therefore all variances) are obtained.

From (4.26), all moments of the form $\mu_{m-1,m:m:n}^{(R_1,\cdots,R_{m-1},0)}$, $m = 2, 3, \cdots, n$, can be determined, since only the single moments, which have already been computed, are needed to calculate them. Eq. (4.25) can then be used to obtain $\mu_{m-1,m:m:n}^{(R_1,\cdots,R_m)}$ for $R_m = 1, 2, \cdots$. Then, using (4.24), all moments of the form

$\mu_{j-1,j:m:n}^{(R_1,\cdots,R_{j-1},0,R_{j+1},\cdots,R_m)}$ $(j < m)$ can be obtained, and using (4.23), all moments of the form $\mu_{j-1,j:m:n}^{(R_1,\cdots,R_m)}$, $j < m$, $R_j \geq 1$, can be calculated. From this point, using (4.26) and (4.25), we can obtain all moments of the form $\mu_{m-2,m:m:n}^{(R_1,\cdots,R_m)}$ and, subsequently, using (4.24) and (4.23), all moments of the form $\mu_{j-2,j:m:n}^{(R_1,\cdots,R_m)}$, $j < m$, can be determined. Continuing this way, all the desired product moments (and therefore all covariances) can be obtained.

4.4 The Pareto Distribution and Truncated Forms

We will present in this section complete recurrence relations satisfied by the single and product moments of progressively Type-II right censored order statistics from the doubly truncated Pareto distribution established by Aggarwala and Balakrishnan (1998d). From these results, we can obtain the corresponding results for the standard Pareto distribution, the left truncated Pareto distribution, and the right truncated Pareto distribution.

The probability density function of the doubly truncated Pareto distribution is given by

$$f(x) = \frac{1}{P-Q}\nu x^{-v-1}, \ 1 \leq \ Q_1 < x < P_1, \ \nu > 0,$$

where $Q = 1-Q_1^{-\nu}$ and $P = 1-P_1^{-\nu}$; here, $1-P$ is the proportion of truncation on the right of the standard Pareto distribution and Q is the proportion of truncation on the left.

The cumulative distribution function of the doubly truncated Pareto distribution is

$$F(x) = \frac{1}{P-Q}\left(Q_1^{-v} - x^{-\nu}\right),$$

and the characterizing differential equation for this distribution is given by

$$f(x) = \frac{\nu}{x}\left[\frac{1-Q}{P-Q} - F(x)\right] = \left(\frac{1-Q}{P-Q} - 1\right)\frac{\nu}{x} + [1 - F(x)]\frac{\nu}{x}.$$

This relation can be used to establish recurrence relations for the single and product moments of progressively Type-II right censored order statistics from the doubly truncated Pareto distribution.

4.4.1 Recurrence Relations for Single Moments

Theorem 4.18 *For $k \geq 0$,*

$$\mu_{1:1:1}^{(0)^{(k)}} = \begin{cases} \frac{\nu}{(k-\nu)(P-Q)}\left(P_1^{k-\nu} - Q_1^{k-\nu}\right), & k \neq \nu \\ \frac{\nu}{P-Q}\left[\ln(P_1) - \ln(Q_1)\right], & k = \nu. \end{cases} \tag{4.27}$$

Remark 4.16 If we let $P \to 1$, this moment exists only for $\nu > k$, in which case its value becomes $\frac{\nu Q_1^{k-\nu}}{(\nu-k)(1-Q)}$.

Theorem 4.19 *For $n \geq 2$ and $k > 0$,*

$$\mu_{1:1:n}^{(n-1)^{(k)}} = \frac{nv}{k-n\nu}\left\{\left(\frac{1-P}{P-Q}\right)\mu_{1:1:n-1}^{(n-2)^{(k)}} - \left(\frac{1-Q}{P-Q}\right)Q_1^k\right\}. \tag{4.28}$$

Theorem 4.20 *For $2 \leq m \leq n-1$, $R_1 \geq 1$ and $k > 0$,*

$$\begin{aligned}
&\mu_{1:m:n}^{(R_1,\cdots,R_m)^{(k)}} \\
&= \frac{\nu}{k-\nu(R_1+1)}\left\{\left(\frac{1-P}{P-Q}\right)\left[\frac{n(n-R_1-1)}{n-1}\mu_{1:m-1:n-1}^{(R_1+R_2,R_3,\cdots,R_m)^{(k)}}\right.\right. \\
&+ \left.\frac{n}{n-1}R_1\mu_{1:m:n-1}^{(R_1-1,R_2,\cdots,R_m)^{(k)}}\right] \\
&+ \left.(n-R_1-1)\,\mu_{1:m-1:n}^{(R_1+R_2+1,R_3,\cdots,R_m)^{(k)}} - n\left(\frac{1-Q}{P-Q}\right)Q_1^k\right\}.
\end{aligned} \tag{4.29}$$

Theorem 4.21 *For $2 \leq m \leq n$, $R_1 = 0$ and $k > 0$,*

$$\begin{aligned}
\mu_{1:m:n}^{(0,R_2,\cdots,R_m)^{(k)}} &= \frac{\nu}{k-\nu}\left\{n\left(\frac{1-P}{P-Q}\right)\mu_{1:m-1:n-1}^{(R_2,R_3,\cdots,R_m)^{(k)}}\right. \\
&+ \left.(n-1)\,\mu_{1:m-1:n}^{(R_2+1,R_3,\cdots,R_m)^{(k)}} - n\left(\frac{1-Q}{P-Q}\right)Q_1^k\right\}.
\end{aligned} \tag{4.30}$$

Theorem 4.22 *For $2 \leq i \leq m-1$, $m \leq n-1$, $R_i \geq 1$ and $k > 0$,*

$$\begin{aligned}
\mu_{i:m:n}^{(R_1,\cdots,R_m)^{(k)}} &= \frac{\nu}{k-\nu(R_1+1)}\left\{\left(\frac{1-P}{P-Q}\right)\right. \\
&\times \left[\frac{A(n,i)}{A(n-1,i-1)}\mu_{i:m-1:n-1}^{(R_1,\cdots,R_{i-1},R_i+R_{i+1},R_{i+2},\cdots,R_m)^{(k)}}\right. \\
&- \frac{A(n,i-1)}{A(n-1,i-2)}\mu_{i-1:m-1:n-1}^{(R_1,\cdots,R_{i-2},R_{i-1}+R_i,R_{i+1},\cdots,R_m)^{(k)}} \\
&+ \left.\frac{A(n,i-1)}{A(n-1,i-1)}R_i\mu_{i:m:n-1}^{(R_1,\cdots R_{i-1},R_i-1,R_{i+1},\cdots,R_m)^{(k)}}\right] \\
&+ (n-R_1-\cdots-R_i-i)\,\mu_{i:m-1:n}^{(R_1,\cdots,R_{i-1},R_i+R_{i+1}+1,R_{i+2},\cdots,R_m)^{(k)}} \\
&- (n-R_1-\cdots-R_{i-1}-i+1) \\
&\times \left.\mu_{i-1:m-1:n}^{(R_1,\cdots,R_{i-2},R_{i-1}+R_i+1,R_{i+1},\cdots,R_m)^{(k)}}\right\},
\end{aligned} \tag{4.31}$$

where $A(p,q)$ is as defined earlier in (4.2).

Theorem 4.23 *For $2 \le i \le m-1$, $m \le n$, $R_i = 0$ and $k > 0$,*

$$\begin{aligned}
\mu_{i:m:n}^{(R_1,\cdots R_{i-1},0,R_{i+1},\cdots,R_m)^{(k)}} &= \frac{\nu}{k-\nu}\left\{\left(\frac{1-P}{P-Q}\right)\frac{A(n,i-1)}{A(n-1,i-2)}\right.\\
&\times \left[\mu_{i:m-1:n-1}^{(R_1,\cdots,R_{i-1},R_{i+1},R_{i+2},\cdots,R_m)^{(k)}}\right.\\
&- \left.\mu_{i-1:m-1:n-1}^{(R_1,\cdots,R_{i-1},R_{i+1},\cdots,R_m)^{(k)}}\right]\\
&+ (n-R_1-\cdots-R_{i-1}-i)\\
&\times \mu_{i:m-1:n}^{(R_1,\cdots,R_{i-1},R_{i+1}+1,R_{i+2},\cdots,R_m)^{(k)}}\\
&- (n-R_1-\cdots-R_{i-1}-i+1)\\
&\times \left.\mu_{i-1:m-1:n}^{(R_1,\cdots,R_{i-2},R_{i-1}+1,R_{i+1},\cdots,R_m)^{(k)}}\right\}.
\end{aligned} \tag{4.32}$$

Theorem 4.24 *For $2 \le m \le n-1$, $R_m \ge 1$ and $k > 0$,*

$$\begin{aligned}
\mu_{m:m:n}^{(R_1,\cdots,R_m)^{(k)}} &= \frac{\nu}{k-\nu(R_m+1)}\left\{\left(\frac{1-P}{P-Q}\right)\right.\\
&\times \left[-\frac{A(n,m-1)}{A(n-1,m-2)}\mu_{m-1:m-1:n-1}^{(R_1,\cdots,R_{m-2},R_{m-1}+R_m)^{(k)}}\right.\\
&+ \left.\frac{A(n,m-1)}{A(n-1,m-1)}R_m\mu_{m:m:n-1}^{(R_1,\cdots R_{m-1},R_m-1)^{(k)}}\right]\\
&- \left.(n-R_1-\cdots-R_{m-1}-m+1)\,\mu_{m-1:m-1:n}^{(R_1,\cdots,R_{m-2},R_{m-1}+R_m+1)^{(k)}}\right\}.
\end{aligned} \tag{4.33}$$

Theorem 4.25 *For $2 \le m \le n$, $R_m = 0$ and $k > 0$,*

$$\begin{aligned}
\mu_{m:m:n}^{(R_1,\cdots,R_{m-1},0)^{(k)}} &= \frac{\nu}{k-\nu}\left\{-(n-R_1-\cdots-R_{m-1}-m+1)\right.\\
&\times \mu_{m-1:m-1:n}^{(R_1,\cdots,R_{m-2},R_{m-1}+1)^{(k)}}\\
&+ \left.\left(\frac{1-P}{P-Q}\right)\frac{A(n,m-1)}{A(n-1,m-2)}\left[P_1^k-\mu_{m-1:m-1:n-1}^{(R_1,\cdots,R_{m-2},R_{m-1})^{(k)}}\right]\right\}.
\end{aligned} \tag{4.34}$$

Remark 4.17 If we let $P \to 1$ in the last two relations, we must also have $\nu(R_m + 1) > k$ for existence.

Remark 4.18 The relations presented in this subsection are complete in the sense that they will enable one to compute all the single moments of progressively Type-II right censored order statistics from the doubly, right ($Q \to 0$) and left ($P \to 1$) truncated Pareto distributions for all sample sizes and all censoring schemes. The recursive algorithm will be presented in detail in Section 4.4.3.

4.4.2 Recurrence Relations for Product Moments

Adopting steps similar to those in Section 4.2.2 for the product moments of progressively Type-II right censored order statistics from the standard exponential distribution, we may obtain the following recurrence relations for the product moments of progressively Type-II right censored order statistics from the doubly truncated Pareto distribution.

Theorem 4.26 *For $1 \le i < j \le m-1$, $m \le n-1$ and $R_j \ge 1$,*

$$
\begin{aligned}
&\mu_{i,j:m:n}^{(R_1,\cdots,R_m)} \\
&= \frac{\nu}{1-\nu(R_j+1)} \\
&\times \left\{ \left(\frac{1-P}{P-Q} \right) \left[\frac{A(n,j)}{A(n-1,j-1)} \mu_{i,j:m-1:n-1}^{(R_1,\cdots,R_{j-1},R_j+R_{j+1},R_{j+2},\cdots,R_m)} \right.\right. \\
&- \frac{A(n,j-1)}{A(n-1,j-2)} \mu_{i,j-1:m-1:n-1}^{(R_1,\cdots,R_{j-2},R_{j-1}+R_j,R_{j+1},\cdots,R_m)} \\
&+ \left. \frac{A(n,j-1)}{A(n-1,j-1)} R_j \mu_{i,j:m:n-1}^{(R_1,\cdots,R_{j-1},R_j-1,R_{j+1},\cdots,R_m)} \right] \\
&+ (n-R_1-\cdots-R_j-j)\, \mu_{i,j:m-1:n}^{(R_1,\cdots,R_{j-1},R_j+R_{j+1}+1,R_{j+2},\cdots,R_m)} \\
&- \left. (n-R_1-\cdots-R_{j-1}-j+1)\, \mu_{i,j-1:m-1:n}^{(R_1,\cdots,R_{j-2},R_{j-1}+R_j+1,R_{j+1},\cdots,R_m)} \right\}.
\end{aligned}
\tag{4.35}
$$

Theorem 4.27 *For $1 \le i < j \le m-1$, $m \le n$ and $R_j = 0$,*

$$
\begin{aligned}
\mu_{i,j:m:n}^{(R_1,\cdots R_{j-1},0,R_{j+1},\cdots,R_m)} &= \frac{\nu}{1-\nu} \left\{ \left(\frac{1-P}{P-Q} \right) \frac{A(n,j-1)}{A(n-1,j-2)} \right. \\
&\times \left[\mu_{i,j:m-1:n-1}^{(R_1,\cdots,R_{j-1},R_{j+1},\cdots,R_m)} \right. \\
&- \left. \mu_{i,j-1:m-1:n-1}^{(R_1,\cdots,R_{j-1},R_{j+1},\cdots,R_m)} \right]
\end{aligned}
$$

$$
\begin{aligned}
&+ \quad (n - R_1 - \cdots - R_{j-1} - j) \\
&\times \quad \mu_{i,j:m-1:n}^{(R_1,\cdots,R_{j-1},R_{j+1}+1,R_{j+2},\cdots,R_m)} \\
&- \quad (n - R_1 - \cdots - R_{j-1} - j + 1) \\
&\times \quad \mu_{i,j-1:m-1:n}^{(R_1,\cdots,R_{j-2},R_{j-1}+1,R_{j+1},\cdots,R_m)} \Big\}.
\end{aligned}
\tag{4.36}
$$

Theorem 4.28 *For* $1 \le i \le m-1$, $m \le n-1$ *and* $R_m \ge 1$,

$$
\begin{aligned}
\mu_{i,m:m:n}^{(R_1,\cdots,R_m)} &= \frac{\nu}{1-\nu\,(R_m+1)} \left\{ \left(\frac{1-P}{P-Q} \right) \right. \\
&\times \left[- \frac{A\,(n,m-1)}{A\,(n-1,m-2)} \mu_{i,m-1:m-1:n-1}^{(R_1,\cdots,R_{m-2},R_{m-1}+R_m)} \right. \\
&+ \left. \frac{A\,(n,m-1)}{A\,(n-1,m-1)} R_m \, \mu_{i,m:m:n-1}^{(R_1,\cdots R_{m-1},R_m-1)} \right] \\
&- \left. (n - R_1 - \cdots - R_{m-1} - m + 1)\, \mu_{i,m-1:m-1:n}^{(R_1,\cdots,R_{m-2},R_{m-1}+R_m+1)} \right\}.
\end{aligned}
\tag{4.37}
$$

Theorem 4.29 *For* $1 \le i \le m-1$, $m \le n$ *and* $R_m = 0$,

$$
\begin{aligned}
\mu_{i,m:m:n}^{(R_1,\cdots,R_{m-1},0)} &= \frac{\nu}{1-\nu} \left\{ \left(\frac{1-P}{P-Q} \right) \frac{A\,(n,m-1)}{A\,(n-1,m-2)} \right. \\
&\times \left[P_1 \mu_{i:m-1:n-1}^{(R_1,\cdots,R_{m-1})} - \mu_{i,m-1:m-1:n-1}^{(R_1,\cdots,R_{m-2},R_{m-1})} \right] \\
&- \left. (n - R_1 - \cdots - R_{m-1} - m + 1)\, \mu_{i,m-1:m-1:n}^{(R_1,\cdots,R_{m-2},R_{m-1}+1)} \right\}.
\end{aligned}
\tag{4.38}
$$

Remark 4.19 If we let $P \to 1$ in the last two relations, we must also have $\nu\,(R_m+1) > k$ for existence.

Remark 4.20 Using these recurrence relations, we can obtain all the product moments of progressively Type-II right censored order statistics from the doubly truncated Pareto distribution for all sample sizes and all censoring schemes $(R_1, \cdots, R_m)$. The recursive algorithm is described in the next subsection.

Remark 4.21 If we set $R_1 = \cdots = R_m = 0$ and consequently $m = n$, all the above relations reduce to the corresponding results for the single and product moments of usual order statistics from the doubly truncated Pareto distribution established by Balakrishnan and Joshi (1982).

4.4.3 Recursive Algorithm

Using the recurrence relations developed in Sections 4.4.1 and 4.4.2, the means, variances and covariances of all progressively Type-II right censored order statistics can be readily computed as follows:

Setting $k = 1$ and $k = 2$, (4.27) will give us the values of $\mu_{1:1:1}^{(0)}$ and $\mu_{1:1:1}^{(0)^{(2)}}$, respectively. From these values, we can recursively compute $\mu_{1:1:n}^{(n-1)}$ and $\mu_{1:1:n}^{(n-1)^{(2)}}$ for $n = 2, 3, \cdots$ using (4.28). Thus, all the first and second moments with $m = 1$ for all sample sizes n can be obtained. Next, using (4.30), we can determine all moments of the form $\mu_{1:2:n}^{(0,n-2)}$ and $\mu_{1:2:n}^{(0,n-2)^{(2)}}$, $n = 2, 3, \cdots$. Eq. (4.29) can then be used to obtain $\mu_{1:2:n}^{(R_1,R_2)}$ and $\mu_{1:2:n}^{(R_1,R_2)^{(2)}}$ for $R_1 = 1, 2, \cdots$ and $n \geq 3$. Now, (4.34) can be used to obtain $\mu_{2:2:n}^{(n-2,0)}$ and $\mu_{2:2:n}^{(n-2,0)^{(2)}}$ for all n, and (4.33) can be used next to obtain, for $R_1, R_2 = 1, 2, \cdots$ and $n \geq 3$, all moments of the form $\mu_{2:2:n}^{(R_1,R_2)}$ and $\mu_{2:2:n}^{(R_1,R_2)^{(2)}}$. This process can be continued until all the desired first and second order moments (and therefore all variances) are obtained.

From (4.38), all moments of the form $\mu_{m-1,m:m:n}^{(R_1,\cdots,R_{m-1},0)}$, $m = 2, 3, \cdots, n$, can be determined, since only the single moments, which have already been computed, are needed to calculate them. Eq. (4.37) can then be used to obtain $\mu_{m-1,m:m:n}^{(R_1,\cdots,R_m)}$ for $R_m = 1, 2, \cdots$. Then, using (4.36), all moments of the form $\mu_{j-1,j:m:n}^{(R_1,\cdots,R_{j-1},0,R_{j+1},\cdots,R_m)}$ $(j < m)$ can be obtained, and using (4.35), all moments of the form $\mu_{j-1,j:m:n}^{(R_1,\cdots,R_m)}$, $j < m$, $R_j \geq 1$, can be calculated. From this point, using (4.39) and (4.37), we can obtain all moments of the form $\mu_{m-2,m:m:n}^{(R_1,\cdots,R_m)}$ and using (4.36) and (4.35), all moments of the form $\mu_{j-2,j:m:n}^{(R_1,\cdots,R_m)}$, $j < m$, can be obtained. Continuing this way, all the desired product moments (and therefore all covariances) can be determined.

4.5 The Power Function Distribution and Truncated Forms

You have probably already noticed that the recurrence relations for the truncated forms of the exponential and Pareto distribution look very similar, as do the characterizing differential equations for the two distributions. In fact, they look identical except for a factor of $\frac{\nu}{x}$. The similarities between the Pareto and power function distributions in terms of their characterizing differential equations are just as stark, and as such, rather than presenting the complete set of recurrence relations, we will simply mention the nature of change that will take place in the recurrence relations already presented for the Pareto distribution in the last section.

The probability density function of the doubly truncated power function distribution is given by

$$f(x) = \frac{1}{P-Q} a x^{a-1}, \; 0 \le Q_1 < x < P_1 \le 1, \; a > 0,$$

where $Q = Q_1^a$ and $P = P_1^a$; here, $1-P$ is the proportion of truncation on the right of the standard power function distribution and Q is the proportion of truncation on the left.

The cumulative distribution function of the doubly truncated power function distribution is

$$F(x) = \frac{1}{P-Q}\left(x^a - Q_1^a\right),$$

and the characterizing differential equation for this distribution is given by

$$f(x) = \frac{a}{x}\left[\frac{Q}{P-Q} + F(x)\right] = \left(1 + \frac{Q}{P-Q}\right)\frac{a}{x} - [1 - F(x)]\frac{a}{x}.$$

This differential equation differs from that of the doubly truncated Pareto distribution only in terms of the constant term and the sign of the constant multiplying $[1 - F(x)]$. Therefore, the recurrence relations already presented for the doubly truncated Pareto distribution will be altered in the following ways:

1. ν in the relations for the Pareto distribution will be replaced by a for the power function distribution.

2. $\frac{1-P}{P-Q}$ in the relations for the Pareto distribution will be replaced by $\frac{P}{P-Q}$ for the power function distribution.

3. Terms not multiplied by $\frac{1-P}{P-Q}$ in the relations for the Pareto distribution will be multiplied by -1 for the power function distribution. Note that in converting terms such as $\frac{1-Q}{P-Q} \cdot Q_1^k$, upon using the fact that $\frac{1-Q}{P-Q} = \frac{1-P}{P-Q} + 1$, and writing the term as $\frac{1-P}{P-Q} \cdot Q_1^k + Q_1^k$, it will change to $\frac{P}{P-Q} \cdot Q_1^k - Q_1^k$ by Rules 2 and 3.

4. The terms like $\frac{\nu}{k-\nu(R_i+1)}$ will be replaced with terms like $\frac{a}{k+a(R_i+1)}$.

Furthermore, since the support of this distribution is always finite, regardless of the truncation parameters Q_1 and P_1, there are no special additional restrictions on any of the parameters to ensure existence of the moments.

Thus, for example, the following relations hold for moments of progressively Type-II right censored order statistics from the power function distribution:

For $2 \le i \le m-1$, $m \le n-1$, $R_i \ge 1$ and $k > 0$,

$$\mu_{i:m:n}^{(R_1,\cdots,R_m)^{(k)}}$$

$$= \frac{a}{k + a(R_i + 1)}$$

$$
\begin{aligned}
\times & \left\{ \left(\frac{P}{P-Q}\right) \left[\frac{A(n,i)}{A(n-1,i-1)} \mu_{i:m-1:n-1}^{(R_1,\cdots,R_{i-1},R_i+R_{i+1},R_{i+2},\cdots,R_m)^{(k)}} \right.\right. \\
- & \frac{A(n,i-1)}{A(n-1,i-2)} \mu_{i-1:m-1:n-1}^{(R_1,\cdots,R_{i-2},R_{i-1}+R_i,R_{i+1},\cdots,R_m)^{(k)}} \\
+ & \left. \frac{A(n,i-1)}{A(n-1,i-1)} R_i \, \mu_{i:m:n-1}^{(R_1,\cdots R_{i-1},R_i-1,R_{i+1},\cdots,R_m)^{(k)}} \right] \\
- & (n-R_1-\cdots-R_i-i)\, \mu_{i:m-1:n}^{(R_1,\cdots,R_{i-1},R_i+R_{i+1}+1,R_{i+2},\cdots,R_m)^{(k)}} \\
+ & \left. (n-R_1-\cdots-R_{i-1}-i+1)\, \mu_{i-1:m-1:n}^{(R_1,\cdots,R_{i-2},R_{i-1}+R_i+1,R_{i+1},\cdots,R_m)^{(k)}} \right\};
\end{aligned}
$$

For $1 \le i \le m-1$, $m \le n$ and $R_m = 0$,

$$
\begin{aligned}
\mu_{i,m:m:n}^{(R_1,\cdots,R_{m-1},0)} &= \frac{a}{1+a} \left\{ \left(\frac{P}{P-Q}\right) \frac{A(n,m-1)}{A(n-1,m-2)} \right. \\
&\times \left[P_1 \mu_{i:m-1:n-1}^{(R_1,\cdots,R_{m-1})} - \mu_{i,m-1:m-1:n-1}^{(R_1,\cdots,R_{m-2},R_{m-1})} \right] \\
&+ \left. (n-R_1-\cdots-R_{m-1}-m+1)\, \mu_{i,m-1:m-1:n}^{(R_1,\cdots,R_{m-2},R_{m-1}+1)} \right\};
\end{aligned}
$$

For $2 \le m \le n-1$, $R_1 \ge 1$ and $k > 0$,

$$
\begin{aligned}
& \mu_{1:m:n}^{(R_1,\cdots,R_m)^{(k)}} \\
&= \frac{a}{k+a(R_1+1)} \left\{ \frac{P}{P-Q} \left[\frac{n(n-R_1-1)}{n-1} \mu_{1:m-1:n-1}^{(R_1+R_2,R_3,\cdots,R_m)^{(k)}} \right.\right. \\
&\quad \left. + \frac{n}{n-1} R_1 \mu_{1:m:n-1}^{(R_1-1,R_2,\cdots,R_m)^{(k)}} \right] \\
&\quad \left. -(n-R_1-1)\mu_{1:m-1:n}^{(R_1+R_2+1,R_3,\cdots,R_m)^{(k)}} - n\left(\frac{P}{P-Q}\, Q_1^k - Q_1^k\right) \right\}.
\end{aligned}
$$

The starting values for the recursive algorithm in this case are given by

$$
\mu_{1:1:1}^{(0)^{(k)}} = \begin{cases} \frac{a}{(k+a)(P-Q)} \left(P_1^{k+a} - Q_1^{k+a}\right), & k \ne -a \\ \frac{a}{P-Q} \left[\ln(P_1) - \ln(Q_1)\right], & k = -a,\ Q_1 > 0. \end{cases}
$$

The recursive algorithm for the computation of the single and product moments in this case is virtually identical to the one presented in the last section for the doubly truncated Pareto distribution.

Remark 4.22 If we set $R_1 = \cdots = R_m = 0$ and consequently $m = n$, these relations will readily reduce to the corresponding results for the single and product moments of usual order statistics from the doubly truncated power function distribution established by Balakrishnan and Joshi (1981).

Remark 4.23 It needs to be mentioned that although all the results in this chapter are presented for real k, we will be primarily interested only in the values of k which are positive integers as we will usually require only the means, variances, and so on, for practical applications of these results.

5

Alternative Computational Methods

5.1 Introduction

Clever transformations and efficient recursive algorithms are often useful in obtaining moments and for establishing mathematical properties of progressively Type-II right censored order statistics from a number of distributions, as we have already seen. However, such elegant methods are not possible for all distributions that may be of interest to a practitioner. For this reason, alternative methods for computing moments of progressively censored order statistics must be sought. In this chapter, we present two such methods for the computation of moments of progressively Type-II right censored order statistics. The first applies to an arbitrary continuous distributions for which the moments of *usual* order statistics are known, and the second applies specifically to symmetric distributions for which the moments of progressively Type-II right censored order statistics from the corresponding folded distribution are known. These two methods will further enhance our repertoire of distributions that we can consider as models for lifetime data, and hence will make the use of progressive censoring in real-life situations much more attractive. Finally, we present some first-order approximations to the means, variances and covariances of progressively Type-II right censored order statistics based on Taylor series expansions. These expressions will be used later in Chapter 6 in order to develop and illustrate first-order approximations to the best linear unbiased estimators of location and scale parameters of any distribution of interest.

5.2 Formulas in Terms of Moments of Usual Order Statistics

Anyone who has wrestled with problems involving order statistics from distributions such as the normal, extreme value or log-normal will be aware of the computational complexities associated with even finding single and product moments of *usual* order statistics from these distributions. Quite often, these moments must be obtained through extensive numerical integration methods (to high accuracy, too); see, for example, Tietjen, Kahaner and Beckman (1977), Balakrishnan and Chan (1992), Harter and Balakrishnan (1996), Balakrishnan and Chen (1997, 1999) and Balakrishnan and Rao (1998a,b). To consider generalizing one step further by finding explicit expressions for moments of progressively Type-II right censored order statistics from these distributions is clearly an ambitious task at the very least.

A method of obtaining the single and product moments of progressively Type-II right censored order statistics from a sample of size n from an arbitrary continuous distribution, provided the mean vector and variance-covariance matrix of the usual order statistics from a sample of size n from that distribution are known, was given by Thomas and Wilson (1972). Although the method requires somewhat extensive computation, it is a useful method of obtaining moments of progressively Type-II right censored order statistics in the absence of other more elegant methods like those already seen in Chapter 3. This method will be used extensively in Chapter 10 when we discuss the determination of optimal censoring schemes for the estimation of various parameters of interest from a number of distributions including the ones mentioned earlier in this section.

Thomas and Wilson's method of obtaining the single and product moments of progressively Type-II right censored order statistics from arbitrary continuous distributions works as follows.

Suppose we denote the usual order statistics from a sample of size n from the distribution of interest by $Z_{i:n}, i = 1, 2, \cdots, n$. Further, suppose we denote the m progressively Type-II right censored order statistics from the sample of size n obtained with the censoring scheme $(R_1, R_2, \cdots, R_m)$ by $Z_{j:m:n}, j = 1, 2, \cdots, m$. Then, each progressively Type-II right censored order statistic corresponds to some usual order statistic from the original n items on test, i.e., $Z_{j:m:n} = Z_{K_j:n}$ where the rank of $Z_{j:m:n}$, K_j, can take on the values $K_{j-1}+1, K_{j-1}+2, \cdots, j + R_1 + R_2 + \cdots + R_{j-1}$ for $j = 2, 3, \cdots, m$ and $K_1 = 1$. Thomas and Wilson (1972) have shown that the joint probability mass function of the rank vector can be written as

$$P(K_1, \cdots, K_m) = P(K_1) \prod_{i=2}^{m} P(K_i \mid K_1, \cdots, K_{i-1}) \text{ with } P(K_1 = 1) = 1,$$

where

$$P(K_i \mid K_1, \cdots, K_{i-1}) = \frac{\binom{n-K_i}{\sum_{j=1}^{i-1} R_j + i - K_i}}{\binom{n-K_{i-1}}{\sum_{j=1}^{i-1} R_j + i - K_{i-1} - 1}}, \quad i = 2, \cdots, m.$$

Thus, if all possible rank vectors (say there are M of them) can be listed for a particular censoring scheme $(R_1, R_2, \cdots, R_m)$, then for each rank vector, we can define an $m \times n$ indicator matrix $\boldsymbol{D}_l, l = 1, 2, \cdots, M$, whose (r, s)-th element is 1 if $s = K_r$ and 0 otherwise, so that $\boldsymbol{Z}_{ps} = \boldsymbol{D}_l \boldsymbol{Z}_u$ for some l, where $\boldsymbol{Z}_{ps}$ is the $m \times 1$ vector of progressively Type-II right censored order statistics and $\boldsymbol{Z}_u$ is the $n \times 1$ vector of usual order statistics from the particular distribution of interest. Then, denoting $E(\boldsymbol{Z}_u)$ by $\boldsymbol{\mu}_u$, we can obtain the means of the progressively Type-II right censored order statistics as

$$E\left(\boldsymbol{Z}_{ps}\right)=\boldsymbol{\mu}=EE\left(\boldsymbol{Z}_{ps}\mid\boldsymbol{D}_l\right)=E\left(\boldsymbol{D}_l\boldsymbol{\mu}_u\right)=\left(\sum_{l=1}^{M}\boldsymbol{D}_l p_l\right)\boldsymbol{\mu}_u, \tag{5.1}$$

where p_l is the probability mass function of the rank vector corresponding to $\boldsymbol{D}_l$, $l=1,2,\cdots,M$. Similarly, denoting the variance-covariance matrix of $\boldsymbol{Z}_u$ by $\boldsymbol{\Sigma}_u$, we can obtain the variance-covariance matrix of the progressively Type-II right censored order statistics as

$$\begin{aligned}
Var\left(\boldsymbol{Z}_{ps}\right) &= \boldsymbol{\Sigma} \\
&= E\left(\boldsymbol{Z}_{ps}\boldsymbol{Z}_{ps}'\right)-\boldsymbol{\mu}\boldsymbol{\mu}' \\
&= EE\left(\boldsymbol{D}_l\boldsymbol{Z}_u\boldsymbol{Z}_u'\boldsymbol{D}_l'\mid\boldsymbol{D}_l\right)-\boldsymbol{\mu}\boldsymbol{\mu}' \\
&= E\left[\boldsymbol{D}_l\left(\boldsymbol{\Sigma}_u+\boldsymbol{\mu}_u\boldsymbol{\mu}_u'\right)\boldsymbol{D}_l'\right]-\boldsymbol{\mu}\boldsymbol{\mu}' \\
&= \sum_{i=1}^{M}\boldsymbol{D}_l\left(\boldsymbol{\Sigma}_u+\boldsymbol{\mu}_u\boldsymbol{\mu}_u'\right)\boldsymbol{D}_l'p_l-\boldsymbol{\mu}\boldsymbol{\mu}'. \qquad (5.2)
\end{aligned}$$

Tables of means, variances and covariances of the usual order statistics are available for various sample sizes for numerous distributions including the standard normal, log-normal and extreme value distributions; see, for example, Harter and Balakrishnan (1996). Hence, using the formulas in (5.1) and (5.2), we will be able to compute the means, variances and covariances of progressively Type-II right censored order statistics from all those distributions.

5.3 Formulas in the Case of Symmetric Distributions

In the usual order statistics literature, a large number of results are available for the special case when the random sample is from a symmetric distribution. See, for example, Arnold, Balakrishnan and Nagaraja (1992, pp. 26, 119, 123-126, 128, 136-137, 173-174, 187), and Balakrishnan and Sultan (1998). In this section, we establish results which will enable one to compute the moments of progressively Type-II right censored order statistics from an arbitrary symmetric distribution when the moments of progressively Type-II right censored order statistics and progressively Type-II left withdrawn order statistics from the corresponding folded distribution are available. The idea of progressive Type-II left withdrawal scheme is introduced in order to arrive at this result.

Let us assume the underlying distribution is symmetric about 0 (without loss of any generality), has probability density function $f(x)$ $(-\infty<x<\infty)$, and has cumulative distribution function $F(x)$ $(-\infty<x<\infty)$. Next, let us denote the probability density function for the corresponding folded distribution by $g(x)=2f(x)$, $0<x<\infty$, and the cumulative distribution function by $G(x)=2F(x)-1$, $0<x<\infty$.

The main results established by Aggarwala and Balakrishnan (1998b) and presented here generalize two classical results for the usual order statistics due to Govindarajulu (1963) which state that the (single or product) moments of order statistics from a symmetric distribution can be obtained from the (single or product) moments of order statistics from the corresponding folded distribution.

5.3.1 Progressive Withdrawal

It is a well-known property of the usual order statistics, $X_{i:n}(i = 1, 2, \cdots, n)$, from a symmetric distribution (symmetric about 0), say $F(x)$, that

$$X_{i:n} \stackrel{d}{=} -X_{n-i+1:n}, \quad i = 1, 2, \cdots, n$$

or that jointly,

$$(X_{1:n}, X_{2:n}, \cdots, X_{n:n}) \stackrel{d}{=} (-X_{n:n}, -X_{n-1:n}, \cdots - X_{1:n});$$

for example, see David (1981) and Arnold, Balakrishnan and Nagaraja (1992). Thus, the negatives of the usual order statistics are once again distributed as the usual order statistics from the same symmetric distribution. It will therefore be natural to see whether a similar result holds for progressively Type-II right censored order statistics from a symmetric distribution as it will facilitate the handling of these random variables.

In the case of the usual order statistics, the result just stated is easily observed by using the facts that $f(x) = f(-x)$, and $F(x) = 1 - F(-x)$ and simply rewriting the marginal density of the i-th order statistic as

$$\begin{aligned} & \frac{n!}{(i-1)!\,(n-i)!} f(x)\,[F(x)]^{i-1}\,[1-F(x)]^{n-i} \\ = & \frac{n!}{\{(n-i+1)-1\}!\,\{n-(n-i+1)\}!} f(-x)\,[F(-x)]^{n-i}\,[1-F(-x)]^{i-1}. \end{aligned}$$

However, in working with the progressively Type-II right censored order statistics, since marginal density functions are not available explicity, we begin with the joint distribution of all m progressively Type-II right censored order statistics.

As before, let us denote the sample of progressively Type-II right censored order statistics of size m obtained from a random sample of size n with progressive censoring scheme $(R_1, R_2, \cdots, R_m)$ from the symmetric distribution by $X_{1:m:n}^{(R_1,R_2,\cdots,R_m)}, X_{2:m:n}^{(R_1,R_2,\cdots,R_m)}, \cdots, X_{m:m:n}^{(R_1,R_2,\cdots,R_m)}$. Then, we consider the following transformation:

$$Z_1 = -X_{m:m:n}^{(R_1,R_2,\cdots,R_m)}, Z_2 = -X_{m-1:m:n}^{(R_1,R_2,\cdots,R_m)}, \cdots, Z_m = -X_{1:m:n}^{(R_1,R_2,\cdots,R_m)}.$$

The Jacobian of this transformation is 1, and the joint density function of $(Z_1, Z_2, \cdots, Z_m)$ is then given by

$$\begin{aligned} h_{1,2,\cdots,m}(z_1, z_2, \cdots z_m) &= c\,[F(z_1)]^{R_1} f(z_1)\,[F(z_2)]^{R_2} f(z_2) \\ &\times \cdots [F(z_m)]^{R_m} f(z_m), \\ & \qquad -\infty < z_1 < z_2 < \cdots < z_m < \infty, \end{aligned} \tag{5.3}$$

where

$$c = n(n - R_1 - 1)(n - R_1 - R_2 - 2) \cdots (n - R_1 - \cdots - R_{m-1} - m + 1). \tag{5.4}$$

As a result, the negatives of the progressively Type-II right censored order statistics are *not* jointly distributed as progressively Type-II right censored order statistics. If we interpret (5.4) in terms of live items on test, we would be withdrawing, at the failure times $z_1, z_2, \cdots, z_m$, items which have failed already (whose exact failure times are unknown). Though this may seem to be an impractical idea, mathematical consideration of this set of random variables will give rise to useful results for progressively Type-II right censored order statistics as shall be seen shortly.

To formalize the above discussion, we introduce the following notation: if m random variables have a joint density given by (5.4), we call them a *progressively Type-II left withdrawn sample* of size m from a random sample of size n with withdrawal scheme $(R_m, R_{m-1}, \cdots, R_1)$ from the distribution with probability density function $f(x)$, and denote the corresponding progressively Type-II left withdrawn order statistics by

$${}^{(R_m,\cdots,R_1)}Z_{1:m:n}, {}^{(R_m,\cdots,R_1)}Z_{2:m:n}, \cdots, {}^{(R_m,\cdots,R_1)}Z_{m:m:n}.$$

Notice that this definition applies to any *arbitrary* continuous distribution, and the constant c is exactly the same as the one in the joint density function of the progressively Type-II right censored order statistics. It is clear that $n = m + R_1 + R_2 + \cdots + R_m$, and that if $R_1 = R_2 = \cdots = R_m = 0$ so that $m = n$, there is no withdrawal in which case the usual order statistics are obtained. We also readily note that the largest progressively Type-II left withdrawn order statistic $\left({}^{(R_m,\cdots,R_1)}Z_{m:m:n}\right)$ is distributed as the largest usual order statistic from a sample of size n $(Z_{n:n})$, since we know that $R_1 + R_2 + \cdots + R_m + m - 1 = n - 1$ items have already failed by this time. Also, if $R_1 = R_2 = \cdots = R_{m-1} = 0$ so that $R_m = n - m$, we are left with the case of conventional Type-II left censoring (hence, the name *progressive Type-II left withdrawal*) wherein the largest m usual order statistics are observed.

5.3.2 Properties of Progressively Type-II Left Withdrawn Order Statistics

The following properties, which parallel those given in Theorems 2.3, 2.4 and 2.5 and established by Aggarwala and Balakrishnan (1998a), hold for the progressively Type-II left withdrawn order statistics from any continuous distribution. For convenience, we will drop the preceding superscript $(R_m, R_{m-1}, \cdots, R_1)$ from the notation of progressively Type-II left withdrawn order statistics here; let us also denote the term $n - R_1 - \cdots - R_{i-1} - i$ by R_i'.

Theorem 5.1
(i) The marginal distribution of $Z_{m-i+1:m:n}$, $1 \leq i \leq m$, is free of $R_i, R_{i+1}, \cdots, R_m$.
(ii) $Z_{m-i+1:m:n}, Z_{m-i+2:m:n}, \cdots, Z_{m:m:n}$ form a progressively Type-II left withdrawn sample of size i from n items placed on test with the withdrawal scheme $\left(R_i', R_{i-1}, \cdots, R_1\right)$, viz., $(n - R_1 - \cdots - R_{i-1} - i, R_{i-1}, \cdots, R_1)$.

Theorem 5.2 *The progressively Type-II left withdrawn order statistics from an arbitrary continuous distribution form a "reverse" Markov chain; that is, given $Z_{i:m:n}$, $Z_{j:m:n}$ $(j < i)$ is independent of $Z_{i+1:m:n}, Z_{i+2:m:n}, \cdots, Z_{m:m:n}$.*

Theorem 5.3 *Given $Z_{m-i+1:m:n} = z_{m-i+1}$, $Z_{1:m:n}, \cdots, Z_{m-i:m:n}$ $(1 \leq i \leq m-1)$ are jointly distributed as a progressively Type-II left withdrawn sample of size $m-i$ from $n - R_1 - R_2 - \cdots - R_i - i$ identically distributed random variables from the density $f(x)$ right-truncated at z_{m-i+1}, that is, with density function $f(z)/F(z_{m-i+1})$ $(z < z_{m-i+1})$, and with progressive withdrawal scheme $(R_m, \cdots, R_{i+1})$.*

Remark 5.1 Again, notice that the converse of Theorem 5.3 will not be true in general, since, even after the observation of the i-th progressively Type-II left withdrawn order statistic, say z_i, we may withdraw items which failed before the time z_i. This is because we only know that the withdrawn items failed at some point since time 0.

Remark 5.2 The proofs of Theorems 5.1, 5.2 and 5.3 parallel those of Theorems 2.3, 2.4 and 2.5 and hence are omitted here.

5.3.3 Moments of Progressively Type-II Right Censored Order Statistics from Symmetric Distributions

Govindarajulu (1963) established the following result for usual order statistics.

Theorem 5.4 [Govindarajulu (1963)] *Let us denote the order statistics from a random sample of size n from a population with cumulative distribution function $F(x)$ and probability density function $f(x)$ symmetric about*

0 *(without loss of generality) by* $X_{1:n}, X_{2:n}, \cdots, X_{n:n}$. *Further, let us denote the order statistics from a random sample of size n from the corresponding folded population (folded at zero) with cumulative distribution function* $G(x) = 2F(x) - 1$ $(x \geq 0)$ *and probability density function* $g(x) = 2f(x)$ $(x \geq 0)$ *by* $Y_{1:n}, Y_{2:n}, \cdots, Y_{n:n}$. *Then, denoting* $E\left(X_{i:n}^k\right)$ *by* $\mu_{i:n}^{(k)}$ *and* $E\left(Y_{i:n}^k\right)$ *by* $\alpha_{i:n}^{(k)}$ *for* $i = 1, 2, \cdots, n$ *and* $k = 1, 2, \cdots,$

$$2^n \mu_{i:n}^{(k)} = \sum_{r=0}^{i-1} \binom{n}{r} \alpha_{i-r:n-r}^{(k)} + (-1)^k \sum_{r=i}^{n} \binom{n}{r} \alpha_{r-i+1:r}^{(k)}. \tag{5.5}$$

This result is established by integrating the marginal probability density function of the i-th order statistic multiplied by x^k over the entire real line, splitting the integral into its positive and negative parts, using the relationships

$$f(x) = f(-x), F(x) = 1 - F(-x), g(x) = 2f(x), G(x) = 2F(x) - 1, \tag{5.6}$$

and then expanding certain terms binomially. A simple and interesting probabilistic proof of this result has also been given by Balakrishnan, Govindarajulu and Balasubramanian (1993), while its generalization to the case when the variables are non-identically distributed has been given by Balakrishnan (1989).

We will generalize this result for the case of progressive Type-II right censoring. As usual, let us denote the sample of progressively Type-II right censored order statistics of size m with censoring scheme $(R_1, R_2, \cdots, R_m)$ from a sample of size n from a symmetric distribution (symmetric about zero) by

$$X_{1:m:n}^{(R_1,R_2,\cdots,R_m)}, X_{2:m:n}^{(R_1,R_2,\cdots,R_m)}, \cdots, X_{m:m:n}^{(R_1,R_2,\cdots,R_m)}.$$

Also, let us denote the sample of progressively Type-II right censored order statistics of size m with censoring scheme $(R_1, R_2, \cdots, R_m)$ from a sample of size n from the distribution folded at zero by

$$Y_{1:m:n}^{(R_1,R_2,\cdots,R_m)}, Y_{2:m:n}^{(R_1,R_2,\cdots,R_m)}, \cdots, Y_{m:m:n}^{(R_1,R_2,\cdots,R_m)},$$

and the sample of progressively Type-II left withdrawn order statistics of size m with withdrawal scheme $(R_m, R_{m-1}, \cdots, R_1)$ from a sample of size n from this folded distribution by

$$^{(R_m,\cdots,R_1)}Y_{1:m:n}, {}^{(R_m,\cdots,R_1)}Y_{2:m:n}, \cdots, {}^{(R_m,\cdots,R_1)}Y_{m:m:n}.$$

Further, let us denote $E\left[\left\{X_{i:m:n}^{(R_1,\cdots,R_m)}\right\}^k\right]$ by $\mu_{i:m:n}^{(R_1,\cdots,R_m)^{(k)}}$, $E\left[\left\{Y_{i:m:n}^{(R_1,\cdots,R_m)}\right\}^k\right]$ by $\alpha_{i:m:n}^{(R_1,\cdots,R_m)^{(k)}}$, and $E\left[\left\{{}^{(R_m,\cdots,R_1)}Y_{i:m:n}\right\}^k\right]$ by ${}^{(R_m,\cdots,R_1)}\alpha_{i:m:n}^{(k)}$.

Finally, the constant c in (5.4) can be rewritten as

$$c = (R_m + 1)(R_m + R_{m-1} + 2) \cdots (R_m + R_{m-1} + \cdots + R_1 + m).$$

We will denote this quantity by $c_{R_m,R_{m-1},\cdots,R_1}$; similarly, we will denote

$$c_{R_m,R_{m-1},\cdots,R_i} = (R_m+1)\,(R_m+R_{m-1}+2)\cdots(R_m+\cdots+R_i+m-i+1)\,.$$

The following result, due to Aggarwala and Balakrishnan (1998b), can now be established.

Theorem 5.5 *For $i=1,2,\cdots,m\le n$, and $k\ge 0$,*

$$\begin{aligned}
&2^n\mu_{i:m:n}^{(R_1,\cdots,R_m)^{(k)}}\\
&= c_{R_i',R_{i-1},\cdots,R_1}\left\{\sum_{j=0}^{i-1}\left\{\left[\sum_{l_j=0}^{R_j}\sum_{l_{j-1}=0}^{R_{j-1}}\cdots\sum_{l_1=0}^{R_1}\binom{R_j}{l_j}\binom{R_{j-1}}{l_{j-1}}\cdots\binom{R_1}{l_1}\right.\right.\right.\\
&\times\left.\left.\frac{1}{c_{l_j,l_{j-1},\cdots,l_1}}\right]\frac{\alpha_{i-j:i-j:n-R_1-\cdots-R_j-j}^{\left(R_{j+1},\cdots,R_{i-1},R_i'\right)^{(k)}}}{c_{R_i',R_{i-1},\cdots,R_{j+1}}}\right\}\\
&+\left.(-1)^k\sum_{l_i=0}^{R_i'}\sum_{l_{i-1}=0}^{R_{i-1}}\cdots\sum_{l_1=0}^{R_1}\binom{R_i'}{l_i}\binom{R_{i-1}}{l_{i-1}}\cdots\binom{R_1}{l_1}\frac{{}^{(l_i,l_{i-1},\cdots,l_1)}\alpha_{1:i:l_1+\cdots+l_i+i}^{(k)}}{c_{l_i,l_{i-1},\cdots,l_1}}\right\},
\end{aligned}\tag{5.7}$$

where $R_i' = n-R_1-\cdots-R_{i-1}-i$.

Proof Using Theorem 2.3, we have $\mu_{i:m:n}^{(R_1,\cdots,R_m)^{(k)}} = \mu_{i:i:n}^{\left(R_1,\cdots,R_{i-1},R_i'\right)^{(k)}}$. Thus, we begin with

$$\begin{aligned}
\mu_{i:m:n}^{(R_1,\cdots,R_m)^{(k)}} &= c_{R_i',R_{i-1},\cdots,R_1}\int\!\!\int_{-\infty<x_1<\cdots<x_i<\infty}\!\!\cdots\int x_i^k f(x_1)\,[1-F(x_1)]^{R_1}\cdots\\
&\times\ f(x_{i-1})\,[1-F(x_{i-1})]^{R_{i-1}}\,f(x_i)\,[1-F(x_i)]^{R_i'}\\
&\quad dx_i\,dx_{i-1}\ \cdots\ dx_1.
\end{aligned}\tag{5.8}$$

Now, the area of integration $-\infty<x_1<\cdots<x_i<\infty$, can be divided into $i+1$ mutually exclusive and collectively exhaustive regions as follows:

$$\begin{aligned}
\{-\infty<x_1<\cdots<x_i<\infty\} &= \{0<x_1<\cdots<x_i\}\\
&\cup\ \{(x_1<0)\cap(0<x_2<\cdots<x_i)\}\\
&\cup\ \{(x_1<x_2<0)\cap(0<x_3<\cdots<x_i)\}\\
&\cdots\\
&\cdots\\
&\cup\ \{(x_1<\cdots<x_{i-1}<0)\cap(0<x_i)\}\\
&\cup\ \{x_1<\cdots<x_i<0\}\,.
\end{aligned}$$

Then, the right hand side of (5.8) can be written as, using (5.6),

$$
\begin{aligned}
& c_{R_i',R_{i-1},\cdots,R_1}\Bigg[\Bigg\{\underset{0<x_1<\cdots<x_i<\infty}{\int\int\cdots\int} x_i^k\,\frac{g(x_1)}{2}\left[\frac{1-G(x_1)}{2}\right]^{R_1}\cdots\frac{g(x_{i-1})}{2} \\
& \times\left[\frac{1-G(x_{i-1})}{2}\right]^{R_{i-1}}\frac{g(x_i)}{2}\left[\frac{1-G(x_i)}{2}\right]^{R_i'} dx_1\cdots dx_i\Bigg\} \\
& +\Bigg\{\int_{-\infty}^{0} f(x_1)\left[1-F(x_1)\right]^{R_1}dx_1 \underset{0<x_2<\cdots<x_i<\infty}{\int\int\cdots\int} x_i^k\frac{g(x_2)}{2}\left[\frac{1-G(x_2)}{2}\right]^{R_2} \\
& \times\cdots\frac{g(x_{i-1})}{2}\left[\frac{1-G(x_{i-1})}{2}\right]^{R_{i-1}}\frac{g(x_i)}{2}\left[\frac{1-G(x_i)}{2}\right]^{R_i'} dx_2\cdots dx_i\Bigg\} \\
& +\Bigg\{\underset{-\infty<x_1<x_2<0}{\int\int} f(x_1)\left[1-F(x_1)\right]^{R_1} f(x_2)\left[1-F(x_2)\right]^{R_2}dx_1dx_2 \\
& \times\underset{0<x_3<\cdots<x_i<\infty}{\int\int\cdots\int} x_i^k\frac{g(x_3)}{2}\left[\frac{1-G(x_3)}{2}\right]^{R_3}\cdots\frac{g(x_{i-1})}{2}\left[\frac{1-G(x_{i-1})}{2}\right]^{R_{i-1}} \\
& \times\frac{g(x_i)}{2}\left[\frac{1-G(x_i)}{2}\right]^{R_i'} dx_3\cdots dx_i\Bigg\} \\
& +\cdots\cdots \\
& +\Bigg\{\underset{-\infty<x_1<\cdots<x_{i-1}<0}{\int\int\cdots\int} f(x_1)\left[1-F(x_1)\right]^{R_1}\cdots f(x_{i-1}) \\
& \times\left[1-F(x_{i-1})\right]^{R_{i-1}}dx_1\cdots dx_{i-1}\int_0^\infty x_i^k\frac{g(x_i)}{2}\left[\frac{1-G(x_i)}{2}\right]^{R_i'}dx_i\Bigg\} \\
& +\Bigg\{\underset{-\infty<x_1<\cdots<x_i<0}{\int\int\cdots\int} x_i^k f(x_1)\left[1-F(x_1)\right]^{R_1}\cdots f(x_{i-1}) \\
& \times\left[1-F(x_{i-1})\right]^{R_{i-1}} f(x_i)\left[1-F(x_i)\right]^{R_i'}dx_1\cdots dx_i\Bigg\}\Bigg]. \qquad (5.9)
\end{aligned}
$$

Now, upon making the substitution $y_p = -x_1, y_{p-1} = -x_2, \cdots, y_1 = -x_p$ in the integrals containing the probability density function $f(\cdot)$, where p is the dimension of the integral, (5.9) becomes

$$
c_{R_i',R_{i-1},\cdots,R_1}\Bigg[\Bigg\{\frac{1}{2^n}\underset{0<x_1<\cdots<x_i<\infty}{\int\int\cdots\int} x_i^k g(x_1)\left[1-G(x_1)\right]^{R_1}\cdots g(x_{i-1})
$$

$$\times \left[1-G\left(x_{i-1}\right)\right]^{R_{i-1}} g\left(x_{i}\right)\left[1-G\left(x_{i}\right)\right]^{R_{i}^{\prime}} dx_{1} \cdots dx_{i}\Bigg\}$$

$$+\Bigg\{\frac{1}{2^{n-R_1-1}}\int_0^\infty f\left(y_1\right)\left[F\left(y_1\right)\right]^{R_1} dy_1 \underset{0<x_2<\cdots<x_i<\infty}{\int\int\cdots\int} x_i^k g\left(x_2\right)\left[1-G\left(x_2\right)\right]^{R_2}$$

$$\times \cdots g\left(x_{i-1}\right)\left[1-G\left(x_{i-1}\right)\right]^{R_{i-1}} g\left(x_{i}\right)\left[1-G\left(x_{i}\right)\right]^{R_{i}^{\prime}} dx_{2} \cdots dx_{i}\Bigg\}$$

$$+\Bigg\{\frac{1}{2^{n-R_1-R_2-2}} \underset{0<y_1<y_2<\infty}{\int\int} f\left(y_1\right)\left[F\left(y_1\right)\right]^{R_2} f\left(y_2\right)\left[F\left(y_2\right)\right]^{R_1} dy_1 dy_2$$

$$\times \underset{0<x_3<\cdots<x_i<\infty}{\int\int\cdots\int} x_i^k g\left(x_3\right)\left[1-G\left(x_3\right)\right]^{R_3} \cdots g\left(x_{i-1}\right)$$

$$\times \left[1-G\left(x_{i-1}\right)\right]^{R_{i-1}} g\left(x_{i}\right)\left[1-G\left(x_{i}\right)\right]^{R_{i}^{\prime}} dx_{3} \cdots dx_{i}\Bigg\}$$

$$+\cdots\cdots$$

$$+\Bigg\{\frac{1}{2^{n-R_1-\cdots-R_{i-1}-i+1}} \underset{0<y_1<\cdots<y_{i-1}<\infty}{\int\int\cdots\int} f\left(y_1\right)\left[F\left(y_1\right)\right]^{R_{i-1}} \cdots f\left(y_{i-1}\right)$$

$$\times \left[F\left(y_{i-1}\right)\right]^{R_1} dy_1 \cdots dy_{i-1} \int_0^\infty x_i^k g\left(x_i\right)\left[1-G\left(x_i\right)\right]^{R_i^{\prime}} dx_i\Bigg\}$$

$$+\Bigg\{ \underset{0<y_1<\cdots<y_i<\infty}{\int\int\cdots\int} (-1)^k y_1^k f\left(y_1\right)\left[F\left(y_1\right)\right]^{R_i^{\prime}} f\left(y_2\right)\left[F\left(y_2\right)\right]^{R_{i-1}} \cdots$$

$$\times f\left(y_i\right)\left[F\left(y_i\right)\right]^{R_1} dy_1 \cdots dy_i\Bigg\}\Bigg].$$

In the above expression, for all terms containing the cumulative distribution function $F(\cdot)$, we may write

$$\left[F\left(y_s\right)\right]^{R_t} = \left[\frac{1+G\left(y_s\right)}{2}\right]^{R_t} = \frac{1}{2^{R_t}} \sum_{l_t=0}^{R_t} \binom{R_t}{l_t} \left[G\left(y_s\right)\right]^{l_t}.$$

Upon making this substitution, interchanging the order of summations and integrations, and simplifying the resulting equation, the relation in (5.7) is obtained. □

Remark 5.3 Notice that, upon setting $R_1 = R_2 = \cdots = R_m = 0$, (5.7) simplifies to (5.5), since $\mu_{i:n}^{(k)} = \mu_{i:n:n}^{(0,0,\cdots,0)^{(k)}} = \mu_{i:i:n}^{(0,0,\cdots,0,n-i)^{(k)}}$.

Similar arguments are used to generalize Govindarajulu's (1963) result for the product moments of usual order statistics from symmetric distributions given by (for $1 \le i < j \le n$)

$$2^n \mu_{i,j:n} = \sum_{l=0}^{i-1} \binom{n}{l} \alpha_{i-l,j-l:n-l} - \sum_{l=i}^{j-1} \binom{n}{l} \alpha_{l-i+1:l} \alpha_{j-l:n-l} + \sum_{l=j}^{n} \binom{n}{l} \alpha_{l-j+1,l-i+1:l}, \tag{5.10}$$

where $\mu_{i,j:n} = E(X_{i:n}X_{j:n})$ is the (i,j)-th product moment of the order statistics from the symmetric distribution, and $\alpha_{i,j:n} = E(Y_{i:n}Y_{j:n})$ is the (i,j)-th product moment of the order statistics from the corresponding folded distribution.

Thus, for the progressive Type-II censored and withdrawn order statistics described earlier in this section, let us denote $E\left[X_{i:m:n}^{(R_1,\cdots,R_m)} X_{j:m:n}^{(R_1,\cdots,R_m)}\right]$ by $\mu_{i,j:m:n}^{(R_1,\cdots,R_m)}$, $E\left[Y_{i:m:n}^{(R_1,\cdots,R_m)} Y_{j:m:n}^{(R_1,\cdots,R_m)}\right]$ by $\alpha_{i,j:m:n}^{(R_1,\cdots,R_m)}$, and $E\left[{}^{(R_m,\cdots,R_1)}Y_{i:m:n} \times {}^{(R_m,\cdots,R_1)}Y_{j:m:n}\right]$ by ${}^{(R_m,\cdots,R_1)}\alpha_{i,j:m:n}$.

The result which generalizes the relation in (5.10) for the product moments of progressively Type-II right censored order statistics from symmetric distributions, also due to Aggarwala and Balakrishnan (1998b), is as follows.

Theorem 5.6 *For* $1 \le i < j \le m \le n$,

$$\begin{aligned}
&2^n \mu_{i,j:m:n}^{(R_1,\cdots,R_m)} \\
&= c_{R_j',R_{j-1},\cdots,R_1} \left\{ \sum_{l=0}^{i-1} \left\{ \left[\sum_{k_l=0}^{R_l} \sum_{k_{l-1}=0}^{R_{l-1}} \cdots \sum_{k_1=0}^{R_1} \binom{R_l}{k_l} \binom{R_{l-1}}{k_{l-1}} \cdots \binom{R_1}{k_1} \right. \right. \right. \\
&\times \left. \frac{1}{c_{k_l,k_{l-1},\cdots,k_1}} \right] \left. \frac{\alpha_{i-l,j-l:j-l:n-R_1-\cdots-R_l-l}^{\left(R_{l+1},\cdots R_{j-1},R_j'\right)}}{c_{R_j',R_{j-1},\cdots,R_{\ell+1}}} \right\} \\
&- \sum_{l=i}^{j-1} \left\{ \left[\sum_{k_l=0}^{R_l} \sum_{k_{l-1}=0}^{R_{l-1}} \cdots \sum_{k_1=0}^{R_1} \binom{R_l}{k_l} \binom{R_{l-1}}{k_{l-1}} \cdots \binom{R_1}{k_1} \right. \right. \\
&\times \left. \frac{{}^{(k_l,k_{l-1},\cdots,k_1)}\alpha_{l-i+1:l:k_1+\cdots+k_l+l}}{c_{k_l,k_{l-1},\cdots,k_1}} \right] \left. \frac{\alpha_{j-l:j-l:n-R_1-\cdots-R_l-l}^{\left(R_{l+1},\cdots R_{j-1},R_j'\right)}}{c_{R_j',R_{j-1},\cdots,R_{l+1}}} \right\} \\
&+ \left. \sum_{k_j=0}^{R_j'} \sum_{k_{j-1}=0}^{R_{j-1}} \cdot\cdot \sum_{k_1=0}^{R_1} \binom{R_j'}{k_j} \binom{R_{j-1}}{k_{j-1}} \cdot\cdot \binom{R_1}{k_1} \frac{{}^{(k_j,\cdots,k_1)}\alpha_{1,j-i+1:j:k_1+\cdots+k_j+j}}{c_{k_j,\cdots,k_1}} \right\}
\end{aligned} \tag{5.11}$$

Remark 5.4 The proof of this theorem is quite similar to that of Theorem 5.5; in this case, using Theorem 2.3, we have $\mu_{i,j:m:n}^{(R_1,\cdots,R_m)} = \mu_{i,j:j:n}^{\left(R_1,\cdots R_{j-1},R_j'\right)}$ and the area of integration, $-\infty < x_1 < \cdots < x_j < \infty$, is divided into $j+1$ mutually exclusive and collectively exhaustive regions.

Remark 5.5 Theorems 5.5 and 5.6 can and will be used in Chapter 6 to obtain single and product moments of progressively Type-II right censored order statistics from the Laplace distribution.

5.4 Other Relations for Moments

In this section, we give additional relations for *arbitrary* continuous distribution which will enable us to obtain the moments of progressively Type-II left withdrawn order statistics if the moments of the progressively Type-II right censored order statistics are known, and vice versa. These results were established by Aggarwala and Balakrishnan (1998b).

Theorem 5.7 *For* $1 \le i \le m \le n$,

$$\begin{aligned} \mu_{i:m:n}^{(R_1,\cdots,R_m)^{(k)}} &= c_{R_i',R_{i-1},\cdots,R_1} \sum_{j_i=0}^{R_i'} \sum_{j_{i-1}=0}^{R_{i-1}} \cdots \sum_{j_1=0}^{R_1} \binom{R_i'}{j_i} \binom{R_{i-1}}{j_{i-1}} \cdots \binom{R_1}{j_1} \\ &\times (-1)^{j_1+\cdots+j_i} \left[\frac{\mu_{j_1+\cdots+j_i+i:j_1+\cdots+j_i+i}^{(k)}}{c_{j_1,\cdots,j_i}} \right]. \end{aligned} \tag{5.12}$$

Proof Using the fact that $\mu_{i:m:n}^{(R_1,\cdots,R_m)} = \mu_{i:i:n}^{\left(R_1,\cdots,R_{i-1},R_i'\right)}$ and expanding the terms containing $1 - F(\cdot)$ in (5.8) binomially, interchanging the order of integrations and summations and simplifying, we have

$$\begin{aligned} \mu_{i:m:n}^{(R_1,\cdots,R_m)^{(k)}} &= \mu_{i:i:n}^{\left(R_1,\cdots,R_{i-1},R_i'\right)} \\ &= c_{R_i',R_{i-1},\cdots,R_1} \sum_{j_i=0}^{R_i'} \sum_{j_{i-1}=0}^{R_{i-1}} \cdots \sum_{j_1=0}^{R_1} \binom{R_i'}{j_i} \binom{R_{i-1}}{j_{i-1}} \cdots \binom{R_1}{j_1} \\ &\times (-1)^{j_1+\cdots+j_i} \left[\frac{{}^{(j_1,\cdots,j_i)}\mu_{i:i:j_1+\cdots+j_i+i}^{(k)}}{c_{j_1,\cdots,j_i}} \right]. \end{aligned}$$

Since the largest progressively Type-II left withdrawn order statistic from a sample of size n is also the largest usual order statistic from a sample of size n, the relation in (5.12) then readily follows. □

Remark 5.6 Theorem 5.7 provides another method for obtaining single moments of progressively Type-II right censored order statistics based on the knowledge of single moments of the largest usual order statistics for varying sample sizes.

Remark 5.7 In the special case when there is no censoring, this result reduces to a relation given by Srikantan (1962) for the usual order statistics which is as follows:

$$\mu_{i:n}^{(k)} = \sum_{j=i}^{n} \binom{n}{j}\binom{j-1}{i-1}(-1)^{j-i}\mu_{j:j}^{(k)};$$

see also David (1981) and Arnold, Balakrishnan and Nagaraja (1992).

Proceeding in a manner similar to that of Theorem 5.7, we obtain the following result for the product moments of progressively Type-II right censored order statistics.

Theorem 5.8 *For* $1 \le i < j \le m \le n$,

$$\begin{aligned}\mu_{i,j:m:n}^{(R_1,\cdots,R_m)} &= c_{R_j^{'},R_{j-1},\cdots,R_1} \sum_{l_j=0}^{R_j^{'}} \sum_{l_{j-1}=0}^{R_{j-1}} \cdots \sum_{l_1=0}^{R_1} \binom{R_j^{'}}{l_j}\binom{R_{j-1}}{l_{j-1}} \cdots \binom{R_1}{l_1} \\ &\times \quad (-1)^{l_1+\cdots+l_j}\left[\frac{{}^{(l_1,\cdots,l_j)}\mu_{i,j:j:l_1+\cdots+l_j+j}}{c_{l_1,\cdots,l_j}}\right]. \end{aligned} \tag{5.13}$$

Thus, if we know the product moments of progressively Type-II left withdrawn order statistics from an arbitrary continuous distribution, we can determine all the single and product moments of the progressively Type-II right censored order statistics.

Remark 5.8 For the special case when there is no censoring, (5.13) reduces to the following identity for usual order statistics:

$$\mu_{i,j:n} = \sum_{l=0}^{n-j}\sum_{k=0}^{l}(-1)^{l+k}\frac{n!}{(n-j-l)!\,(l-k)!\,(k+j)!}\mu_{i,j:k+j}.$$

This identity for usual order statistics is well-known; see, for example, Arnold and Balakrishnan (1989), and Balakrishnan and Sultan (1998).

The following two theorems give the converse results of Theorems 5.7 and 5.8.

Theorem 5.9 *For* $1 \le i \le m \le n$,

$$\begin{aligned}{}^{(R_m,R_{m-1},\cdots,R_1)}\mu_{i:m:n}^{(k)} &= c_{R_{m-i+1}^{'},R_{m-i},\cdots,R_1} \sum_{j_{m-i+1}=0}^{R_{m-i+1}^{'}} \sum_{j_{m-i}=0}^{R_{m-i}} \cdots \sum_{j_1=0}^{R_1} \\ &\times \quad (-1)^{j_1+\cdots+j_{m-i+1}}\binom{R_{m-i+1}^{'}}{j_{m-i+1}}\binom{R_{m-i}}{j_{m-i}} \cdots \binom{R_1}{j_1} \\ &\times \quad \left[\frac{\mu_{1:j_1+\cdots+j_{m-i+1}+m-i+1}^{(k)}}{c_{j_1,\cdots,j_{m-i+1}}}\right]. \end{aligned} \tag{5.14}$$

Proof From Theorem 5.1, we have

$$^{(R_m,\cdots,R_1)}\mu_{i:m:n}^{(k)} = ^{\left(R'_{m-i+1},R_{m-i},\cdots,R_1\right)}\mu_{1:m-i+1:n}^{(k)}.$$

Thus, we begin with

$$\begin{aligned} ^{(R_m,\cdots,R_1)}\mu_{i:m:n}^{(k)} &= c_{R'_{m-i+1},R_{m-i},\cdots,R_1} \int\int_{-\infty<x_1<\cdots<x_{m-i+1}<\infty}\cdots\int x_1^k f(x_1) \\ &\times \ [F(x_1)]^{R'_{m-i+1}} f(x_2)\,[F(x_2)]^{R_{m-i}} \cdots f(x_{m-i+1}) \\ &\times \ [F(x_{m-i+1})]^{R_1}\, dx_1 \cdots dx_{m-i+1}. \end{aligned}$$

Now, writing $F(x) = 1 - [1 - F(x)]$, expanding these terms binomially, interchanging the order of summations and integrations, and realizing that the first progressively Type-II right censored order statistic is also the first usual order statistic from a sample of size n, we obtain the relation in (5.14). □

Remark 5.9 Theorem 5.9 provides a method for obtaining single moments of progressively Type-II left withdrawn order statistics based on the knowledge of single moments of the smallest usual order statistics for varying sample sizes.

Remark 5.10 If we consider the case of no censoring, (5.14) simplifies to a result for usual order statistics due to Srikantan (1962) given by

$$\mu_{i:n}^{(k)} = \sum_{j=n-i+1}^{n} \binom{n}{j}\binom{j-1}{n-i}(-1)^{j-n+i-1}\,\mu_{1:j}^{(k)};$$

see also David (1981) and Arnold, Balakrishnan and Nagaraja (1992).

Following lines similar to that of Theorem 5.9, we may establish the following theorem for the product moments of progressively Type-II left withdrawn order statistics.

Theorem 5.10 *For* $1 \le i < j \le m \le n$,

$$\begin{aligned} ^{(R_m,R_{m-1},\cdots,R_1)}\mu_{i,j:m:n} &= c_{R'_{m-i+1},R_{m-i},\cdots,R_1} \sum_{l_{m-i+1}=0}^{R'_{m-i+1}} \sum_{l_{m-i}=0}^{R_{m-i}} \cdots \sum_{l_1=0}^{R_1} \\ &\times \ (-1)^{l_1+\cdots+l_{m-i+1}} \binom{R'_{m-i+1}}{l_{m-i+1}}\binom{R_{m-i}}{l_{m-i}}\cdots\binom{R_1}{l_1} \\ &\times \ \left[\frac{\mu_{1,j-i+1:m-i+1:l_1+\cdots+l_{m-i+1}+m-i+1}^{(l_{m-i+1},l_{m-i},\cdots,l_1)}}{c_{l_1,\cdots,l_{m-i+1}}}\right]. \end{aligned} \tag{5.15}$$

Remark 5.11 For the special case of no censoring, (5.15) reduces to the following identity for the usual order statistics:

$$\mu_{i,j:n} = \sum_{l=0}^{i-1} \sum_{k=0}^{l} (-1)^{l+k} \frac{n!}{(i-l-1)!\,(l-k)!\,(n-i+k-1)!} \mu_{k+1,j-i+k+1:n-i+k+1}.$$

We now have a number of different methods for obtaining the exact moments of progressively Type-II right censored order statistics when explicit expressions are not readily available. These will be useful when we discuss exact linear estimation of parameters of different lifetime distributions when the observed samples are progressively censored (Chapter 6). They will also be used in our discussions on the determination of optimal censoring schemes for the estimation of parameters of interest (Chapter 10).

Though these methods will enable the exact determination of the single and product moments, implementation of them for large values of n will be very difficult (and even if possible, will involve serious rounding errors). For this reason, we present in the next section some first-order approximations for the means, variances and covariances of the progressively Type-II right censored order statistics. These approximate expressions will also be utilized later in Chapter 6 to develop first-order approximations to the best linear unbiased estimators of location and scale parameters.

5.5 First-Order Approximations to the Moments

Suppose the progressively Type-II right censored order statistics have come from the *Uniform*(0,1) distribution. For convenience in notation, let us denote them by $U_{1:m:n}^{(R_1,\cdots,R_m)}, U_{2:m:n}^{(R_1,\cdots,R_m)}, \cdots, U_{m:m:n}^{(R_1,\cdots,R_m)}$. Then, from (1.3), we readily have their joint density function to be

$$f(u_1, u_2, \cdots, u_m) = c \prod_{i=1}^{m} (1-u_i)^{R_i}, \quad 0 < u_1 < \cdots < u_m < 1, \tag{5.16}$$

where c is as given there. Earlier in Chapter 2, we established that the random variables

$$V_i = \frac{1 - U_{m-i+1:m:n}^{(R_1,\cdots,R_m)}}{1 - U_{m-i:m:n}^{(R_1,\cdots,R_m)}}, \ i = 1, \cdots, m-1, \text{ and } V_m = 1 - U_{1:m:n}^{(R_1,\cdots,R_m)} \tag{5.17}$$

are all mutually independent, and further that

$$W_i = V_i^{i+R_m+R_{m-1}+\cdots+R_{m-i+1}}, \quad i = 1, 2, \cdots, m \tag{5.18}$$

are all independently and identically distributed as *Uniform*(0,1).

Recall that from (5.17), we can readily write

$$U_{i:m:n}^{(R_1,\cdots,R_m)} \stackrel{d}{=} 1 - \prod_{j=m-i+1}^{m} V_j, \quad i = 1, 2, \cdots, m, \tag{5.19}$$

where V_j's are independently distributed as $Beta(j + \sum_{k=m-j+1}^{m} R_k, 1)$. Using this result, we obtained the following explicit expressions for means, variances and covariances of progressively Type-II right censored order statistics from the *Uniform*(0,1) distribution:

$$E\left(U_{i:m:n}^{(R_1,\cdots,R_m)}\right) = \Pi_i = 1 - b_i, \quad i = 1, 2, \cdots, m, \tag{5.20}$$

$$\mathrm{Var}\left(U_{i:m:n}^{(R_1,\cdots,R_m)}\right) = k_i b_i, \qquad i = 1, 2, \cdots, m, \tag{5.21}$$

and

$$\mathrm{Cov}\left(U_{i:m:n}^{(R_1,\cdots,R_m)}, U_{j:m:n}^{(R_1,\cdots,R_m)}\right) = k_i b_j, \qquad 1 \le i < j \le m, \tag{5.22}$$

where, for $i = 1, 2, \cdots, m$,

$$\begin{aligned} k_i &= \prod_{k=1}^{i} \left\{ \frac{m-k+2+R_k+R_{k+1}+\cdots+R_m}{m-k+3+R_k+R_{k+1}+\cdots+R_m} \right\} \\ &\quad - \prod_{k=1}^{i} \left\{ \frac{m-k+1+R_k+R_{k+1}+\cdots+R_m}{m-k+2+R_k+R_{k+1}+\cdots+R_m} \right\} \\ &= \prod_{j=m-i+1}^{m} \gamma_j - \prod_{j=m-i+1}^{m} \alpha_j \ \text{ from (2.25) and (2.26)} \end{aligned} \tag{5.23}$$

and

$$\begin{aligned} b_i &= \prod_{k=1}^{i} \left\{ \frac{m-k+1+R_k+R_{k+1}+\cdots+R_m}{m-k+2+R_k+R_{k+1}+\cdots+R_m} \right\} \\ &= \prod_{j=m-i+1}^{m} \alpha_j \ \text{ from (2.25) and (2.26).} \end{aligned} \tag{5.24}$$

Observe that, if we set $R_1 = \cdots = R_m = 0$ so that there is no progressive censoring, then the above expressions reduce to $E(U_{i:n}) = \frac{i}{n+1}$, $\mathrm{Var}(U_{i:n}) = \frac{i(n-i+1)}{(n+1)^2(n+2)}$ and $\mathrm{Cov}(U_{i:n}, U_{j:n}) = \frac{i(n-j+1)}{(n+1)^2(n+2)}$ which are well-known results on order statistics from the *Uniform*(0,1) distribution; see, for example, David (1981, pp. 35–36) and Arnold, Balakrishnan and Nagaraja (1992, pp. 19–20).

We shall now use these expressions to get first-order approximations to the means, variances and covariances of progressively Type-II right censored order statistics from an arbitrary continuous distribution $F(\cdot)$. From the inverse probability integral transformation, we readily have the relationship

$$Y_{i:m:n}^{(R_1,\cdots,R_m)} \stackrel{d}{=} F^{-1}\left(U_{i:m:n}^{(R_1,\cdots,R_m)}\right), \tag{5.25}$$

where $F^{-1}(\cdot)$ is the inverse cumulative distribution function of the lifetime distribution from which the progressively censored sample has come. Expanding the function on the right hand side of (5.25) in a Taylor series around $E(U_{i:m:n}^{(R_1,\cdots,R_m)}) = \Pi_i$ and then taking expectations and retaining the first term, we obtain the approximation

$$E\left(Y_{i:m:n}^{(R_1,\cdots,R_m)}\right) \simeq F^{-1}(\Pi_i), \quad i = 1, 2, \cdots, m, \tag{5.26}$$

where Π_i is as given in (5.20). Proceeding similarly, we obtain the approximations

$$\text{Var}\left(Y_{i:m:n}^{(R_1,\cdots,R_m)}\right) \simeq \left\{F^{-1^{(1)}}(\Pi_i)\right\}^2 k_i b_i, \quad i = 1, 2, \cdots, m, \tag{5.27}$$

and

$$\text{Cov}\left(Y_{i:m:n}^{(R_1,\cdots,R_m)}, Y_{j:m:n}^{(R_1,\cdots,R_m)}\right) \simeq F^{-1^{(1)}}(\Pi_i) F^{-1^{(1)}}(\Pi_j) k_i b_j, \ i < j, \tag{5.28}$$

where k_i and b_i are as given in (5.23) and (5.24), respectively, and $F^{-1^{(1)}}(u) = \frac{d}{du}F^{-1}(u)$. This type of a Taylor series approximation for the usual order statistics was given by David and Johnson (1954) wherein it has been shown that the Taylor series terms [in (5.26)] beyond the first-order are of order $1/n$ thus yielding a large-sample approximation. As a matter of fact, if we set $R_1 = \cdots = R_m = 0$, the above approximations reduce to the first terms in the corresponding formulas given by David and Johnson (1954). If more precision is needed in the approximations, then more terms could be retained in the Taylor series expansion.

6

Linear Inference

6.1 One-Parameter (Scale) Models

6.1.1 Introduction

We have to this point developed a number of mathematical results for progressively Type-II censored order statistics, some of which are quite elegant and intuitively pleasing, and others which may be classified by many as "the reason why mathematics is so unpopular." However, all of the results obtained thus far enjoy real and practical applications. One such application has already been considered, namely, computer simulation of progressively censored order statistics. We will consider in this chapter the problem of linear inference for the parameters of lifetime distributions when observed samples are progressively censored. Many results presented will be for *general* progressively Type-II censored samples which were introduced in Section 1.5. To obtain results for progressively Type-II right censored order statistics, we would simply set $r = 0$.

Consider an arbitrary "standard" continuous distribution $F(x)$. Then, if the single and product moments of the progressively Type-II censored order statistics from $F(\cdot)$ are known, the best linear unbiased estimators for scale parameters may be obtained following the steps in, for example, Arnold, Balakrishnan and Nagaraja (1992, pp. 171–173). We will briefly outline the results here.

If we are to consider the linear transformation $\boldsymbol{Y} = \theta \boldsymbol{X}$, where the vector $\boldsymbol{X}$ represents a vector of progressively Type-II censored order statistics from the standard distribution $F(x)$, then the best linear unbiased estimator of θ is obtained by minimizing with respect to θ the *generalized variance* $Q(\theta) = (\boldsymbol{Y} - \theta\boldsymbol{\mu})' \boldsymbol{\Sigma}^{-1} (\boldsymbol{Y} - \theta\boldsymbol{\mu})$, where $\boldsymbol{\mu}$ is the mean vector of $\boldsymbol{X}$ and $\boldsymbol{\Sigma}$ is the variance-covariance matrix of $\boldsymbol{X}$. The minimum occurs at

$$\theta^* = \frac{\boldsymbol{\mu}'\boldsymbol{\Sigma}^{-1}}{\boldsymbol{\mu}'\boldsymbol{\Sigma}^{-1}\boldsymbol{\mu}}\boldsymbol{Y}. \tag{6.1}$$

This is the *best linear unbiased estimator* of θ. The variance of θ^* is then easily obtained as

$$Var(\theta^*) = \frac{\theta^2}{\boldsymbol{\mu}'\boldsymbol{\Sigma}^{-1}\boldsymbol{\mu}}. \tag{6.2}$$

Various linear estimators for progressively censored samples have been discussed by a number of authors, including Mann (1969a), Thomas and Wilson

(1972), and Cacciari and Montanari (1987). Many of these results involve numerical analysis and approximations for specific distributions such as the extreme value and Weibull. In this chapter, we present many closed form results for best linear unbiased estimators, as well as a few scenarios where modern computing resources are necessary to perform the required calculations. We also present first-order approximations to these best linear unbiased estimators.

6.1.2 The Exponential Distribution

For a general progressively Type-II censored sample from the standard exponential distribution, let us denote the corresponding general progressively Type-II censored order statistics by ${}_rX_{r+1:m:n}, {}_rX_{r+2:m:n}, \cdots, {}_rX_{m:m:n}$.

Now, let us represent the general progressively Type-II censored order statistics with the same censoring scheme from the *Exponential* (θ) distribution, that is, with probability density function

$$f(y;\theta) = \frac{1}{\theta}e^{-y/\theta}, \; y \geq 0, \; \theta > 0,$$

by ${}_rY_{r+i:m:n}$, $i = 1, 2, \cdots, m-r$. Using the expressions for the means, variances, and covariances of ${}_rX_{r+1:m:n}, {}_rX_{r+2:m:n}, \cdots, {}_rX_{m:m:n}$ obtained in Section 3.4.2, we see that the $(m-r) \times (m-r)$ variance-covariance matrix is of the special form $\sigma_{ij} = s_i t_j$ (where $t_j = 1$ for all j) which can be inverted explicitly as given in Graybill (1983, pp. 198). As discussed in Arnold, Balakrishnan and Nagaraja (1992, pp. 174-175), the general form for the inverse of a non-singular $k \times k$ symmetric matrix (c_{ij}), where $c_{ij} = a_i b_j \; (i \leq j)$, is given by $(c^{i,j})$ where

$$c^{i,j} = \begin{cases} -\left(a_{i+1}b_i - a_i b_{i+1}\right)^{-1}, \; j = i+1, \; i = 1, \cdots, k-1 \\ \frac{a_{i+1}b_{i-1} - a_{i-1}b_{i+1}}{(a_i b_{i-1} - a_{i-1}b_i)(a_{i+1}b_i - a_i b_{i+1})}, \; i = j = 2, \; \cdots, \; k-1 \\ a_2\left[a_1\left(a_2 b_1 - a_1 b_2\right)\right]^{-1}, \; i = j = 1 \\ b_{k-1}\left[b_k\left(a_k b_{k-1} - a_{k-1}b_k\right)\right]^{-1}, \; i = j = k \\ 0, \; j > i+1. \end{cases} \tag{6.3}$$

We can then obtain the exact best linear unbiased estimator (BLUE) of θ and its variance, using (6.1) and (6.2). Explicit expressions of the BLUE and its variance, as given in Balakrishnan and Sandhu (1996), are as follows:

$$\begin{aligned} \theta^* &= \frac{1}{m-r-1+\frac{\alpha^2_{r+1:n}}{\beta_{r+1:n}}} \\ &\quad \times \left[\sum_{j=r+2}^{m}\left(R_j+1\right)\left({}_rY_{j:m:n} - {}_rY_{r+1:m:n}\right) + \left(\frac{\alpha_{r+1:n}}{\beta_{r+1:n}}\right){}_rY_{r+1:m:n}\right] \end{aligned} \tag{6.4}$$

and

$$Var\left(\theta^*\right) = \frac{\theta^2}{m-r-1+\frac{\alpha^2_{r+1:n}}{\beta_{r+1:n}}}, \tag{6.5}$$

where, as in Section 3.4.2,

$$\alpha_{r+1:n} = \sum_{i=1}^{r+1} \frac{1}{n-i+1}, \quad \beta_{r+1:n} = \sum_{i=1}^{r+1} \frac{1}{(n-i+1)^2}.$$

Remark 6.1 It is of interest to observe that the precision of the BLUE of in (6.5) depends only on r, m and n, and not on the progressive censoring scheme $(R_{r+1}, \cdots, R_m)$.

The BLUE of θ in (6.4) can also be derived directly by adopting the approach of Balakrishnan and Rao (1997b). First of all, from Theorem 3.4 we have

$$\begin{aligned}
Z_{r+1} &= {}_rY_{r+1:m:n}/\theta \\
&= {}_rX_{r+1:m:n}, \\
Z_{r+2} &= (n-r-R_{r+1}-1)({}_rY_{r+2:m:n} - {}_rY_{r+1:m:n})/\theta \\
&= (n-r-R_{r+1}-1)({}_rX_{r+2:m:n} - {}_rX_{r+1:m:n}), \\
\cdots & \quad \cdots \\
Z_m &= (n-r-R_{r+1}-\cdots-R_{m-1}-m+r+1)({}_rY_{m:m:n} - {}_rY_{m-1:m:n})/\theta \\
&= (n-r-R_{r+1}-\cdots-R_{m-1}-m+r+1)({}_rX_{m:m:n} - {}_rX_{m-1:m:n})
\end{aligned}$$

to be independent random variables, and that $Z_{r+2}, \cdots, Z_m$ to be distributed as standard exponential. Consequently, we have

$$\begin{aligned}
E(Z_{r+1}) &= \sum_{i=1}^{r+1} \frac{1}{n-i+1} = \alpha_{r+1:n}, \ E(Z_i) = 1 \quad (r+2 \le i \le m), \\
\text{Var}(Z_{r+1}) &= \sum_{i=1}^{r+1} \frac{1}{(n-i+1)^2} = \beta_{r+1:n}, \ \text{Var}(Z_i) = 1 \quad (r+2 \le i \le m),
\end{aligned}$$

and

$$\text{Cov}(Z_i, Z_j) = 0 \quad (r+1 \le i \ne j \le m).$$

Let us now take an arbitrary linear estimator of θ as

$$\theta^* = \sum_{i=r+1}^{m} a_i \, {}_rY_{i:m:n} = \theta \sum_{i=r+1}^{m} a'_i Z_i \tag{6.6}$$

from which we readily find

$$E(\theta^*) = \theta \left\{ a'_{r+1}\alpha_{r+1:n} + \sum_{i=r+2}^{m} a'_i \right\} \tag{6.7}$$

and

$$\text{Var}(\theta^*) = \theta^2 \left\{ a'^2_{r+1}\beta_{r+1:n} + \sum_{i=r+2}^{m} a'^2_i \right\}. \tag{6.8}$$

If we require the estimator θ^* in (6.6) to be unbiased for θ, then (6.7) gives the unbiasedness condition as

$$a'_{r+1}\alpha_{r+1:n} + \sum_{i=r+2}^{m} a'_i = 1. \tag{6.9}$$

So, for the derivation of BLUE, we need to minimize the variance in (6.8) subject to the unbiasedness condition in (6.9). For this purpose, let us use the Lagrangian method and consider

$$S = a'^2_{r+1}\beta_{r+1:n} + \sum_{i=r+2}^{m} a'^2_i - 2\lambda\left\{a'_{r+1}\alpha_{r+1:n} + \sum_{i=r+2}^{m} a'_i - 1\right\},$$

where λ is the Lagrangian multiplier. Minimizing S with respect to $a'_{r+1}, a'_{r+2}, \cdots, a'_m$, we obtain the equations

$$\begin{aligned} &a'_{r+1}\beta_{r+1:n} - \lambda\alpha_{r+1:n} = 0, \\ &a'_i - \lambda = 0 \quad \text{for } i = r+2, \cdots, m, \end{aligned} \tag{6.10}$$

which gives

$$a'_{r+1} = \lambda\,\frac{\alpha_{r+1:n}}{\beta_{r+1:n}} \quad \text{and} \quad a'_i = \lambda \quad (i = r+2, \cdots, m).$$

The unbiasedness condition in (6.9) then yields the optimal choice of the Lagrangian multiplier λ as

$$\lambda = 1\Big/\left\{m - r - 1 + \frac{\alpha^2_{r+1:n}}{\beta_{r+1:n}}\right\}.$$

Eq. (6.10) gives the coefficients a'_i's as

$$a'_{r+1} = \frac{\alpha_{r+1:n}}{\beta_{r+1:n}\left\{m - r - 1 + \frac{\alpha^2_{r+1:n}}{\beta_{r+1:n}}\right\}} \text{ and } a'_{r+2} = \cdots = a'_m = \frac{1}{m - r - 1 + \frac{\alpha^2_{r+1:n}}{\beta_{r+1:n}}}$$

from which we obtain the coefficients a_i's as

$$\begin{aligned} a_{r+1} &= a'_{r+1} - (n - r - R_{r+1} - 1)a'_{r+2} \\ &= \frac{1}{m - r - 1 + \frac{\alpha^2_{r+1:n}}{\beta_{r+1:n}}}\left\{\frac{\alpha_{r+1:n}}{\beta_{r+1:n}} - (n - r - R_{r+1} - 1)\right\}, \\ a_{r+2} &= (n - r - R_{r+1} - 1)a'_{r+2} - (n - r - R_{r+1} - R_{r+2} - 2)a'_{r+3} \\ &= (R_{r+2} + 1)\Big/\left(m - r - 1 + \frac{\alpha^2_{r+1:n}}{\beta_{r+1:n}}\right), \\ \cdots & \quad \cdots \end{aligned}$$

$$
\begin{aligned}
\cdots \quad & \quad \cdots \\
a_{m-1} &= (R_{m-1}+1)\Big/\left(m-r-1+\frac{\alpha_{r+1:n}^2}{\beta_{r+1:n}}\right), \\
a_m &= (n-r-R_{r+1}-\cdots-R_{m-1}-m+r+1)a'_m \\
&= (R_m+1)\Big/\left(m-r-1+\frac{\alpha_{r+1:n}^2}{\beta_{r+1:n}}\right).
\end{aligned}
$$

These coefficients, when substituted in (6.6), give the BLUE of θ exactly as presented in (6.4). Furthermore, we obtain from (6.8)

$$
\begin{aligned}
\mathrm{Var}(\theta^*) &= \theta^2 \frac{1}{\left\{m-r-1+\frac{\alpha_{r+1:n}^2}{\beta_{r+1:n}}\right\}^2}\left\{\frac{\alpha_{r+1:n}^2}{\beta_{r+1:n}}+(m-r-1)\right\} \\
&= \theta^2\Big/\left\{m-r-1+\frac{\alpha_{r+1:n}^2}{\beta_{r+1:n}}\right\},
\end{aligned}
$$

which agrees with (6.5).

Remark 6.2 For the special case of progressive Type-II right censoring (viz., $r=0$), we obtain from (6.4) and (6.5) that

$$
\theta^* = \frac{1}{m}\sum_{i=1}^{m}(R_i+1)Y_{i:m:n} \quad \text{and} \quad \mathrm{Var}(\theta^*) = \frac{\theta^2}{m}.
$$

In Chapter 7, we will see that this also is the maximum likelihood estimator of θ. In this case, using the property of spacings established in Theorem 2.7, it can be shown that $2m\theta^*/\theta$ is distributed as chi-square with $2m$ degrees of freedom. Also, Chandrasekar, Leo Alexander and Balakrishnan (1999) have recently shown that a minimum risk equivariant estimator of θ^ℓ is given by

$$
\frac{\Gamma(m+\ell)}{\Gamma(m+2\ell)}\left\{\sum_{i=1}^{m}(R_i+1)Y_{i:m:n}\right\}^\ell.
$$

Remark 6.3 Suppose k general progressively Type-II censored samples are available from the *Exponential(θ)* distribution, with the i-th censored sample denoted by

$$
{}_{r_i}Y_{i;r_i+1:m_i:n_i},\ {}_{r_i}Y_{i;r_i+2:m_i:n_i},\ \cdots,\ {}_{r_i}Y_{i;m_i:m_i:n_i}
$$

and the corresponding general progressive censoring scheme by $(R_{i,r_i+1}, R_{i,r_i+2}, \cdots, R_{i,m_i})$ for $i=1,2,\cdots,k$. Then, Balakrishnan, Cramer, Kamps and Schenk (1999) have derived the BLUE of θ to be

$$\theta^* = \frac{1}{\sum_{i=1}^{k}\left(m_i - r_i - 1 + \frac{\alpha^2_{r_i+1:n_i}}{\beta_{r_i+1:n_i}}\right)}$$
$$\times\left[\sum_{i=1}^{k}\sum_{j=r_i+2}^{m_i}(R_{i,j}+1)\left({}_{r_i}Y_{i;j:m_i:n_i} - {}_{r_i}Y_{i;r_i+1:m_i:n_i}\right)\right.$$
$$\left.+\sum_{i=1}^{k}\frac{\alpha_{r_i+1:n_i}}{\beta_{r_i+1:n_i}}\,{}_{r_i}Y_{i;r_i+1:m_i:n_i}\right]$$

and its variance as

$$\operatorname{Var}(\theta^*) = \theta^2 \Big/ \sum_{i=1}^{k}\left(m_i - r_i - 1 + \frac{\alpha^2_{r_i+1:n_i}}{\beta_{r_i+1:n_i}}\right),$$

where

$$\alpha_{r_i+1:n_i} = \sum_{j=1}^{r_i+1}\frac{1}{(n_i-j+1)} \quad \text{and} \quad \beta_{r_i+1:n_i} = \sum_{j=1}^{r_i+1}\frac{1}{(n_i-j+1)^2}\,.$$

6.1.3 The Uniform Distribution

For a general progressively Type-II censored sample from the *Uniform*(0, 1) distribution, let us denote the corresponding general progressively Type-II censored order statistics by ${}_rU_{r+1:m:n}, {}_rU_{r+2:m:n}, \cdots, {}_rU_{m:m:n}$.

In Section 3.4.3, we found that

$$E\left({}_rU_{r+i:m:n}\right) = 1 - \prod_{j=m-i+1}^{m}\alpha_j,\ i = 1,\cdots,m-r,$$
$$Var\left({}_rU_{r+i:m:n}\right) = \left[\prod_{j=m-i+1}^{m}\alpha_j\right]\left[\prod_{j=m-i+1}^{m}\gamma_j - \prod_{j=m-i+1}^{m}\alpha_j\right],$$
$$i = 1,\cdots,m-r$$

and

$$Cov\left({}_rU_{r+i:m:n},\ {}_rU_{r+k:m:n}\right) = \left[\prod_{j=m-i+1}^{m}\alpha_j\right]\left[\prod_{j=m-k+1}^{m}\gamma_j - \prod_{j=m-k+1}^{m}\alpha_j\right],$$
$$k < i,\ i = 2,\cdots,m-r,$$

where α_j and γ_j, $j = r+1, \cdots, m$, are as defined earlier in Section 3.4.3.

Upon making use of the above expressions, we see that the $(m-r)\times(m-r)$ variance-covariance matrix is again of the special form $\sigma_{ik} = s_i t_k$ which can be inverted explicitly using (6.3). The elements of the symmetric tri-diagonal inverted matrix $(c^{i,j})$ are as follows:

$$\begin{aligned}
c^{i,i} &= \frac{\gamma_{m-i}\gamma_{m-i+1} - \alpha_{m-i}\alpha_{m-i+1}}{\beta_{m-i}\beta_{m-i+1}\prod_{j=m-i+1}^{m}\alpha_j\gamma_j}, \quad i = 1, 2, \cdots, m-r-1, \\
c^{m-r,m-r} &= \frac{1}{\beta_{r+1}\prod_{j=r+1}^{m}\alpha_j\prod_{j=r+2}^{m}\gamma_j}, \\
c^{i,i+1} &= \frac{-1}{\beta_{m-i}\prod_{m-i+1}^{m}\alpha_j\gamma_j}, \quad i = 1, 2, \cdots, m-r-1, \\
c^{i,j} &= 0 \quad \text{otherwise},
\end{aligned} \tag{6.11}$$

where β_j, $j = r+1, \cdots, m$, are as defined earlier in Section 3.4.3.

Now, let us represent the general progressively Type-II censored order statistics with the same censoring scheme from the $Uniform\,(0, \theta)$ distribution by ${}_rY_{r+i:m:n}$, $i = 1, 2, \cdots, m-r$. Using (6.11), (6.1) and (6.2), we can obtain the exact best linear unbiased estimator (BLUE) of θ. Explicit expressions of the BLUE and its variance are as follows [see Aggarwala and Balakrishnan (1998a)]:

$$\begin{aligned}
\theta^* &= \left\{ \sum_{i=1}^{m-r-1} \frac{\gamma_{m-i}\gamma_{m-i+1} - \alpha_{m-i}\alpha_{m-i+1} - \beta_{m-i+1} - \alpha_{m-i+1}\gamma_{m-i+1}\beta_{m-i}}{\beta_{m-i}\beta_{m-i+1}\prod_{j=m-i+1}^{m}\alpha_j\gamma_j} \right. \\
&\quad \left. \times\, {}_rY_{r+i:m:n} + \frac{(1-\alpha_{r+1})}{\beta_{r+1}\alpha_{r+1}\prod_{j=r+2}^{m}\alpha_j\gamma_j}\, {}_rY_{m:m:n} \right\} \\
&\quad \times \left\{ \sum_{i=1}^{m-r-1} \frac{\gamma_{m-i}\gamma_{m-i+1} - \alpha_{m-i}\alpha_{m-i+1} - \beta_{m-i+1} - \alpha_{m-i+1}\gamma_{m-i+1}\beta_{m-i}}{\beta_{m-i}\beta_{m-i+1}\prod_{j=m-i+1}^{m}\alpha_j\gamma_j} \right. \\
&\quad \left. \times \left(1 - \prod_{j=m-i+1}^{m}\alpha_j\right) + \frac{(1-\alpha_{r+1})\left(1 - \prod_{j=r+1}^{m}\alpha_j\right)}{\beta_{r+1}\alpha_{r+1}\prod_{j=r+2}^{m}\alpha_j\gamma_j} \right\}^{-1}
\end{aligned} \tag{6.12}$$

and

$$\begin{aligned}
&Var\,(\theta^*) \\
&= \theta^2 \left\{ \sum_{i=1}^{m-r-1} \frac{\gamma_{m-i}\gamma_{m-i+1} - \alpha_{m-i}\alpha_{m-i+1} - \beta_{m-i+1} - \alpha_{m-i+1}\gamma_{m-i+1}\beta_{m-i}}{\beta_{m-i}\beta_{m-i+1}\prod_{j=m-i+1}^{m}\alpha_j\gamma_j} \right. \\
&\quad \left. \times \left(1 - \prod_{j=m-i+1}^{m}\alpha_j\right) + \frac{(1-\alpha_{r+1})\left(1 - \prod_{j=r+1}^{m}\alpha_j\right)}{\beta_{r+1}\alpha_{r+1}\prod_{j=r+2}^{m}\alpha_j\gamma_j} \right\}^{-1}.
\end{aligned} \tag{6.13}$$

Remark 6.4 It is of interest to note that the precision of θ^* in (6.13) does depend on the censoring scheme used, unlike in the one-parameter exponential case seen earlier in Section 6.1.2.

Remark 6.5 In the special case when there is no censoring, (that is, $m = n,\ r = 0,\ R_1 = \cdots = R_n = 0$), the following expressions are obtained:

$$\alpha_i = \frac{i}{i+1},\ \gamma_i = \frac{i+1}{i+2},\ \beta_i = \frac{1}{(i+2)(i+1)},\ i = 1, 2, \cdots, n, \tag{6.14}$$

so that

$$\begin{aligned} c^{i,i} &= 2(n+1)(n+2),\ i = 1, 2, \cdots, n, \\ c^{i,i+1} &= -(n+1)(n+2),\ i = 1, 2, \cdots, n-1, \end{aligned}$$

and we have

$$\theta^* = \frac{n+1}{n} Y_{n:n},\ Var(\theta^*) = \frac{\theta^2}{n(n+2)},$$

which is a well-known result; see, for example, Arnold, Balakrishnan and Nagaraja (1992).

6.1.4 The Pareto Distribution

For a progressively Type-II right censored sample from the standard $Pareto(\nu)$ distribution, let us denote the corresponding progressively Type-II censored order statistics by $X_{1:m:n},\ X_{2:m:n}, \cdots,\ X_{m:m:n}$.

Now, let us represent the progressively Type-II right censored order statistics with the same censoring scheme from the $Pareto(\nu, \theta)$ distribution, that is, with probability density function

$$f(w; \nu, \theta) = \frac{\nu}{\theta}\left(\frac{w}{\theta}\right)^{-\nu-1},\ w > \theta,\ \nu > 0,\ \theta > 0$$

by $W_{i:m:n},\ i = 1, 2, \cdots, m$. Using the expressions for the means, variances, and covariances of $X_{1:m:n},\ X_{2:m:n}, \cdots, X_{m:m:n}$ obtained in (2.36), we see that the $m \times m$ variance-covariance matrix is again of the special form $\sigma_{ij} = s_i t_j$ which can be inverted explicitly using (6.3). The elements of the symmetric tri-diagonal inverted variance-covariance matrix of $\boldsymbol{X}$, $\left(c^{i,j}\right)$, are given by

$$\begin{aligned} c^{i,i} &= \frac{\gamma_i \gamma_{i+1} - \beta_i^2 \beta_{i+1}^2}{\left(\prod_{k=1}^{i} \gamma_k\right)\left(\gamma_i - \beta_i^2\right)\left(\gamma_{i+1} - \beta_{i+1}^2\right)}, i = 1, 2, ..., m-1, \\ c^{m,m} &= \frac{1}{\left(\prod_{k=1}^{m-1} \gamma_k\right)\left(\gamma_m - \beta_m^2\right)}, \\ c^{i,i+1} &= \frac{\beta_{i+1}}{\left(\prod_{k=1}^{i} \gamma_k\right)\left(\beta_{i+1}^2 - \gamma_{i+1}\right)}, i = 1, 2, ..., m-1, \\ c^{i,j} &= 0 \text{ otherwise}, \end{aligned} \tag{6.15}$$

where α_i, β_i, and γ_i, $i = 1, 2, \cdots, m$, are as given in (2.35).

We can, therefore, obtain the exact best linear unbiased estimator (BLUE) of θ and its variance, using (6.1) and (6.2). The explicit expressions for the BLUE and its variance are as follows:

$$\theta^* = \frac{\sum_{i=1}^m \sum_{k=1}^m c^{k,i}\mu_k W_{i:m:n}}{\sum_{j=1}^m \sum_{k=1}^m c^{k,j}\mu_k\mu_j} \tag{6.16}$$

and

$$Var\,(\theta^*) = \frac{\theta^2}{\sum_{j=1}^m \sum_{k=1}^m c^{k,j}\mu_k\mu_j}, \tag{6.17}$$

where $\mu_i = \prod_{k=1}^i \beta_k$ $(i = 1, 2, \cdots, m)$ is the mean of $X_{i:m:n}$; see Aggarwala and Balakrishnan (1998c).

Remark 6.6 The above general expressions, (6.16) and (6.17), for the BLUE of the scale parameter and its variance may be used for the scale-parameter BLUE based on progressively Type-II right censored order statistics (or any order statistics) from *any* distribution which has a corresponding variance-covariance matrix of the order statistics from the standard distribution of the form $\sigma_{i,j} = s_i t_j$. Here, $c^{i,j}$ are then elements of the inverted variance-covariance matrix of the order statistics from the standard distribution and μ_i are elements of the corresponding mean vector.

6.1.5 First-Order Approximation to the BLUE

Let us assume that the progressively Type-II right censored sample $Y_{1:m:n}^{(R_1,\cdots,R_m)}$, $Y_{2:m:n}^{(R_1,\cdots,R_m)}, \cdots, Y_{m:m:n}^{(R_1,\cdots,R_m)}$ has arisen from a lifetime distribution belonging to the scale-parameter family, that is, with density function $f(y;\theta) = \frac{1}{\theta}\, f(\frac{y}{\theta})$. In this case, either from the generalized least-squares theory [see Lloyd (1952)] or through Lagrangian multiplier method [see Balakrishnan and Rao (1997b, 1999)], one can get the BLUE of θ as

$$\theta^* = \left(\frac{\boldsymbol{\mu}^T\boldsymbol{\Sigma}^{-1}}{\boldsymbol{\mu}^T\boldsymbol{\Sigma}^{-1}\boldsymbol{\mu}}\right)\boldsymbol{Y} \tag{6.18}$$

and its variance as

$$\mathrm{Var}(\theta^*) = \theta^2/(\boldsymbol{\mu}^T\boldsymbol{\Sigma}^{-1}\boldsymbol{\mu}). \tag{6.19}$$

In the above formulas, $\boldsymbol{Y}$ denotes the observed vector of progressively Type-II right censored order statistics $(Y_{1:m:n}^{(R_1,\cdots,R_m)}, \cdots, Y_{m:m:n}^{(R_1,\cdots,R_m)})^T$, $\boldsymbol{X} = \boldsymbol{Y}/\theta$ denotes the vector of corresponding standardized order statistics, $\boldsymbol{\mu} = E(\boldsymbol{X})$, and $\boldsymbol{\Sigma} = \mathrm{Var}(\boldsymbol{X})$. The usage of the BLUE in (6.18) becomes somewhat restrictive due to the complications in determining the variance-covariance matrix $\boldsymbol{\Sigma}$ for any given lifetime distribution F and also in getting $\boldsymbol{\Sigma}^{-1}$ for large m.

We, therefore, seek to derive a first-order approximation to the BLUE of θ in (6.18) by making use of the approximate expressions of $\boldsymbol{\mu}$ and $\boldsymbol{\Sigma}$ presented earlier in Section 5.5. To this end, we first note from (5.27) and (5.28) that the first-order approximation to the variance-covariance matrix $\boldsymbol{\Sigma} = ((\Sigma_{i,j}))$ has the special form

$$\Sigma_{i,j} = \beta_i \gamma_j, \qquad 1 \le i \le j \le m, \tag{6.20}$$

where

$$\beta_i = F^{-1^{(1)}}(\Pi_i) k_i \quad \text{and} \quad \gamma_j = F^{-1^{(1)}}(\Pi_j) b_j, \tag{6.21}$$

where k_i and b_i are as given in (5.23) and (5.24), respectively, and $F^{-1^{(1)}}(u) = \frac{d}{du} F^{-1}(u)$. Then, from (6.3), $\boldsymbol{\Sigma}^{-1} = ((\Sigma^{i,j}))$ is a symmetric tri-diagonal matrix given by

$$\Sigma^{i,j} = \begin{cases} \dfrac{k_2}{\{F^{-1^{(1)}}(\Pi_1)\}^2 \; k_1(k_2 b_1 - k_1 b_2)}, & i = j = 1, \\[2ex] \dfrac{k_{i+1} b_{i-1} - k_{i-1} b_{i+1}}{\{F^{-1^{(1)}}(\Pi_i)\}^2 \; (k_i b_{i-1} - k_{i-1} b_i)(k_{i+1} b_i - k_i b_{i+1})}, & i = j = 2, \cdots, m-1, \\[2ex] \dfrac{b_{m-1}}{\{F^{-1^{(1)}}(\Pi_m)\}^2 \; b_m(k_m b_{m-1} - k_{m-1} b_m)}, & i = j = m, \\[2ex] \dfrac{-1}{F^{-1^{(1)}}(\Pi_i) F^{-1^{(1)}}(\Pi_{i+1})(k_{i+1} b_i - k_i b_{i+1})}, & i = j-1 = 1, \cdots, m-1, \\[2ex] 0, & \text{otherwise.} \end{cases} \tag{6.22}$$

Upon using the expressions in (5.26) and (6.22), we obtain the first-order approximation to the BLUE of θ in (6.18) as [Balakrishnan and Rao (1997a)]

$$\begin{aligned} \theta^* = \frac{1}{\Delta} \Bigg[& \left\{ \frac{F^{-1}(\Pi_1) k_2}{\{F^{-1^{(1)}}(\Pi_1)\}^2 \; k_1(k_2 b_1 - k_1 b_2)} \right. \\ & \left. - \frac{F^{-1}(\Pi_2)}{F^{-1^{(1)}}(\Pi_1) F^{-1^{(1)}}(\Pi_2) \; (k_2 b_1 - k_1 b_2)} \right\} Y_{1:m:n}^{(R_1,\cdots,R_m)} \\ & + \sum_{i=2}^{m-1} \left\{ \frac{-F^{-1}(\Pi_{i-1})}{F^{-1^{(1)}}(\Pi_{i-1}) F^{-1^{(1)}}(\Pi_i)(k_i b_{i-1} - k_{i-1} b_i)} \right. \\ & - \frac{F^{-1}(\Pi_{i+1})}{F^{-1^{(1)}}(\Pi_i) F^{-1^{(1)}}(\Pi_{i+1})(k_{i+1} b_i - k_i b_{i+1})} \\ & \left. + \frac{F^{-1}(\Pi_i)(k_{i+1} b_{i-1} - k_{i-1} b_{i+1})}{\{F^{-1^{(1)}}(\Pi_i)\}^2 \; (k_i b_{i-1} - k_{i-1} b_i)(k_{i+1} b_i - k_i b_{i+1})} \right\} Y_{i:m:n}^{(R_1,\cdots,R_m)} \end{aligned}$$

$$+\left\{\frac{-F^{-1}(\Pi_{m-1})}{F^{-1^{(1)}}(\Pi_{m-1})F^{-1^{(1)}}(\Pi_m)(k_m b_{m-1}-k_{m-1}b_m)} + \frac{F^{-1}(\Pi_m)b_{m-1}}{\{F^{-1^{(1)}}(\Pi_m)\}^2\, b_m(k_m b_{m-1}-k_{m-1}b_m)}\right\} Y_{m:m:n}^{(R_1,\cdots,R_m)}\Bigg],$$

(6.23)

where

$$\begin{aligned}\Delta \;=\; & \frac{\{F^{-1}(\Pi_1)\}^2\, k_2}{\{F^{-1^{(1)}}(\Pi_1)\}^2\, k_1(k_2b_1-k_1b_2)} \\ & + \sum_{i=2}^{m-1} \frac{\{F^{-1}(\Pi_i)\}^2\,(k_{i+1}b_{i-1}-k_{i-1}b_{i+1})}{\{F^{-1^{(1)}}(\Pi_i)\}^2\,(k_ib_{i-1}-k_{i-1}b_i)(k_{i+1}b_i-k_ib_{i+1})} \\ & + \frac{\{F^{-1}(\Pi_m)\}^2\, b_{m-1}}{\{F^{-1^{(1)}}(\Pi_m)\}^2\, b_m(k_mb_{m-1}-k_{m-1}b_m)} \\ & - 2\sum_{i=1}^{m-1} \frac{F^{-1}(\Pi_i)F^{-1}(\Pi_{i+1})}{F^{-1^{(1)}}(\Pi_i)F^{-1^{(1)}}(\Pi_{i+1})(k_{i+1}b_i-k_ib_{i+1})}\,. \end{aligned} \qquad (6.24)$$

Furthermore, the first-order approximation to the variance of the BLUE of θ in (6.19) is given by

$$\mathrm{Var}(\theta^*) = \theta^2/\Delta, \qquad (6.25)$$

where Δ is as given in (6.24).

Example 6.1 Nelson (1982, p. 228, Table 6.1) reported data on times to breakdown of an insulating fluid in an accelerated test conducted at various test voltages. For illustrating the method of estimation developed in this section, let us consider the following progressively Type-II right censored sample of size $m = 8$ generated from the $n = 19$ observations recorded at 34 kilovolts in Nelson's (1982) Table 6.1, as given by Viveros and Balakrishnan (1994):

Table 6.1: Progressively censored sample generated from the times to breakdown data on insulating fluid tested at 34 kilovolts by Nelson (1982)

i	1	2	3	4	5	6	7	8
$y_{i:8:19}$	0.19	0.78	0.96	1.31	2.78	4.85	6.50	7.35
R_i	0	0	3	0	3	0	0	5

In this case, let us assume a scale-parameter exponential distribution with density function

$$f(y;\theta) = \frac{1}{\theta}\, e^{-y/\theta}, \qquad y \geq 0,\ \theta > 0 \qquad (6.26)$$

as the time-to-breakdown distribution. Then, we note that the standardized variable $X = Y/\theta$ has a standard exponential distribution in which case

$$F(x) = 1 - e^{-x}, \; F^{-1}(u) = -\ln(1-u) \text{ and } F^{-1^{(1)}}(u) = 1/(1-u).$$

From (6.23), we now determine the BLUE of θ to be

$$\begin{aligned} \theta^* &= (0.12305 \times 0.19) + (0.12305 \times 0.78) + (0.47255 \times 0.96) \\ &\quad + (0.12458 \times 1.31) + (0.47153 \times 2.78) + (0.12798 \times 4.85) \\ &\quad + (0.12808 \times 6.50) + (0.82641 \times 7.35) \\ &= 9.57. \end{aligned}$$

Similarly, from (6.24) we determine

$$\text{Var}(\theta^*) = 0.11713\, \theta^2$$

so that we obtain the standard error of the estimate θ^* to be

$$\text{SE}(\theta^*) = \theta^*(0.11713)^{1/2} = 3.28.$$

It is of interest to mention here that in this case the exact BLUE of θ (which is also the MLE) is given by $\frac{1}{m}\sum_{i=1}^{m}(R_i+1)Y_{i:m:n}^{(R_1,\cdots,R_m)}$ and its exact variance is θ^2/m; see Remark 6.2. These results yield the exact BLUE of θ to be 9.09 and its standard error to be 3.21. Observe the closeness of these values to the corresponding values obtained above from the first-order approximations even though we have n and m to be small, viz., $n = 19$ and $m = 8$.

6.2 Two-Parameter (Location-Scale) Models

6.2.1 Introduction

Consider again an arbitrary "standard" continuous distribution $F(x)$. Suppose now that we believe our progressively censored observations to be represented by the linear transformation $\boldsymbol{Y} = \mu\mathbf{1} + \sigma\boldsymbol{X}$, where the vector $\boldsymbol{X}$ represents a vector of progressively Type-II censored order statistics from the standard distribution $F(x)$ (which we will assume to be of length p); then the best linear unbiased estimators of μ and σ will be obtained by minimizing the generalized variance $Q(\boldsymbol{\theta}) = (\boldsymbol{Y} - \boldsymbol{A\theta})'\,\boldsymbol{\Sigma}^{-1}(\boldsymbol{Y} - \boldsymbol{A\theta})$ with respect to $\boldsymbol{\theta}$ where $\boldsymbol{\theta} = (\mu, \sigma)'$, $\boldsymbol{A}$ is the $p \times 2$ matrix $(\mathbf{1}, \boldsymbol{\mu})$, $\mathbf{1}$ is the $p \times 1$ vector with components all 1's, $\boldsymbol{\mu}$ is the mean vector of $\boldsymbol{X}$, and $\boldsymbol{\Sigma}$ is the variance-covariance matrix of $\boldsymbol{X}$. The minimum occurs when

$$\mu^* = -\boldsymbol{\mu}'\boldsymbol{\Gamma Y} \quad \text{and} \quad \sigma^* = \mathbf{1}'\boldsymbol{\Gamma Y}, \tag{6.27}$$

where $\boldsymbol{\Gamma} = \boldsymbol{\Sigma}^{-1}(\mathbf{1}\boldsymbol{\mu}' - \boldsymbol{\mu}\mathbf{1}')\boldsymbol{\Sigma}^{-1}/\Delta$ and $\Delta = \left(\mathbf{1}'\boldsymbol{\Sigma}^{-1}\mathbf{1}\right)\left(\boldsymbol{\mu}'\boldsymbol{\Sigma}^{-1}\boldsymbol{\mu}\right) - \left(\mathbf{1}'\boldsymbol{\Sigma}^{-1}\boldsymbol{\mu}\right)^2$.

From these expressions, variances and covariance of the estimators are readily obtained as

$$\begin{aligned} Var\left(\mu^{*}\right) &= \sigma^{2}\boldsymbol{\mu}'\boldsymbol{\Sigma}^{-1}\boldsymbol{\mu}/\Delta, \\ Var\left(\sigma^{*}\right) &= \sigma^{2}\mathbf{1}'\boldsymbol{\Sigma}^{-1}\mathbf{1}/\Delta, \\ Cov\left(\mu^{*},\sigma^{*}\right) &= -\sigma^{2}\boldsymbol{\mu}'\boldsymbol{\Sigma}^{-1}\mathbf{1}/\Delta. \end{aligned} \tag{6.28}$$

6.2.2 The Exponential Distribution

Consider the two-parameter exponential distribution with probability density function

$$f\left(y;\mu,\sigma\right) = \frac{1}{\sigma}e^{-(y-\mu)/\sigma},\ y \geq \mu,\ \sigma > 0.$$

Let us denote the general progressively Type-II censored sample from this distribution by ${}_{r}Y_{r+i:m:n}$, $i = 1, 2, \cdots, m-r$. Then, again using the moments for the standard exponential distribution obtained in Section 3.4.2 and the formulas in (6.27), we derive, after much algebraic simplification, the BLUEs of μ and σ to be [Balakrishnan and Sandhu (1996)]

$$\mu^{*} = {}_{r}Y_{r+1:m:n} - \frac{\alpha_{r+1:n}}{m-r-1}\sum_{i=r+2}^{m}\left(R_i+1\right)\left({}_{r}Y_{i:m:n} - {}_{r}Y_{r+1:m:n}\right) \tag{6.29}$$

and

$$\sigma^{*} = \frac{1}{m-r-1}\sum_{i=r+2}^{m}\left(R_i+1\right)\left({}_{r}Y_{i:m:n} - {}_{r}Y_{r+1:m:n}\right), \tag{6.30}$$

respectively. Further, upon using (6.28), the variances and covariance of these estimators are given by Balakrishnan and Sandhu (1996) as

$$Var\left(\mu^{*}\right) = \sigma^{2}\left(\frac{\alpha_{r+1}^{2}}{m-r-1} - \beta_{r+1:n}\right), \tag{6.31}$$

$$Var\left(\sigma^{*}\right) = \frac{\sigma^{2}}{m-r-1} \tag{6.32}$$

and

$$Cov\left(\mu^{*},\sigma^{*}\right) = -\sigma^{2}\frac{\alpha_{r+1:n}}{m-r-1}\,. \tag{6.33}$$

Remark 6.7 By writing σ^{*} in (6.30) equivalently as

$$\begin{aligned} \sigma^{*} &= \frac{1}{m-r-1}\sum_{j=2}^{m-r}\left(n-r-R_{r+1}-\cdots-R_{r+j-1}-j+1\right) \\ &\quad \times\left({}_{r}Y_{r+j:m:n} - {}_{r}Y_{r+j-1:m:n}\right) \end{aligned}$$

and making use of the fact that the spacings $(n - r - R_{r+1} - \cdots - R_{r+j-1} - j + 1)({}_rY_{r+j:m:n} - {}_rY_{r+j-1:m:n})/\sigma$, $j = 2, \cdots, m$, are independent standard exponential random variables, we have $2\,(m - r - 1)\,\sigma^*/\sigma$ to be distributed exactly as a central chi-square random variable with $2\,(m - r - 1)$ degrees of freedom. This fact can be used to develop confidence intervals or tests of hypothesis about σ. However, no such exact distributional result can be obtained for the pivotal quantity $(\mu^* - \mu)\,/\sigma^*$ in this general case; for the special case $r = 0$, this quantity simplifies to $\frac{Y_{1:m:n} - \mu}{\sigma^*} - \frac{1}{n}$. Using the fact that $2\,(Y_{1:m:n} - \mu)\,/\sigma^*$ has a chi-square distribution with 2 degrees of freedom and that the random variables $2\,(m - 1)\,\sigma^*/\sigma$ and $2\,(Y_{1:m:n} - \mu)\,/\sigma^*$ are statistically independent, we find that $\frac{\mu^* - \mu}{\sigma^*} + \frac{1}{n}$ has an F-distribution with $(2{,}2m-2)$ degrees of freedom. This pivotal quantity can therefore be used to perform tests of inference concerning μ; see Viveros and Balakrishnan (1994).

Remark 6.8 It is important to note here that the precision of the BLUEs of μ and σ, as given by (6.31)–(6.33), depend only on r, m, and n, and not on the progressive censoring scheme $(R_{r+1}, \cdots, R_m)$.

Example 6.2 For the purpose of illustration, let us consider Nelson's data (1982, p. 228, Table 6.1) which gives observations on times to breakdown of an insulating fluid in an accelerated test conducted at various test voltages. In analyzing the complete data, Nelson assumed a scaled Weibull distribution for the times to breakdown (from the 90% confidence interval [0.459, 1.381] that he determined for the shape parameter, it is quite clear that an exponential model is also appropriate for this data). For the purpose of illustrating the methods of inference presented here, we generated the following general progressively Type-II censored sample from the $n = 19$ observations (the smallest observation was censored and three stages of progressive censoring were introduced).

i	1	2	3	4	5	6	7	8
y_i	–	0.78	0.96	1.31	2.78	4.85	6.50	7.35
R_i	–	0	3	0	3	0	0	5

In Example 6.1, we assumed a one-parameter (scaled) exponential distribution for the data at hand, and determined the BLUE of θ and its standard error as $\theta^* = 9.09$ and $SE\,(\theta^*) = 3.21$. If, instead, we assume a two-parameter (location-scale) exponential distribution for the data at hand, we get from (6.29) and (6.30) that $\mu^* = -0.274$ and $\sigma^* = 9.743$, and their standard errors to be $SE\,(\mu^*) = \sigma^*\sqrt{0.0078} = 0.861$ and $SE\,(\sigma^*) = \sigma^*/\sqrt{6} = 3.978$. From these values, we calculate the BLUE of the expected time to breakdown to be $\mu^* + \sigma^* = 9.469$ with its standard error being 3.735. We may note that the standard error in this case is larger than in the one-parameter case. (The negative value of μ^* should probably indicate here that the two-parameter model is not quite appropriate).

Example 6.3 In Example 3.2, a general progressively Type-II censored sample from the exponential distribution with the following censoring scheme was generated (with $\mu = 25$ and $\sigma = 10$):

$$\begin{aligned} n &= 50;\ r = 5;\ R_6 = 2;\ R_7 = 1;\ R_8 = 0;\ R_9 = 2;\ R_{10} = 1;\ R_{11} = 3; \\ R_{12} &= 2;\ R_{13} = 3;\ R_{14} = 4;\ R_{15} = 3;\ R_{16} = 0;\ R_{17} = 2;\ R_{18} = 1;\ R_{19} = 2; \\ R_{20} &= 4. \end{aligned}$$

The observed sample (of size 15) was:

25.99609, 26.17323, 26.55884, 26.65558, 27.32842, 27.52826, 28.58114, 28.58350, 28.68850, 29.09515, 29.17521, 29.47387, 29.61337, 33.44267, 35.74206.

The BLUEs are obtained as $\mu^* = 24.66292$ and $\sigma^* = 10.54061$ with standard errors $SE\,(\mu^*) = 0.65082$ and $SE\,(\sigma^*) = 2.81710$. These estimated values are very close to the true values of $\mu = 25$ and $\sigma = 10$.

The BLUEs of μ and σ in (6.29) and (6.30) can also be derived directly by adopting the approach of Balakrishnan and Rao (1997b). As in Section 6.1.2, let us define

$$\begin{aligned} Z_{r+1} &= ({}_rY_{r+1:m:n} - \mu)/\sigma, \\ Z_{r+2} &= (n - r - R_{r+1} - 1)({}_rY_{r+2:m:n} - {}_rY_{r+1:m:n})/\sigma, \\ \cdots & \quad \cdots \\ \cdots & \quad \cdots \\ Z_m &= (n - r - R_{r+1} - \cdots - R_{m-1} - m + r + 1)({}_rY_{m:m:n} - {}_rY_{m-1:m:n})/\sigma. \end{aligned}$$

As before, we have Z_i's to be independent random variables and their means, variances and covariances are exactly as given in Section 6.1.2. Let us now take arbitrary linear estimators of μ and σ as

$$\mu^* = \sum_{i=r+1}^{m} a_i\ {}_rY_{i:m:n} = a'_{r+1}\ {}_rY_{r+1:m:n} + \sigma \sum_{i=r+2}^{m} a'_i Z_i \tag{6.34}$$

and

$$\sigma^* = \sum_{i=r+1}^{m} b_i\ {}_rY_{i:m:n} = b'_{r+1}\ {}_rY_{r+1:m:n} + \sigma \sum_{i=r+2}^{m} b'_i Z_i. \tag{6.35}$$

We then readily find

$$\begin{aligned} E(\mu^*) &= a'_{r+1}(\mu + \sigma\alpha_{r+1:n}) + \sigma \sum_{i=r+2}^{m} a'_i \\ &= \mu\ a'_{r+1} + \sigma \left\{ a'_{r+1}\alpha_{r+1:n} + \sum_{i=r+2}^{m} a'_i \right\} \qquad (6.36) \end{aligned}$$

and

$$\begin{aligned} E(\sigma^*) &= b'_{r+1}(\mu + \sigma\alpha_{r+1:n}) + \sigma \sum_{i=r+2}^{m} b'_i \\ &= \mu\, b'_{r+1} + \sigma \left\{ b'_{r+1}\alpha_{r+1:n} + \sum_{i=r+2}^{m} b'_i \right\}. \end{aligned} \tag{6.37}$$

If we require the estimators μ^* and σ^* in (6.34) and (6.35) to be unbiased for μ and σ, respectively, then (6.36) and (6.37) give the unbiasedness conditions as

$$a'_{r+1} = 1,\ \alpha_{r+1:n} + \sum_{i=r+2}^{m} a'_i = 0,\ b'_{r+1} = 0,\ \sum_{i=r+2}^{m} b'_i = 1. \tag{6.38}$$

If we now wish to minimize Var(μ^*)+ Var(σ^*) subject to the unbiasedness conditions in (6.38), we can use the Lagrangian method and consider

$$\begin{aligned} S &= \beta_{r+1:n} + \sum_{i=r+2}^{m} a_i'^2 + \sum_{i=r+2}^{m} b_i'^2 - 2\lambda_1 \left\{ \alpha_{r+1:n} + \sum_{i=r+2}^{m} a'_i \right\} \\ &\qquad -2\lambda_2 \left\{ \sum_{i=r+2}^{m} b'_i - 1 \right\}, \end{aligned}$$

where λ_1 and λ_2 are the Lagrangian multipliers. Minimizing S with respect to a'_i and b'_i $(i = r+2, \cdots, m)$, we obtain the equations

$$a'_i - \lambda_1 = 0 \ \text{ and } \ b'_i - \lambda_2 = 0 \ \text{ for } \ i = r+2, \cdots, m, \tag{6.39}$$

which gives, from the unbiasedness conditions in (6.38),

$$a'_i = -\frac{\alpha_{r+1:n}}{m-r-1} \ \text{ and } \ b'_i = \frac{1}{m-r-1} \ \text{ for } \ i = r+2, \cdots, m.$$

We then obtain the coefficients a_i's and b_i's as

$$\begin{aligned} a_{r+1} &= a'_{r+1} - (n - r - R_{r+1} - 1)a'_{r+2} \\ &= 1 + (n - r - R_{r+1} - 1)\, \frac{\alpha_{r+1:n}}{m-r-1}, \\ a_{r+2} &= (n - r - R_{r+1} - 1)a'_{r+2} - (n - r - R_{r+1} - R_{r+2} - 2)a'_{r+3} \\ &= -(R_{r+2} + 1)\, \frac{\alpha_{r+1:n}}{m-r-1}, \\ \cdots & \quad \cdots \\ \cdots & \quad \cdots \\ a_m &= (n - m - R_{r+1} - \cdots - R_{m-1} + 1)a'_m \\ &= -(R_m + 1)\, \frac{\alpha_{r+1:n}}{m-r-1}, \end{aligned}$$

and

$$\begin{aligned}
b_{r+1} &= b'_{r+1} - (n-r-R_{r+1}-1)b'_{r+2} \\
&= -\frac{(n-r-R_{r+1}-1)}{m-r-1}, \\
b_{r+2} &= (n-r-R_{r+1}-1)b'_{r+2} - (n-r-R_{r+1}-R_{r+2}-2)b'_{r+3} \\
&= \frac{R_{r+2}+1}{m-r-1}, \\
\cdots & \quad \cdots \\
\cdots & \quad \cdots \\
b_m &= (n-m-R_{r+1}-\cdots-R_{m-1}+1)b'_m \\
&= \frac{R_m+1}{m-r-1}.
\end{aligned}$$

These coefficients, when substituted in (6.34) and (6.35), give the BLUEs of μ and σ exactly as presented in (6.29) and (6.30), respectively. Furthermore, the expressions of variances and covariance of these estimators agree with those in (6.31) – (6.33). This establishes the fact that the best linear unbiased estimators are "trace-efficient linear unbiased estimators." Instead of minimizing the trace $\mathrm{Var}(\mu^*)+\mathrm{Var}(\sigma^*)$, if we choose to minimize the determinant $\mathrm{Var}(\mu^*)\,\mathrm{Var}(\sigma^*)$ - $(\mathrm{Cov}(\mu^*,\sigma^*))^2$, we will find that the estimators so derived are once again the best linear unbiased estimators; this states the fact that the best linear unbiased estimators are "determinant-efficient linear unbiased estimators". As Balakrishnan and Rao (1999) have proved, these properties are subsumed in the general result that the best linear unbiased estimators possess "complete covariance matrix dominance" in the class of all linear unbiased estimators.

Remark 6.9 For the special case of progressive Type-II right censoring (viz., $r=0$), we obtain from (6.29) and (6.30) that

$$\mu^* = Y_{1:m:n} - \frac{1}{n(m-1)}\sum_{i=2}^{m}(R_i+1)\left(Y_{i:m:n}-Y_{1:m:n}\right)$$

and

$$\sigma^* = \frac{1}{m-1}\sum_{i=2}^{m}(R_i+1)\left(Y_{i:m:n}-Y_{1:m:n}\right).$$

As pointed our earlier in Remark 6.7, in this case exact inference for σ and μ can be developed based on central chi-square and F distributions, respectively. Also, Chandresekar, Leo Alexander and Balakrishnan (1999) have recently shown that a minimum risk equivariant estimator of $(\mu,\sigma)^T$ is given by

$$\left(Y_{1:m:n} - \frac{1}{mn}\sum_{i=2}^{m}(R_i+1)\left(Y_{i:m:n}-Y_{1:m:n}\right),\ \frac{1}{m}\sum_{i=2}^{m}(R_i+1)\left(Y_{i:m:n}-Y_{1:m:n}\right)\right)^T.$$

These authors have also discussed the minimum risk equivariant estimation of $(\mu, \sigma^2)^T$.

Remark 6.10 Recently, Chandrasekar and Balakrishnan (1999) proposed a multiparameter version of Tukey's (1965) linear sensitivity measure [see also Nagaraja (1994)] and showed that for the progressively Type-II right censored sample from the *Exponential*(μ, σ) distribution, this measure of information equals

$$S(\boldsymbol{Y}; \mu, \sigma) = \frac{1}{\sigma^2}\begin{pmatrix} n^2 & m \\ m & m \end{pmatrix}.$$

As Chandrasekar and Balakrishnan (1999) have shown, this is precisely the inverse of the variance-covariance matrix of the BLUEs μ^* and σ^* presented in Remark 6.9.

Remark 6.11 Suppose k general progressively Type-II censored samples are available from the *Exponential*(μ, σ) distribution, with the i-th censored sample denoted by

$$_{r_i}Y_{i;r_i+1:m_i:n_i},\ _{r_i}Y_{i;r_i+2:m_i:n_i}, \cdots,\ _{r_i}Y_{i;m_i:m_i:n_i}$$

and the corresponding general progressive censoring scheme by $(R_{i,r_i+1}, R_{i,r_i+2}, \cdots, R_{i,m_i})$ for $i = 1, 2, \cdots, k$. Then, Balakrishnan, Cramer, Kamps and Schenk (1999) have derived the BLUEs of μ and σ to be

$$\begin{aligned}\mu^* &= \frac{1}{\sum_{i=1}^k \frac{m_i - r_i + 1}{\beta_{r_i+1:n_i}}} \\ &\quad \times \Bigg[\sum_{i=1}^{k} \frac{m_i - r_i + 1}{\beta_{r_i+1:n_i}}\, {}_{r_i}Y_{i;r_i+1:m_i:n_i} - \sum_{i=1}^{k} \frac{\alpha_{r_i+1:n_i}}{\beta_{r_i+1:n_i}} \\ &\quad \times \sum_{j=r_i+2}^{m_i} (R_{i,j}+1)\left({}_{r_i}Y_{i;j:m_i:n_i} - {}_{r_i}Y_{i;r_i+1:m_i:n_i}\right)\Bigg]\end{aligned}$$

and

$$\begin{aligned}\sigma^* &= \frac{1}{\sum_{i=1}^k \frac{m_i - r_i + 1}{\beta_{r_i+1:n_i}}} \sum_{i=1}^{k} \frac{1}{\beta_{r_i+1:n_i}} \\ &\quad \times \sum_{j=r_i+2}^{m_i} (R_{i,j}+1)\left({}_{r_i}Y_{i;j:m_i:n_i} - {}_{r_i}Y_{i;r_i+1:m_i:n_i}\right).\end{aligned}$$

6.2.3 The Uniform Distribution

Exact expressions for the BLUEs of μ and σ in the two-parameter *Uniform*$(\mu, \mu + \sigma)$ distribution can also be obtained, as well as their variances and covariance, using (6.27) and (6.28). Let us denote the general progressively

Type-II censored order statistics from the *Uniform*$(\mu, \mu+\sigma)$ distribution by ${}_rY_{r+1:m:n}, {}_rY_{r+2:m:n}, \cdots, {}_rY_{m:m:n}$.

Letting

$$\begin{aligned}\Delta &= \left\{\sum_{i=1}^{m-r-1} \frac{\gamma_{m-i}\gamma_{m-i+1} - \alpha_{m-i}\alpha_{m-i+1} - \beta_{m-i+1} - \alpha_{m-i+1}\gamma_{m-i+1}\beta_{m-i}}{\beta_{m-i}\beta_{m-i+1}\prod_{j=m-i+1}^{m}\alpha_j\gamma_j}\right.\\ &\quad \left.\times\left(1-\prod_{j=m-i+1}^{m}\alpha_j\right) + \frac{(1-\alpha_{r+1})\left(1-\prod_{j=r+1}^{m}\alpha_j\right)}{\beta_{r+1}\alpha_{r+1}\prod_{j=r+2}^{m}\alpha_j\gamma_j}\right\}\\ &\quad \times\left\{\sum_{i=1}^{m-r-1}\frac{\gamma_{m-i}\gamma_{m-i+1} - \alpha_{m-i}\alpha_{m-i+1} - 2\beta_{m-i+1}}{\beta_{m-i}\beta_{m-i+1}\prod_{j=m-i+1}^{m}\alpha_j\gamma_j}\right.\\ &\quad \left.+\frac{1}{\beta_{r+1}\alpha_{r+1}\prod_{j=r+2}^{m}\alpha_j\gamma_j}\right\}\\ &\quad -\left\{\sum_{i=1}^{m-r-1}\frac{\gamma_{m-i}\gamma_{m-i+1} - \alpha_{m-i}\alpha_{m-i+1} - \beta_{m-i+1} - \alpha_{m-i+1}\gamma_{m-i+1}\beta_{m-i}}{\beta_{m-i}\beta_{m-i+1}\prod_{j=m-i+1}^{m}\alpha_j\gamma_j}\right.\\ &\quad \left.+\frac{1-\alpha_{r+1}}{\beta_{r+1}\alpha_{r+1}\prod_{j=r+2}^{m}\alpha_j\gamma_j}\right\}^2, \qquad (6.40)\end{aligned}$$

where α_i, β_i, and γ_i are exactly as defined earlier in Section 3.4.3, we obtain (after a great deal of algebra), as given in Aggarwala and Balakrishnan (1998a), the following expressions:

$$\begin{aligned}\sigma^*\Delta &= \sum_{i=2}^{m-r-1}\frac{\gamma_{m-i}\gamma_{m-i+1} - \alpha_{m-i}\alpha_{m-i+1} - \beta_{m-i+1} - \alpha_{m-i+1}\gamma_{m-i+1}\beta_{m-i}}{\beta_{m-i}\beta_{m-i+1}\prod_{j=m-i+1}^{m}\alpha_j\gamma_j}\\ &\qquad \times {}_rY_{r+i:m:n}\\ &\quad + \frac{1-\alpha_{r+1}}{\beta_{r+1}\beta_m\alpha_{r+1}\prod_{j=r+2}^{m}\alpha_j\gamma_j}\,{}_rY_{m:m:n}\\ &\quad - \left\{\sum_{i=2}^{m-r-1}\frac{\gamma_{m-i}\gamma_{m-i+1} - \alpha_{m-i}\alpha_{m-i+1} - \beta_{m-i+1} - \alpha_{m-i+1}\gamma_{m-i+1}\beta_{m-i}}{\beta_{m-i}\beta_{m-i+1}\beta_m\prod_{j=m-i+1}^{m}\alpha_j\gamma_j}\right.\\ &\quad \left.+ \frac{1-\alpha_{r+1}}{\beta_{r+1}\beta_m\alpha_{r+1}\prod_{j=r+2}^{m}\alpha_j\gamma_j}\right\}{}_rY_{r+1:m:n} \qquad (6.41)\end{aligned}$$

with

$$\begin{aligned}&Var(\sigma^*)\\ &\quad = \frac{\sigma^2}{\Delta}\left\{\sum_{i=1}^{m-r-1}\frac{\gamma_{m-i}\gamma_{m-i+1} - \alpha_{m-i}\alpha_{m-i+1} - 2\beta_{m-i+1}}{\beta_{m-i}\beta_{m-i+1}\prod_{j=m-i+1}^{m}\alpha_j\gamma_j}\right.\\ &\quad\quad \left.+\frac{1}{\beta_{r+1}\alpha_{r+1}\prod_{j=r+2}^{m}\alpha_j\gamma_j}\right\}, \qquad (6.42)\end{aligned}$$

$$\begin{aligned}
\mu^* \Delta &= \left\{ \sum_{i=2}^{m-r-1} \frac{\gamma_{m-i}\gamma_{m-i+1} - \alpha_{m-i}\alpha_{m-i+1} - \beta_{m-i+1} - \alpha_{m-i+1}\gamma_{m-i+1}\beta_{m-i}}{\beta_{m-i}\beta_{m-i+1}\beta_m \prod_{j=m-i+1}^{m} \alpha_j\gamma_j} \right. \\
&\times \left. \left(1 - \prod_{j=m-i+1}^{m} \alpha_j\right) + \frac{(1-\alpha_{r+1})\left(1 - \prod_{j=r+1}^{m}\alpha_j\right)}{\beta_{r+1}\beta_m\alpha_{r+1}\prod_{j=r+2}^{m}\alpha_j\gamma_j} \right\} {}_rY_{r+1:m:n} \\
&- \left\{ \sum_{i=2}^{m-r-1} \frac{\gamma_{m-i}\gamma_{m-i+1} - \alpha_{m-i}\alpha_{m-i+1} - \beta_{m-i+1} - \alpha_{m-i+1}\gamma_{m-i+1}\beta_{m-i}}{\beta_{m-i}\beta_{m-i+1}\beta_m \prod_{j=m-i+1}^{m} \alpha_j\gamma_j} \right. \\
&\times \left. {}_rY_{r+i:m:n} + \frac{(1-\alpha_{r+1})\, {}_rY_{m:m:n}}{\beta_{r+1}\beta_m\alpha_{r+1}\prod_{j=r+2}^{m}\alpha_j\gamma_j} \right\} (1-\alpha_m) \qquad (6.43)
\end{aligned}$$

with

$$\begin{aligned}
&Var(\mu^*) \\
&= \frac{\sigma^2}{\Delta} \left\{ \sum_{i=1}^{m-r-1} \frac{\gamma_{m-i}\gamma_{m-i+1} - \alpha_{m-i}\alpha_{m-i+1} - \beta_{m-i+1} - \alpha_{m-i+1}\gamma_{m-i+1}\beta_{m-i}}{\beta_{m-i}\beta_{m-i+1} \prod_{j=m-i+1}^{m} \alpha_j\gamma_j} \right. \\
&\times \left. \left(1 - \prod_{j=m-i+1}^{m} \alpha_j\right) + \frac{(1-\alpha_{r+1})\left(1 - \prod_{j=r+1}^{m}\alpha_j\right)}{\beta_{r+1}\alpha_{r+1}\prod_{j=r+2}^{m}\alpha_j\gamma_j} \right\} \qquad (6.44)
\end{aligned}$$

and

$$\begin{aligned}
&Cov(\mu^*, \sigma^*) \\
&= -\frac{\sigma^2}{\Delta} \left\{ \sum_{i=1}^{m-r-1} \frac{\gamma_{m-i}\gamma_{m-i+1} - \alpha_{m-i}\alpha_{m-i+1} - \beta_{m-i+1} - \alpha_{m-i+1}\gamma_{m-i+1}\beta_{m-i}}{\beta_{m-i}\beta_{m-i+1} \prod_{j=m-i+1}^{m} \alpha_j\gamma_j} \right. \\
&+ \left. \frac{1-\alpha_{r+1}}{\beta_{r+1}\alpha_{r+1}\prod_{j=r+2}^{m}\alpha_j\gamma_j} \right\}. \qquad (6.45)
\end{aligned}$$

Remark 6.12 For the case of no censoring, upon using the expressions of α_i, β_i and γ_i in (6.14), the above results simplify to the well-known results

$$\begin{aligned}
\sigma^* &= \frac{n+1}{n-1}\left(Y_{n:n} - Y_{1:n}\right), \quad \mu^* = \frac{1}{n-1}\left(nY_{1:n} - Y_{n:n}\right), \\
Var(\sigma^*) &= \frac{2\sigma^2}{(n+2)(n-1)}, \quad Var(\mu^*) = \frac{n\sigma^2}{(n+1)(n+2)(n-1)}, \\
Cov(\mu^*, \sigma^*) &= \frac{-\sigma^2}{(n+1)(n-1)}.
\end{aligned}$$

6.2.4 The Pareto Distribution

Let us denote the progressively Type-II right censored order statistics from the location-scale shifted $Pareto\,(\nu)$ distribution, that is, with density function

$$f\,(w;\nu,\mu,\sigma) = \frac{\nu}{\sigma}\left(\frac{w-\mu}{\sigma}\right)^{-\nu-1}, \; w > \mu+\sigma, \; \nu > 0, \; -\infty < \mu < \infty, \sigma > 0,$$

by $W_{1:m:n}, W_{2:m:n}, \cdots, W_{m:m:n}$. In other words, $W_{i:m:n} = \mu + \sigma X_{i:m:n}$ where $X_{i:m:n}$, $i = 1, 2, \cdots, m$, are progressively Type-II right censored order statistics from the standard $Pareto\,(\nu)$ distribution. From the expression of the inverted variance-covariance matrix for the standard $Pareto\,(\nu)$ distribution given in (6.15), we can obtain the exact BLUEs of the location and scale parameters, μ and σ, respectively, using (6.27). Letting $E(X_{i:m:n}) = \mu_i, i = 1, 2, ..., m$, we are able to write the explicit expressions of the BLUEs as follows [see in Aggarwala and Balakrishnan (1998c)]:

$$\mu^* = \frac{-1}{\Delta}\sum_{j=1}^{m}\sum_{i=1}^{m}\sum_{l=1}^{m}\sum_{k=1}^{m}\mu_i c^{i,k}\left(\mu_l - \mu_k\right) c^{l,j} W_{j:m:n}, \tag{6.46}$$

$$\sigma^* = \frac{1}{\Delta}\sum_{j=1}^{m}\sum_{i=1}^{m}\sum_{l=1}^{m}\sum_{k=1}^{m} c^{i,k}\left(\mu_l - \mu_k\right) c^{l,j} W_{j:m:n}, \tag{6.47}$$

where

$$\Delta = \left(\sum_{i=1}^{m}\sum_{j=1}^{m} c^{i,j}\right)\left(\sum_{i=1}^{m}\sum_{j=1}^{m}\mu_i\mu_j c^{i,j}\right) - \left(\sum_{i=1}^{m}\sum_{j=1}^{m}\mu_i c^{i,j}\right)^2 \tag{6.48}$$

and $c^{i,j}$ are as presented earlier in (6.15).

Furthermore, using (6.28),

$$Var\,(\mu^*) = \sigma^2\left(\sum_{i=1}^{m}\sum_{j=1}^{m}\mu_i\mu_j c^{i,j}\right)/\Delta, \tag{6.49}$$

$$Var\,(\sigma^*) = \sigma^2\left(\sum_{i=1}^{m}\sum_{j=1}^{m} c^{i,j}\right)/\Delta, \tag{6.50}$$

and

$$Cov\,(\mu^*,\sigma^*) = -\sigma^2\left(\sum_{i=1}^{m}\sum_{j=1}^{m}\mu_i c^{i,j}\right)/\Delta. \tag{6.51}$$

Remark 6.13 The above general expressions for the BLUEs of location and scale parameters and their variances and covariance may be used for BLUEs based on progressively Type-II right censored order statistics from any distribution which has a variance-covariance matrix of the form $\sigma_{i,j} = s_i t_j$. Here, $c^{i,j}$ are the elements of the inverted variance-covariance matrix of the order statistics and μ_i are the elements of the mean vector of $\mathbf{X}$.

Example 6.4 Consider a sample of size $n = 15$, and the following progressive Type-II right censoring scheme: $R_1 = 5, R_2 = 0, R_3 = 2, R_4 = 0, R_5 = 3$ (so that $m = 5$). Using (2.32), a progressively Type-II right censored sample from the *Pareto* (3) distribution with location parameter $\mu = 0$ and scale parameter $\sigma = 5$ was generated. The generated sample was: $5.11073, 5.34932, 5.36434, 5.70137, 5.90067$.

From (6.46), we are able to compute the coefficients of the progressively Type-II right censored order statistics corresponding to the BLUE for μ. These are: 8.00439, -1.18584, -2.71754, -0.94208, -2.15893. Using (6.47), the coefficients corresponding to the BLUE for σ are: –6.84874, 1.15948, 2.65715, 0.92115, 2.11096. Therefore, the estimates of the location and scale parameters based on this sample are

$$\mu^* = 1.87680, \quad \sigma^* = 3.16207,$$

and their standard errors and estimated covariance are

$$SE(\mu^*) = 1.83036, \quad SE(\sigma^*) = 1.79113, \quad \widehat{Cov}(\mu^*, \sigma^*) = -3.27578.$$

It should be noted here that it has been assumed that the shape parameter ($\nu = 3$) is known here. This may not always be the case. Aggarwala and Balakrishnan (1998c) have conducted a sensitivity analysis with respect to ν for this example in the event that this parameter is not known to the practitioner and must be estimated in some way. The following is a table of the BLUEs which would result if different values of ν are assumed:

ν	μ^*	σ^*
2.6	2.34507	2.69475
2.8	2.11118	2.92813
3.0	1.87680	3.16207
3.2	1.64202	3.39645
3.4	1.40691	3.63120

6.2.5 The Laplace Distribution

Using the results presented in Section 5.3.3 for the single and product moments of progressively Type-II right censored order statistics from arbitrary continuous symmetric distributions, one may determine the best linear unbiased estimators (BLUEs) for parameters of scale- and location-scale shifted symmetric distributions when samples are progressively Type-II right censored, provided the single and product moments of progressively Type-II right censored and progressively Type-II left withdrawn order statistics from the corresponding folded distribution are known.

We can, therefore, consider the Laplace (or double exponential) distribution with location parameter μ and scale parameter σ, whose cumulative distribution function is given by

$$F(y) = \begin{cases} \frac{1}{2}e^{(y-\mu)/\sigma}, & x \le \mu, \\ 1 - \frac{1}{2}e^{-(y-\mu)/\sigma}, & x > \mu. \end{cases} \tag{6.52}$$

This distribution has been used to model certain real life-test data [see Bain and Engelhardt (1991)].

Notice that we may not be able to easily write down explicit expressions for the BLUEs of μ and σ, or even the inverse of the variance-covariance matrix, as it is not necessarily in a form, such as $\sigma_{ij} = s_i t_j$, which is easily explicitly inverted. However, as long as the mean vector is known and the variance-covariance matrix is obtainable and invertible, as is the case for the Laplace distribution, we may use the general formulae (6.27) and (6.28) to obtain the required BLUEs and their variances and covariance. Similar statements may be made for scale-shifted distributions, where we would instead use (6.1) and (6.2). The real problem, then, in being able to perform this type of linear inference for general distributions is the determination of single and product moments of progressively Type-II censored order statistics from standard distributions. It is with this realization that methods to obtain moments of progressively Type-II censored order statistics such as those presented in Chapters 4 and 5 become so powerful and important to the practitioner and researcher. It should be noted that although the estimates may be calculated with just the knowledge of the single and product moments of the progressively Type-II censored order statistics, the procedure will be performed more efficiently for distributions in which formulae for BLUEs are explicitly known. Furthermore, as in the case of the exponential distribution, for example, the knowledge of the *distributions* of estimators allows for even more sophisticated inference (like confidence intervals and tests of hypotheses) for the parameters of interest.

Example 6.5 A progressively Type-II right censored sample of size $m = 10$ from a sample of size $n = 20$ from the Laplace distribution in (6.41) with $\mu = 25$ and $\sigma = 5$ was simulated, with censoring scheme $\boldsymbol{R} = (2, 0, 0, 2, 0, 0, 0, 2, 0, 4)$. This sample was simulated using the algorithm given in Section 3.2. The simulated progressively Type-II right censored sample is as follows:

19.21167876, 21.97364262, 23.41776818, 23.66253070, 23.80222832,
24.23017797, 25.62072188, 25.86990938, 26.47997028, 27.55344134.

Using the results for moments of progressively Type-II right censored order statistics from the exponential distribution given in Section 2.2 as well as Theorems 5.9 and 5.10, moments of progressively Type-II left withdrawn order statistics from the exponential distribution were calculated for the censoring scheme $\boldsymbol{R}$. It may be noted here that an alternate method of obtaining these moments of progressively Type-II left withdrawn order statistics is through recursive algorithms. Recurrence relations for moments of progressively Type-II left withdrawn order statistics will be very similar to those for moments of progressively Type-II right censored order statistics for the exponential distribution and its truncated forms established in Chapter 4.

Using (5.7) and (5.6), we are now able to calculate the desired single and product moments, and therefore, variances and covariances, of progressively Type-II right censored order statistics with censoring scheme $\boldsymbol{R}$ from the Laplace distribution; see Aggarwala and Balakrishnan (1999). This was done using Maple V, Release 3, with exact results (fractional form). Using these results and the formulae for best linear unbiased estimators for location and scale parameters given in (6.27), the best linear unbiased estimators for μ and σ were obtained. The coefficients $a_{i:m:n}$ for the BLUE of μ, to 10 decimal places, are given by

$$-0.0215281009,\ 0.0098155747,\ 0.0069570375,\ -0.0901429654,$$
$$0.2068718600,\ 0.0772377616,\ 0.1039355790,\ 0.2099103064,$$
$$0.1371049958,\ 0.3598379513,$$

and the coefficients $b_{i:m:n}$ for the BLUE of σ are given by

$$-0.1310403616,\ -0.0930984026,\ -0.0957542723,\ -0.2299462758,$$
$$0.0918178668,\ -0.0494235976,\ -0.0299327584,\ 0.0453859038,$$
$$0.0594679977,\ 0.4325239001.$$

The variance of the BLUE of μ is $0.0750292328\sigma^2$, the variance of the BLUE of σ is $0.1097132433\sigma^2$, and the covariance of the BLUEs is $0.0108583470\sigma^2$.

Thus, based on the progressively Type-II right censored sample given above, we obtain the BLUEs of μ and σ and their standard errors as

$$\begin{aligned} \mu^* &= 26.26607179, \quad SE(\mu^*) = 0.7233296358, \\ \sigma^* &= 2.640711801, \quad SE(\sigma^*) = 0.2897210562. \end{aligned}$$

Remark 6.14 Using the results given in Govindarajulu (1966), the variances and covariance of the BLUEs for a conventional Type-II right censored sample of size 10 from a sample of size $n = 20$ from a Laplace distribution were also calculated. In this case, the censoring pattern can be written as $(0, 0, 0, 0, 0, 0, 0, 0, 0, 10)$. The variance of the BLUE of μ is $0.0700\sigma^2$, the variance of the BLUE of σ is $0.1095\sigma^2$, and the covariance of the BLUEs is $0.0133\sigma^2$. These variances are only slightly more favorable than the variances given above for the censoring pattern $(2, 0, 0, 2, 0, 0, 0, 2, 0, 4)$. However, for this censoring scheme ($\boldsymbol{R}$), items censored early on may be of use to the experimenter. The question of an optimal censoring pattern in terms of the variances of BLUEs will be addressed further in Chapter 10.

6.2.6 The Extreme Value Distribution

The standard extreme value distribution has probability density function given by

$$f(x) = e^{x} \, exp\{-e^{x}\}, \ -\infty < x < \infty. \tag{6.53}$$

This distribution has been used extensively in modelling failure times of items; see, for example, Nelson (1982) and Viveros and Balakrishnan (1994). In addition, because of its close relation to the Weibull random variable, this distribution will be an important one to consider.

We have not as of yet discussed any mathematical properties of progressively Type-II censored order statistics from the extreme value distribution, nor have we considered the problem of obtaining moments of progressively Type-II censored order statistics from this distribution in order to perform linear inference, for example. This is due in part to the fact that if we even consider the special case of *no censoring* for this distribution (that is, the usual order statistics), we are unable to develop efficient closed-form expressions for moments. However, tables of means, variances and covariances of the usual order statistics for the standard extreme value distribution given in (6.53) are available in Balakrishnan and Chan (1992) for sample sizes up to $n = 30$. Using these moments of the usual order statistics and Thomas and Wilson's (1972) results presented earlier in Section 5.2, we are able to numerically determine the moments of progressively Type-II right censored order statistics from the standard extreme value distribution. Using these values and (6.27), we may perform location-scale linear inference based on progressively Type-II right censored samples from extreme value distributions. As mentioned earlier, this argument can be applied to any distribution for which only the single and product moments of the *usual* order statistics are known.

Example 6.6 As an illustration, we calculated the coefficients of a progressively Type-II right censored sample from the extreme value distribution when the censoring scheme is $(22, 0, 0)$, that is, all censoring is performed immediately following the first observed failure out of $n = 25$ items on test. This was done for a location-scale shifted model. Using the Fortran programming language, we first computed the (single and product) moments of the progressively Type-II right censored order statistics using the moments of the usual order statistics given in Balakrishnan and Chan (1992). We then used (6.27) and obtained the following coefficients for μ^* : 0.0169, 0.0804, 0.9027, and for σ^* : -0.2256, -0.0876, 0.3132. Furthermore, using (6.28), we have $Var(\mu^*) = 0.5613\sigma^2$ and $Var(\sigma^*) = 0.1311\sigma^2$. In comparison, for the conventional Type-II right censoring scheme $(0, 0, 22)$, we can compute (simply using the moments of the usual order statistics) $Var(\mu^*) = 2.9046\sigma^2$ and $Var(\sigma^*) = 0.4847\sigma^2$. These values are much *higher* in the case of conventional censoring than for our selected scheme, and as it turns out, we find that based on a sample size of $n = 25$ and a progressively Type-II right censored sample of size $m = 3$, our selected censoring scheme $(22, 0, 0)$ is the *most efficient* censoring scheme possible if our efficiency criterion is defined by the trace, $Var\,(\mu^*) + Var\,(\sigma^*)$. This will be discussed in more detail in Chapter 10.

6.2.7 First-Order Approximations to the BLUEs

Let us assume that the progressively Type-II censored sample $Y_{1:m:n}^{(R_1,\cdots,R_m)}$, $Y_{2:m:n}^{(R_1,\cdots,R_m)}, \cdots, Y_{m:m:n}^{(R_1,\cdots,R_m)}$ has come from a lifetime distribution belonging to a location-scale parameter family, that is, with density function $f(y;\mu,\sigma) = \frac{1}{\sigma}\, f(\frac{y-\mu}{\sigma})$. As we already saw in Section 6.2.1, using either the generalized least-squares theory [see Lloyd (1952)] or the Lagrangian multiplier method [see Balakrishnan and Rao (1997b, 1999)], one can derive the BLUEs of μ and σ as

$$\mu^* = -\boldsymbol{\alpha}^T\boldsymbol{\Gamma}\boldsymbol{Y} \quad \text{and} \quad \sigma^* = \mathbf{1}^T\boldsymbol{\Gamma}\boldsymbol{Y} \tag{6.54}$$

and their variances and covariance as

$$\begin{aligned} \mathrm{Var}(\mu^*) &= \sigma^2(\boldsymbol{\alpha}^T\boldsymbol{\Sigma}^{-1}\boldsymbol{\alpha})/\Delta, \quad \mathrm{Var}(\sigma^*) = \sigma^2(\mathbf{1}^T\boldsymbol{\Sigma}^{-1}\mathbf{1})/\Delta, \\ \mathrm{Cov}(\mu^*,\sigma^*) &= -\sigma^2(\boldsymbol{\alpha}^T\boldsymbol{\Sigma}^{-1}\mathbf{1})/\Delta. \end{aligned} \tag{6.55}$$

In the above formulas, $\boldsymbol{Y}$ denotes the observed vector of progressively Type-II right censored order statistics $\left(Y_{1:m:n}^{(R_1,\cdots,R_m)}, \cdots, Y_{m:m:n}^{(R_1,\cdots,R_m)}\right)^T$, $\boldsymbol{X} = (\boldsymbol{Y} - \mu\mathbf{1})/\sigma$ denotes the vector of corresponding standardized order statistics, $\mathbf{1}$ is a $m \times 1$ column vector of 1's, $\boldsymbol{\alpha} = E(\boldsymbol{X})$, $\boldsymbol{\Sigma} = \mathrm{Var}(\boldsymbol{X})$, $\boldsymbol{\Gamma}$ is a skew-symmetric matrix given by

$$\boldsymbol{\Gamma} = \frac{\boldsymbol{\Sigma}^{-1}(\mathbf{1}\boldsymbol{\alpha}^T - \boldsymbol{\alpha}\mathbf{1}^T)\boldsymbol{\Sigma}^{-1}}{\Delta}, \tag{6.56}$$

and

$$\Delta = (\boldsymbol{\alpha}^T\boldsymbol{\Sigma}^{-1}\boldsymbol{\alpha})(\mathbf{1}^T\boldsymbol{\Sigma}^{-1}\mathbf{1}) - (\boldsymbol{\alpha}^T\boldsymbol{\Sigma}^{-1}\mathbf{1})^2. \tag{6.57}$$

As in Section 6.1.5, we seek to derive here first-order approximations to the BLUEs of μ and σ in (6.54) by making use of the approximate expressions of $\boldsymbol{\alpha}$ and $\boldsymbol{\Sigma}^{-1}$ presented in (5.26) and (6.22). For this purpose, let us denote

$$\begin{aligned} \Delta_1 &= \frac{\{F^{-1}(\Pi_1)\}^2\; k_2}{\{F^{-1(1)}(\Pi_1)\}^2\; k_1(k_2b_1 - k_1b_2)} \\ &+ \sum_{i=2}^{m-1} \frac{\{F^{-1}(\Pi_i)\}^2\; (k_{i+1}b_{i-1} - k_{i-1}b_{i+1})}{\{F^{-1(1)}(\Pi_i)\}^2\; (k_ib_{i-1} - k_{i-1}b_i)(k_{i+1}b_i - k_ib_{i+1})} \\ &+ \frac{\{F^{-1}(\Pi_m)\}^2\; b_{m-1}}{\{F^{-1(1)}(\Pi_m)\}^2\; b_m(k_mb_{m-1} - k_{m-1}b_m)} \\ &- 2\sum_{i=1}^{m-1} \frac{F^{-1}(\Pi_i)F^{-1}(\Pi_{i+1})}{F^{-1(1)}(\Pi_i)F^{-1(1)}(\Pi_{i+1})(k_{i+1}b_i - k_ib_{i+1})}, \end{aligned} \tag{6.58}$$

$$\begin{aligned}
\Delta_2 &= \frac{k_2}{\{F^{-1(1)}(\Pi_1)\}^2\ k_1(k_2 b_1 - k_1 b_2)} \\
&\quad + \sum_{i=2}^{m-1} \frac{k_{i+1}b_{i-1} - k_{i-1}b_{i+1}}{\{F^{-1(1)}(\Pi_i)\}^2\ (k_i b_{i-1} - k_{i-1}b_i)(k_{i+1}b_i - k_i b_{i+1})} \\
&\quad + \frac{b_{m-1}}{\{F^{-1(1)}(\Pi_m)\}^2\ b_m(k_m b_{m-1} - k_{m-1}b_m)} \\
&\quad - 2\sum_{i=1}^{m-1} \frac{1}{F^{-1(1)}(\Pi_i)F^{-1(1)}(\Pi_{i+1})(k_{i+1}b_i - k_i b_{i+1})}, \qquad (6.59)
\end{aligned}$$

$$\begin{aligned}
\Delta_3 &= \frac{F^{-1}(\Pi_1)k_2}{\{F^{-1(1)}(\Pi_1)\}^2\ k_1(k_2 b_1 - k_1 b_2)} \\
&\quad + \sum_{i=2}^{m-1} \frac{F^{-1}(\Pi_i)(k_{i+1}b_{i-1} - k_{i-1}b_{i+1})}{\{F^{-1(1)}(\Pi_i)\}^2\ (k_i b_{i-1} - k_{i-1}b_i)(k_{i+1}b_i - k_i b_{i+1})} \\
&\quad + \frac{F^{-1}(\Pi_m)b_{m-1}}{\{F^{-1(1)}(\Pi_m)\}^2\ b_m(k_m b_{m-1} - k_{m-1}b_m)} \\
&\quad - \sum_{i=1}^{m-1} \frac{F^{-1}(\Pi_i) + F^{-1}(\Pi_{i+1})}{F^{-1(1)}(\Pi_i)F^{-1(1)}(\Pi_{i+1})(k_{i+1}b_i - k_i b_{i+1})}, \qquad (6.60)
\end{aligned}$$

and

$$\Delta = \Delta_1\Delta_2 - \Delta_3^2. \qquad (6.61)$$

Then, from (6.55) we have the variances and covariance of the BLUEs to be

$$\mathrm{Var}(\mu^*) = \sigma^2\Delta_1/\Delta,\ \mathrm{Var}(\sigma^*) = \sigma^2\Delta_2/\Delta,\ \mathrm{Cov}(\mu^*,\sigma^*) = -\sigma^2\Delta_3/\Delta. \qquad (6.62)$$

Example 6.7 Consider the log-times to breakdown of an insulating fluid in an accelerated test conducted at various test voltages reported by Nelson (1982, p. 228, Table 6.1). In this case, we have the following progressively Type-II censored sample of size $m = 8$ generated from the $n = 19$ observations recorded at 34 kilovolts in Nelson's (1982) Table 6.1, as presented by Viveros and Balakrishnan (1994):

Table 6.2: Progressively censored sample generated from the log-times to breakdown data on insulating fluid tested at 34 kilovolts by Nelson (1982)

i	1	2	3	4	5	6	7	8
$y_{i:8:19}$	−1.6608	−0.2485	−0.0409	0.2700	1.0224	1.5789	1.8718	1.9947
R_i	0	0	3	0	3	0	0	5

Now, let us assume an extreme value distribution with density function

$$f(y;\mu,\sigma) = \frac{1}{\sigma}\, e^{(y-\mu)/\sigma}\, e^{-e^{(y-\mu)/\sigma}}, \qquad -\infty < y < \infty, \tag{6.63}$$

for the distribution of log-times to breakdown. Realize here that this assumption means that we are assuming a two-parameter Weibull distribution for the lifetimes of the units in the original time scale. Then, we have the standardized variable $X = (Y - \mu)/\sigma$ to have the standard extreme value distribution in which case

$$F(x) = 1 - e^{-e^{x}},\ F^{-1}(u) = \ln(-\ln(1-u)) \text{ and } F^{-1^{(1)}}(u) = \frac{-1}{(1-u)\ln(1-u)}\,.$$

From (6.54), we determine the BLUEs of μ and σ to be

$$\begin{aligned} \mu^* &= (-0.09888 \times -1.6608) + (-0.06737 \times -0.2485) \\ &\quad + (-0.00296 \times -0.0409) + (-0.04081 \times 0.2700) + (0.12238 \times 1.0224) \\ &\quad + (0.00760 \times 1.5789) + (0.04516 \times 1.8718) + (1.03488 \times 1.9947) \\ &= 2.456 \end{aligned}$$

and

$$\begin{aligned} \sigma^* &= (-0.15392 \times -1.6608) + (-0.11755 \times -0.2485) + (-0.11670 \times -0.0409) \\ &\quad + (-0.10285 \times 0.2700) + (-0.03942 \times 1.0224) + (-0.07023 \times 1.5789) \\ &\quad + (0.04037 \times 1.8718) + (0.64104 \times 1.9947) \\ &= 1.31377. \end{aligned}$$

From (6.62), we determine

$$\text{Var}(\mu^*) = 0.16442\sigma^2,\ \text{Var}(\sigma^*) = 0.10125\sigma^2,\ \text{Cov}(\mu^*, \sigma^*) = 0.06920\sigma^2$$

so that we obtain the standard errors of the estimates μ^* and σ^* to be

$$\text{SE}(\mu^*) = \sigma^*(0.16442)^{1/2} = 0.53272 \text{and} \text{SE}(\sigma^*) = \sigma^*(0.10125)^{1/2} = 0.41804.$$

It may be noted here that, by employing a numerical maximization method, Viveros and Balakrishnan (1994) determined in this case the maximum likelihood estimates of μ and σ to be 2.222 and 1.026, respectively. Keep in mind that these maximum likelihood estimates are both biased estimates.

6.3 Best Linear Invariant Estimation

While the last two sections dealt with linear unbiased estimators, it will also be possible to derive "efficient" linear estimators without requiring them to be unbiased. For example, in the case of a scale-family of distributions, instead of requiring a linear estimator that has minimum variance in the class of all linear unbiased estimators, we may choose to find the linear estimator that has minimum mean square error. Such estimators, studied by Mann (1969b), are called *best linear invariant estimators* (BLIEs).

For the purpose of illustration, let us consider the scale-parameter exponential distribution discussed in Section 6.1.2. Once again, let us assume that we have observed a general progressively Type-II censored sample

$$ {}_rY_{r+1:m:n},\ {}_rY_{r+2:m:n}, \cdots,\ {}_rY_{m:m:n} $$

with the censoring scheme $(R_{r+1}, \cdots, R_m)$. Let Z_i's be as defined in Section 6.1.2. Let us now consider a general linear estimator of θ as

$$ \theta^{**} = \sum_{i=r+1}^{m} d_i\ {}_rY_{i:m:n} = \theta \sum_{i=r+1}^{m} d_i' Z_i \tag{6.64} $$

from which we readily find

$$ E(\theta^{**}) = \theta \left\{ d_{r+1}' \alpha_{r+1:n} + \sum_{i=r+2}^{m} d_i' \right\} \tag{6.65} $$

and

$$ \mathrm{Var}(\theta^{**}) = \theta^2 \left\{ d_{r+1}'^2 \beta_{r+1:n} + \sum_{i=r+2}^{m} d_i'^2 \right\}. \tag{6.66} $$

Then, the mean square error of the estimator θ^{**} in (6.64) is

$$ MSE(\theta^{**}) = \theta^2 \left[d_{r+1}'^2 \beta_{r+1:n} + \sum_{i=r+2}^{m} d_i'^2 + \left\{ d_{r+1}' \alpha_{r+1:n} + \sum_{i=r+2}^{m} d_i' - 1 \right\}^2 \right]. \tag{6.67} $$

Minimizing (6.67) with respect to d_i', we obtain the equations

$$ d_{r+1}' \beta_{r+1:n} + \left\{ d_{r+1}' \alpha_{r+1:n} + \sum_{i=r+2}^{m} d_i' - 1 \right\} \alpha_{r+1:n} = 0, $$

$$ d_i' + \left\{ d_{r+1}' \alpha_{r+1:n} + \sum_{i=r+2}^{m} d_i' - 1 \right\} = 0 \quad \text{for} \quad i = r+2, \cdots, m. \tag{6.68} $$

Solving (6.68), we obtain

$$ d_{r+1}' = \frac{1}{m - r + \frac{\alpha_{r+1:n}^2}{\beta_{r+1:n}}} \cdot \frac{\alpha_{r+1:n}}{\beta_{r+1:n}} \quad \text{and} \quad d_i' = \frac{1}{m - r + \frac{\alpha_{r+1:n}^2}{\beta_{r+1:n}}} \quad (r+2 \le i \le m) $$

from which we obtain the coefficients d_i's as

$$\begin{aligned} d_{r+1} &= d'_{r+1} - (n-r-R_{r+1}-1)d'_{r+2} \\ &= \frac{1}{m-r+\frac{\alpha^2_{r+1:n}}{\beta_{r+1:n}}}\left\{\frac{\alpha_{r+1:n}}{\beta_{r+1:n}} - (n-r-R_{r+1}-1)\right\}, \\ d_{r+2} &= (n-r-R_{r+1}-1)d'_{r+2} - (n-r-R_{r+1}-R_{r+2}-2)d'_{r+3} \\ &= (R_{r+2}+1)\Big/\left(m-r+\frac{\alpha^2_{r+1:n}}{\beta_{r+1:n}}\right), \\ \cdots & \quad \cdots \\ \cdots & \quad \cdots \\ d_{m-1} &= (R_{m-1}+1)\Big/\left(m-r+\frac{\alpha^2_{r+1:n}}{\beta_{r+1:n}}\right), \\ d_m &= (n-r-R_{r+1}-\cdots-R_{m-1}-m+r+1)d'_m \\ &= (R_m+1)\Big/\left(m-r+\frac{\alpha^2_{r+1:n}}{\beta_{r+1:n}}\right). \end{aligned}$$

Upon substituting these coefficients in (6.64), we obtain the BLIE of θ as

$$\begin{aligned} \theta^{**} = & \frac{1}{m-r+\frac{\alpha^2_{r+1:n}}{\beta_{r+1:n}}}\left[\sum_{j=r+2}^{m}(R_j+1)\left({}_rY_{j:m:n} - {}_rY_{r+1:m:n}\right)\right. \\ & \left. + \left(\frac{\alpha_{r+1:n}}{\beta_{r+1:n}}\right) {}_rY_{r+1:m:n}\right]; \end{aligned} \tag{6.69}$$

its mean square error then follows from (6.67) as

$$MSE(\theta^{**}) = \frac{\theta^2}{m-r+\frac{\alpha^2_{r+1:n}}{\beta_{r+1:n}}} . \tag{6.70}$$

Upon comparing this with $Var(\theta^*)$ in (6.5), we obtain the relative efficiency of the BLIE to the BLUE of θ as

$$100\left(1+\frac{1}{m-r-1+\frac{\alpha^2_{r+1:n}}{\beta_{r+1:n}}}\right)\% .$$

Proceeding similarly, we can derive the BLIE of θ for any scale-family of distributions.

In the case of location-scale family of distributions, Mann (1969b) showed that the BLIEs of μ and σ are given by

$$\mu^{**} = \mu^* - \left(\frac{V_3}{1+V_2}\right)\sigma^* \quad \text{and} \quad \sigma^{**} = \frac{\sigma^*}{1+V_2} , \tag{6.71}$$

respectively, where μ^* and σ^* are the BLUEs of μ and σ as given in (6.27), and $\mathrm{Var}(\sigma^*) = \sigma^2 V_2$ and $\mathrm{Cov}(\mu^*, \sigma^*) = \sigma^2 V_3$ as given in (6.28).

From (6.71), one can then determine the BLIEs of μ and σ for any location-scale distribution. One can also present first-order approximations to these BLIEs using results presented earlier in Section 6.2.7.

7

Likelihood Inference: Type-I and Type-II Censoring

7.1 Introduction

In Chapter 6, we considered linear inference based on progressively Type-II censored order statistics. Linear inference is popular because, in addition to having many desirable properties which we associate with good estimators, linear estimators have a very simple form, viz. the estimators are linear combinations of observed data values. As a result, these estimators are often quite simple to interpret from a practitioner's point of view. Furthermore, we are able to very easily calculate the variances and covariance of linear estimators. This is not always true for other types of estimators, such as maximum likelihood and moment estimators. However, the linear inference that we have discussed applies only to scale- or location-scale families of distributions. If a new parameter, such as a shape parameter or a threshold parameter that can not be written as a simple shift from the standard distribution, is introduced, other methods of estimating these parameters need to be considered.

Anyone who has taken a mathematical statistics course will be inclined to consider using maximum likelihood estimation when faced with the joint probability distribution of a number of random variables and the problem of estimating parameter(s) from that joint distribution; maximum likelihood estimation has an intuitively pleasing foundation, and generally, differentiation of likelihood functions is straightforward. However, we often forget that the resulting equations to be solved for the unknown parameters are not quite so straightforward! High-speed computers have certainly helped to ease the problem of solving systems of equations; yet, likelihood functions are not always the classic, smooth functions which we would like to see, and it is sometimes not as straightforward as running a simple algorithm. Furthermore, the actual estimators obtained are sometimes difficult to interpret practically. (This is not to say that problems arising from maximum likelihood estimation are not worthwhile to pursue from a mathematical point of view, of course!) Nevertheless, maximum likelihood estimation is often the most feasible method to use when doing statistical inference, as the only information required to obtain maximum likelihood estimators (MLEs) is the joint distribution of the observed values. Furthermore, MLEs do have very desirable properties, particularly in the asymptotic case. Often, when expressions for MLEs are difficult to work with, the estimators are "linearized"; see, for example, Thomas and Wilson (1972). From a practical point of view, whenever two or more methods of attacking a problem are possible, it is a good idea to be able to compare the different methods before making a process decision.

Earlier in Chapter 1, we briefly discussed the genesis of progressive censoring, and research in this area which has been carried out thus far. Much of this previous work is in maximum likelihood estimation, which seems like a good starting point, as the joint distribution of progressively censored order statistics is readily obtained. In addition, the method of maximum likelihood is tractable in Type-I as well as the Type-II censoring scenarios, as the number of observed failures in a life-test need not be known in advance. In this chapter, we will present many of these previous results, as well as some more recent and newly developed results for maximum likelihood estimation when a progressive censoring scheme (Type-I or Type-II) has been applied to a random sample of items. The results presented here will not be exhaustive in terms of different parameter combinations, distributions and censoring scenarios; but, the reader will secure a very good idea of the various likelihood methods which can be considered, and then applying many of the methods presented here to other distributions will be rather straightforward.

7.2 General Continuous Distributions

We are already well acquainted with the likelihood function to be maximized when a general progressively Type-II censored sample based on n independent units with identical lifetime distributions from an arbitrary continuous distribution $F(x)$ is observed. Recall the likelihood function to be maximized will be

$$\begin{aligned} L(\boldsymbol{\theta}) &= c^* \; [F(x_{r+1};\boldsymbol{\theta})]^r \prod_{i=r+1}^{m} f(x_i;\boldsymbol{\theta}) \, [1 - F(x_i;\boldsymbol{\theta})]^{R_i}, \\ & \qquad x_{r+1} < x_{r+2} < \cdots < x_m, \end{aligned} \tag{7.1}$$

where

$$c^* = \frac{n!}{r!(n-r-1)!}(n - R_{r+1} - r - 1)\cdots(n - R_{r+1} - \cdots - R_{m-1} - m + 1)$$

and $\boldsymbol{\theta}$ is the vector of parameters to be estimated.

In the case of general progressively Type-I censored samples, where again n independent and identical units are put on test, times $T_0, T_1, \cdots, T_m$ are fixed such that observation begins at time T_0 ($T_0 = 0$ if the sample is a progressively Type-I *right* censored sample), at which time R_0 surviving units are withdrawn from experimentation (provided R_0 units are still surviving), and continues such that at times $T_1, T_2, \cdots, T_m$, the number of units removed are $R_1, R_2, \cdots, R_m$, respectively, provided there are surviving units available to be removed at these various censoring times. Here, the experimenter fixes the values $T_0, T_1, T_2, \cdots, T_m, R_0, R_1, R_2, \cdots, R_{m-1}$ (recall that R_m is the random variable corresponding to whatever number of units remain at time T_m); however, it is also possible to view all of the values R_i, $i = 0, \cdots, m$, as random variables. As already alluded to, it is possible that some value(s) of $R_i, i = 0, 1, \cdots, m-1$, are actually less than anticipated, and in these cases, experimentation could even end before time T_m since all experimental units have failed before that time. (For example, if R_j is pre-set to 5, but at time T_j, $j < m$, of experimentation, there are actually only 4 live units remaining, experimentation will end at time T_j.) The likelihood function which we will now present may be modified accordingly, assuming the censoring pattern $(R_0, R_1, R_2, \cdots, R_{m-1})$ is chosen independently of the parameter vector $\boldsymbol{\theta}$. Practically, this should rarely happen, as one reason to employ progressive censoring schemes is to reduce the amount of time required to perform an experiment, implying that the time T_m is much less than what is believed apriori to be a moderately long product lifetime. One more point to keep in mind in the Type-I scenario is that it is also possible to pre-fix R_m. In the case where more than R_m items are still surviving at time T_m, the experiment will continue to run until all remaining items have failed. The likelihood function to be maximized will not change.

We are now able to write down the likelihood function to be maximized when a general progressively Type-I censored sample is observed. If k_1 items fail before observation begins (i.e. before time T_0), and k_2 items are observed to fail during experimentation, and the resulting observed failure times are $x_{k_1+1}, x_{k_2+1}, \cdots, x_{k_1+k_2}$, the likelihood function to be maximized is given by

$$L(\boldsymbol{\theta}) = C\left[F(T_0;\boldsymbol{\theta})\right]^{k_1} \prod_{i=k_1+1}^{k_1+k_2} f(x_i;\boldsymbol{\theta}) \prod_{i=0}^{m} \left[1 - F(T_i;\boldsymbol{\theta})\right]^{R_i}, \qquad (7.2)$$

where C is the normalizing constant independent of θ. Notice that this likelihood function is similar to the likelihood function given in (7.1) for progressively Type-II censored samples with k_1 replacing r, k_2 replacing $m-r$ and T_0 replacing x_{r+1}. However, in the Type-I case, we now have two distinct sets of times: censoring times and failure times, i.e. T_i here do not coincide with x_i, as in the case of Type-II censoring. Therefore, if we develop likelihood results for general progressively Type-I censored samples, they will be readily modified, and often simplified, to results for general progressively Type-II censored samples if, in addition to the above modifications, T_i are replaced by x_i.

Remark 7.1 It is worthwhile to note that when choosing a censoring scheme in the Type-I scenario, we need not ensure that $R_1 + R_2 + \cdots + R_m + m = n$. However, we will have $k_1 + k_2 + R_0 + R_1 + \cdots + R_m = n$, where the quantities on the left-hand side of this equality are not all constants.

Remark 7.2 A variant of the Type-I censoring scheme may be to remove some fixed percentage of live units at various times during the experiment, thereby eliminating the need to pre-set a censoring pattern.

Remark 7.3 It is this type of scheme that was considered in the original query posed to the journal *Technometrics* regarding the broken lamps (see Section 1.2). Assuming the time of accidental loss of a unit to be independent of the parameter vector to be estimated, the above likelihood function may be used to find likelihood estimates, with T_i $(i \geq 0)$ being replaced by the breakage time(s). Kaplan and Meier (1958) have also discussed this situation in human subjects.

A number of authors, including Viveros and Balakrishnan (1994), have approximated Type-I censored samples, where observation can only begin after time $t = 1$ second, by which time 3 units have already failed, using Type-II methods, which are usually more straightforward and tractable than Type-I methods. When designing a life-test experiment, the practitioner should also keep this feature of Type-II censoring in mind. In general, if it is practically feasible to employ a Type-II censoring scheme, so that the censoring times and failure times coincide, the resulting analysis will be both simpler and usually more complete.

7.3 Specific Continuous Distributions

7.3.1 The Normal Distribution

It is fitting to begin our discussion of likelihood estimation with the first results on progressive censoring ever published in a widely read journal, viz. the results given by Cohen (1963), and, incidentally, part of the eventual response given by Cohen (1966) to the query posed in *Technometrics*. Cohen first considered a progressively Type-I right censored sample from a $Normal\ (\mu, \sigma^2)$ distribution, i.e., the failure-time distribution of the n independent items on test is

$$f(y) = \frac{1}{\sigma\sqrt{2\pi}} e^{-(y-\mu)^2/(2\sigma^2)}, \quad -\infty < x < \infty, \ -\infty < \mu < \infty, \ \sigma > 0. \qquad (7.3)$$

Denoting the observed (ordered) progressively censored sample by y_i, $i = 1, \cdots, k_2$, the log-likelihood function to be maximized for the location and scale parameters will be (recall that for a progressively Type-I right censored sample, $k_1 = R_0 = T_0 = 0$)

$$\ln L(\mu, \sigma) = constant - n \ln \sigma - \frac{1}{2} \sum_{i=1}^{k_2} \left(\frac{y_i - \mu}{\sigma} \right)^2 + \sum_{i=1}^{m} R_i \ln (1 - F_i),$$

where

$$F_i = \Phi\left(\xi_i\right), \quad \xi_i = \frac{T_i - \mu}{\sigma}$$

and $\Phi\left(\cdot\right)$ is the cumulative distribution function of the standard normal ($\mu = 0$, $\sigma = 1$) distribution. To maximize the log-likelihood, the score equations to be solved are

$$\begin{aligned} \frac{\sum_{i=1}^{k_2} y_i}{k_2} &= \overline{y} = \mu - \sigma \sum_{i=1}^{m} \frac{R_i}{k_2} Z_i, \\ \frac{\sum_{i=1}^{k_2} \left(y_i - \overline{y}\right)^2}{k_2} &= \sigma^2 \left[1 - \sum_{i=1}^{m} \frac{R_i}{k_2} \xi_i Z_i - \sum_{i=1}^{m} \left(\frac{R_i}{k_2} Z_i\right)^2\right], \end{aligned}$$

where $Z_i = \frac{\varphi(\xi_i)}{1-\Phi(\xi_i)}$ and $\varphi\left(\cdot\right)$ is the probability density function of the standard normal distribution. Cohen (1963) has discussed the numerical solutions of these equations in detail. Now, many mathematical computer packages contain algorithms to solve the above system of equations efficiently. Note that maximum likelihood estimation for progressively Type-II right censored samples will result in the same system of equations, with the censoring times, T_i, replaced by the failure times, y_i.

Remark 7.4 In determining the variances and covariance of the MLEs, one could examine (at least asymptotically) the asymptotic variance-covariance matrix,

$$\begin{pmatrix} -E\left(\frac{\partial^2 lnL}{\partial \mu^2}\right) & -E\left(\frac{\partial^2 lnL}{\partial \mu \partial \sigma}\right) \\ -E\left(\frac{\partial^2 lnL}{\partial \mu \partial \sigma}\right) & -E\left(\frac{\partial^2 lnL}{\partial \sigma^2}\right) \end{pmatrix}^{-1}.$$

7.3.2 The Exponential Distribution

In Chapter 6, best linear unbiased estimation for the one- and two-parameter exponential distributions was considered. (Remember, BLUEs can only be discussed for Type-II censoring). Recall that we denoted the general progressively Type-II censored order statistics from the distribution of interest by ${}_rY_{r+i:m:n}$, $i = 1, 2, \cdots, m - r$. For the one-parameter exponential distribution, the MLE of the scale parameter θ does not exist in an explicit form and has to be determined from the likelihood equation by a numerical method. The equation to be solved numerically for θ is [see Balakrishnan and Sandhu (1996)]

$${}_rY_{r+1:m:n} \frac{re^{-{}_rY_{r+1:m:n}/\theta}}{1 - e^{-{}_rY_{r+1:m:n}/\theta}} = -\theta\left(m - r\right) \sum_{i=r+1}^{m} \left(R_i + 1\right) {}_rY_{i:m:n}. \qquad (7.4)$$

In the special case when $r = 0$, however, the MLE of θ is given explicitly by

$$\widehat{\theta} = \frac{1}{m} \left[nY_{1:m:n} + \sum_{j=1}^{m} \left(R_j + 1\right)\left(Y_{j:m:n} - Y_{1:m:n}\right)\right] = \frac{1}{m} \sum_{j=1}^{m} \left(R_j + 1\right) Y_{j:m:n} \qquad (7.5)$$

which becomes identical with the BLUE of θ given in (6.4).

Remark 7.5 Cohen (1963) derived the maximum likelihood estimator for θ when the observed sample is a progressively Type-I right censored sample. The estimator is

$$\widehat{\theta} = \frac{\sum_{i=1}^{k_2} y_i + \sum_{i=1}^{m} R_i T_i}{k_2} = \frac{\textit{Total Accumulated Life}}{k_2},$$

and it reduces to (7.5) when we adjust it for Type-II samples, viz. T_i are replaced by y_i, and k_2, the number of observed failures times, is replaced by m.

Remark 7.6 Suppose k progressively Type-II right censored samples are available from the *Exponential*(θ) distribution, with the i-th censored sample denoted by

$$Y_{i;1:m_i:n_i}, Y_{i;2:m_i:n_i}, \cdots, Y_{i;m_i:m_i:n_i}$$

and the corresponding progressive censoring scheme by $(R_{i,1}, R_{i,2}, \cdots, R_{i,m_i})$ for $i = 1, 2, \cdots, k$. Then, Balakrishnan, Cramer, Kamps and Schenk (1999) have derived the MLE of θ to be

$$\hat{\theta} = \frac{1}{m_+} \sum_{i=1}^{k} \sum_{j=1}^{m_i} (R_{i,j} + 1) Y_{i;j:m_i:n_i}^{(R_{i,1},\cdots,R_{i,m_i})},$$

where $m_+ = \sum_{i=1}^{k} m_i$. These authors have also shown that $\hat{\theta}$ is distributed as gamma with shape parameter m_+ and scale parameter θ/m_+; furthermore, $\hat{\theta}$ is an unbiased estimator of θ with $\mathrm{Var}(\hat{\theta}) = \theta^2/m_+$, is sufficient and strongly consistent for θ, and is asymptotically normally distributed.

We will now denote the general progressively Type-II censored order statistics from the two-parameter exponential distribution (location parameter μ, scale parameter σ) by ${}_rY_{r+i:m:n}$, $i = 1, \cdots, m - r$. The log-likelihood function based on this general progressively Type-II censored sample is given by [see Balakrishnan and Sandhu (1996)]

$$\begin{aligned} \ln L &= (\textit{Constant}) - (m-r)\ln\sigma + r\ln\left(1 - e^{-\frac{{}_rY_{r+1:m:n}-\mu}{\sigma}}\right) \\ &\quad - \sum_{i=r+1}^{m} (R_i + 1)\left(\frac{{}_rY_{i:m:n} - \mu}{\sigma}\right). \end{aligned} \tag{7.6}$$

For the special case when $r = 0$, the log-likelihood function in (7.6) is monotonically increasing in μ so that the MLE of μ is $\widehat{\mu} = Y_{1:n}$. In this case, the MLE of σ is given by

$$\widehat{\sigma} = \frac{1}{m} \sum_{i=2}^{m} (R_i + 1)\left(Y_{i:m:n} - Y_{1:m:n}\right).$$

It may be noted that the MLEs $\widehat{\mu}$ and $\widehat{\sigma}$ are both biased.

For the case when $r > 0$, eventhough it initially appears from (7.6) that the MLEs of μ and σ may not exist explicitly, it turns out that the MLEs of μ and σ take on simple linear forms. To see this, we have the likelihood equations for μ and σ from (7.6) to be

$$\frac{\partial \ln L}{\partial \mu} = -\frac{r}{\sigma}\left(\frac{e^{-({}_rY_{r+1:m:n}-\mu)/\sigma}}{1-e^{-({}_rY_{r+1:m:n}-\mu)/\sigma}}\right) + \frac{1}{\sigma}\sum_{i=r+1}^{m}(R_i+1) = 0 \tag{7.7}$$

and

$$\begin{aligned}\frac{\partial \ln L}{\partial \sigma} &= \frac{m-r}{\sigma} - \frac{r}{\sigma}\left(\frac{{}_rY_{r+1:m:n}-\mu}{\sigma}\right)\left(\frac{e^{-({}_rY_{r+1:m:n}-\mu)/\sigma}}{1-e^{-({}_rY_{r+1:m:n}-\mu)/\sigma}}\right) - \frac{(n-r)\,\mu}{\sigma^2} \\ &\quad + \frac{1}{\sigma^2}\sum_{i=r+1}^{m}(R_i+1)\,{}_rY_{i:m:n} = 0. \end{aligned} \tag{7.8}$$

Eq. (7.7) simplifies to

$$r\frac{e^{-({}_rY_{r+1:m:n}-\mu)/\sigma}}{1-e^{-({}_rY_{r+1:m:n}-\mu)/\sigma}} = n-r$$

which immediately yields the MLE of μ to be

$$\widehat{\mu} = {}_rY_{r+1:m:n} + \widehat{\sigma}\ln\left(1-\frac{r}{n}\right). \tag{7.9}$$

Eq. (7.8), in conjunction with the MLE of μ in (7.9), reduces to

$$-\frac{m-r}{\sigma} - \frac{n-r}{\sigma^2}{}_rY_{r+1:m:n} + \frac{1}{\sigma^2}\sum_{i=r+1}^{m}(R_i+1)\,{}_rY_{i:m:n} = 0$$

which readily yields the MLE of σ to be

$$\widehat{\sigma} = \frac{1}{m-r}\sum_{i=r+2}^{m}(R_i+1)\left({}_rY_{i:m:n} - {}_rY_{r+1:m:n}\right). \tag{7.10}$$

Remark 7.7 The derivation of the MLEs of μ and σ in (7.9) and (7.10) for the general progressively Type-II censored sample case generalizes the results of Cohen for the special case of the progressively Type-II right-censored sample ($r = 0$). It is also of interest to mention here that the results of Kambo (1978) for the case of doubly Type-II censored samples may be deduced as a special case from (7.9) and (7.10).

Remark 7.8 Comparing the MLEs in (7.9) and (7.10) with the BLUEs in (6.29) and (6.30), it is quite clear that the BLUEs are simply the MLEs adjusted for their bias. As a result, the expressions of the variances and covariance of the BLUEs presented in (6.31) – (6.33) may be used to derive expressions for the variances and covariance of the MLEs of μ and σ given in (7.9) and (7.10). These quantities turn out to be

$$Var\,(\widehat{\mu}) = \left\{ \beta_{r+1} + \left[\frac{\ln\left(1 - \frac{r}{n}\right)}{m - r} \right]^2 (m - r - 1) \right\} \sigma^2$$

and

$$Var\,(\widehat{\sigma}) = \frac{m - r - 1}{(m - r)^2} \sigma^2,$$

respectively.

Remark 7.9 Due to the equivalence of the MLEs and the BLUEs in this case, the inference developed in Section 6.2.2 for the parameters μ and σ using pivotal quantities based on the BLUEs will be the same as those based on the MLEs. For the special case of progressively Type-II right censored samples, this has been noted by Viveros and Balakrishnan (1994).

Remark 7.10 For large m, it is clear that the MLE $\widehat{\sigma}$ in (7.10) approaches the BLUE σ^* in (6.30). Furthermore, since

$$\alpha_{r+1} = \sum_{i=1}^{r+1} \frac{1}{n - i + 1} = \sum_{i=1}^{n} \frac{1}{i} - \sum_{i=1}^{n-r-1} \frac{1}{i}$$

and

$$\lim_{N \to \infty} \left(\sum_{k=1}^{N} \frac{1}{k} - \ln N \right) = \gamma \quad \text{(Euler's constant)}$$

[see Abramowitz and Stegun (1965, p. 68)], the MLE $\widehat{\mu}$ in (7.9) becomes identical with the BLUE μ^* in (6.29) for large m (and, therefore, large n).

Example 7.1 Earlier in Example 6.2, we considered Nelson's data on times to breakdown, where a one-parameter exponential model seemed appropriate. The one-parameter BLUE in that case was found to be $\theta^* = 9.110$ with a standard error of 3.221. Using (7.4), the MLE of θ is found numerically to be $\widehat{\theta} = 9.111$. Note the closeness in the two estimates.

Example 7.2 In Example 6.3, a general progressively Type-II censored sample from the two-parameter exponential distribution was generated with $\mu = 25$ and $\sigma = 10$. The BLUEs in this case were obtained earlier as $\mu^* = 24.66292$ and $\sigma^* = 10.54061$ with standard errors $SE\,(\mu^*) = 0.65082$ and $SE\,(\sigma^*) = 2.81710$. Using (7.9) and (7.10), the MLE's are obtained as $\widehat{\mu} = 24.95956$ and $\widehat{\sigma} = 9.83791$ with standard errors $SE\,(\widehat{\mu}) = 10.54061\sqrt{.00360} = 0.61102$ and $SE\,(\widehat{\sigma}) = 2.62929$. In both cases, the estimated values are very close to the corresponding actual values. Furthermore, the two methods of estimation, BLUE and MLE, yield very similar estimates and standard errors.

Remark 7.11 Suppose we have progressively Type-II right censored samples of size m each from *Exponential*(μ_1, σ_1) and *Exponential*(μ_2, σ_2) distributions, obtained from n units on test with the same progressive censoring scheme $(R_1, \cdots, R_m)$. Let $(\hat{\mu}_1, \hat{\sigma}_1)$ and $(\hat{\mu}_2, \hat{\sigma}_2)$ be the corresponding MLEs obtained from the two progressively censored samples. Then, Balakrishnan and Mi (1999) have recently established an *order-preserving property* of these MLEs. Specifically, they have proved that if $\mu_1 \leq \mu_2$ and $\sigma_1 \leq \sigma_2$, then $\hat{\mu}_1 \leq_{\text{st}} \hat{\mu}_2$ and $\hat{\sigma}_1 \leq_{\text{st}} \hat{\sigma}_2$, where $\leq_{\text{st}}$ denotes the usual stochastic ordering between random variables.

Remark 7.12 Suppose k progressively Type-II right censored samples are available from an *Exponential*(μ, σ) distribution, with the i-th censored sample denoted by

$$Y_{i;1:m_i:n_i}, Y_{i;2:m_i:n_i}, \cdots, Y_{i;m_i:m_i:n_i}$$

and the corresponding progressive censoring scheme by $(R_{i,1}, \cdots, R_{i,m_i})$ for $i = 1, 2, \cdots, k$. Then, Balakrishnan, Cramer, Kamps and Schenk (1999) have derived the MLEs of μ and σ to be

$$\hat{\mu} = \min\left\{Y_{1;1:m_i:n_i}^{(R_{1,1},\cdots,R_{1,m_1})}, \cdots, Y_{k;1:m_k:n_k}^{(R_{k,1},\cdots,R_{k,m_k})}\right\}$$

and

$$\hat{\sigma} = \frac{1}{m_+} \sum_{i=1}^{k} \sum_{j=1}^{m_i} (R_{i,j} + 1) \left(Y_{i;j:m_i:n_i}^{(R_{i,1},\cdots,R_{i,m_i})} - \hat{\mu}\right),$$

where $m_+ = \sum_{i=1}^{k} m_i$. These authors have further shown that $\hat{\mu}$ and $\hat{\sigma}$ are independent, $(\hat{\mu}, \hat{\sigma})$ is complete sufficient for (μ, σ), $\hat{\sigma}$ is distributed as gamma with shape parameter $m_+ - 1$ and scale parameter σ / m_+, and $\hat{\mu}$ is distributed as *Exponential*$(\mu, \sigma/n_+)$ where $n_+ = \sum_{i=1}^{k} n_i$. Balakrishnan, Cramer, Kamps and Schenk (1999) have also shown that $\tilde{\sigma} = \frac{m_+}{m_+ - 1}\,\hat{\sigma}$ and $\tilde{\mu} = \hat{\mu} - \frac{\tilde{\sigma}}{n_+}$ are UMVUEs (uniformly minimum variance unbiased estimators) of σ and μ, respectively, with

$$\text{Var}(\tilde{\sigma}) = \frac{\sigma^2}{m_+ - 1}, \ \text{Var}(\tilde{\mu}) = \frac{m_+}{(m_+ - 1)n_+^2}\,\sigma^2 \text{ and } \text{Cov}(\tilde{\mu}, \tilde{\sigma}) = -\frac{\sigma^2}{(m_+ - 1)n_+}.$$

They have further discussed the maximum likelihood estimation of μ and σ based on k general progressively Type-II censored samples.

7.3.3 The Weibull Distribution

In one of series of papers on likelihood estimation under progressive censoring that appeared in *Technometrics*, Cohen (1975) considered both maximum likelihood and modified maximum likelihood estimation for the three-parameter [location(μ) - scale(σ) - shape(θ)] Weibull distribution with cumulative distribution function

$$F(y) = 1 - e^{-(y-\mu)^{\theta}/\sigma}, \quad y > \mu, \quad -\infty < \mu < \infty, \quad \sigma > 0, \quad \theta > 0.$$

The Weibull distribution is commonly used to model failure-time distributions, due to desirable properties such as a positive, increasing hazard rate, and as such, many of the original papers published on progressive censoring considered the Weibull distribution as the failure time distribution of interest. These include Cohen (1966, 1975), Mann (1969a, 1971), Ringer and Sprinkle (1972), Thomas and Wilson (1972), Gibbons and Vance (1983), Cacciari and Montanari (1987) and Montanari and Cacciari (1988), who used progressive Type-II right censoring in order to study the aging process of an insulated cable.

The log-likelihood function for a progressively Type-I right censored sample from this distribution is given by

$$\begin{aligned}\ln L &= (Constant) + k_2 \ln\theta - k_2 \ln\sigma + (\theta - 1)\sum_{i=1}^{k_2} \ln(y_i - \mu) \\ &\quad - \frac{1}{\sigma}\left[\sum_{i=1}^{k_2} (y_i - \mu)^{\theta} + \sum_{i=1}^{m} R_i (T_i - \mu)^{\theta}\right],\end{aligned}$$

where y_i is the observed (ordered) progressively censored sample. Again, the results which we will present for progressively Type-I right censored samples may be adapted to Type-II scenarios and simplified in the usual way.

From inspection of the likelihood function, we see that we may consider the three cases $\theta < 1$, $\theta = 1$, and $\theta > 1$ separately. If it is known that $\theta = 1$, then we have simply a location-scale exponential distribution, whose estimators we have already derived in the last section. If $\theta < 1$, then we see that the likelihood function becomes infinite as $\mu \to y_1$, and therefore $\widehat{\mu} = y_1$ [in fact, as Cohen (1975) has pointed out, $\widehat{\mu} = y_1 - \frac{\eta}{2}$ where η is the unit of precision of measurements made.] The additional score equations to be solved for σ and θ in this case are given by

$$\begin{aligned}\frac{\partial \ln L}{\partial \sigma} &= 0 = -\frac{k_2}{\sigma} + \frac{1}{\sigma^2}\left[\sum_{i=1}^{k_2} (y_i - \mu)^{\theta} + \sum_{i=1}^{m} R_i (T_i - \mu)^{\theta}\right], \\ \frac{\partial \ln L}{\partial \theta} &= 0 = \frac{k_2}{\theta} + \sum_{i=1}^{k_2} \ln(y_i - \mu) \\ &\quad - \frac{1}{\sigma}\left[\sum_{i=1}^{k_2} (y_i - \mu)^{\theta} \ln(y_i - \mu) + \sum_{i=1}^{m} R_i (T_i - \mu)^{\theta} \ln(T_i - \mu)\right].\end{aligned}$$

We can now eliminate σ between these two equations, obtaining a single equation as

$$\frac{\sum_{i=1}^{k_2}(y_i-\mu)^\theta \ln(y_i-\mu)+\sum_{i=1}^{m} R_i (T_i-\mu)^\theta \ln(T_i-\mu)}{\sum_{i=1}^{k_2}(y_i-\mu)^\theta+\sum_{i=1}^{m} R_i (T_i-\mu)^\theta}$$
$$-\frac{1}{\theta}-\frac{1}{k_2}\sum_{i=1}^{k_2}\ln(y_i-\mu)=0.$$

When μ or its estimate is known, as in the present case and the two-parameter case discussed by Cohen (1966), one needs to solve only this equation for θ and then obtain the estimate for σ by substitution back into the original equation, $\frac{\partial \ln L}{\partial \sigma}=0$, given above.

Finally, for the case when $\theta > 1$, one must solve a system of three equations, consisting of the two equations $\frac{\partial \ln L}{\partial \sigma}=0$ and $\frac{\partial \ln L}{\partial \theta}=0$ given above, along with the equation

$$\begin{aligned}\frac{\partial \ln L}{\partial \mu}=0 &= \frac{\theta}{\sigma}\left[\sum_{i=1}^{k_2}(y_i-\mu)^{\theta-1}+\sum_{i=1}^{m}R_i(T_i-\mu)^{\theta-1}\right]\\ &\quad -(\theta-1)\sum_{i=1}^{k_2}(y_i-\mu)^{-1}.\end{aligned}$$

Again, simplification by eliminating the parameter σ is possible. Cohen (1975) has discussed the numerical solutions to these equations in more detail, using y_1 as an initial approximation for $\widehat{\mu}$. Again, many computer algorithms available today will have suitable methods for solving this system of equations directly.

A method of likelihood estimation known as *modified maximum likelihood estimation* is possible for cases in which the location parameter also serves as a threshold parameter. Cohen (1975) has discussed this method of estimation for the three-parameter Weibull distribution. In this method, the last equation in the system of equations mentioned above, $\frac{\partial \ln L}{\partial \mu}=0$, is replaced by the equation $E(Y_1)=y_1$. For the three-parameter Weibull distribution, this becomes

$$\mu+\left(\frac{\sigma}{n}\right)^{1/\theta}\Gamma\left(1+\frac{1}{\theta}\right)=y_1.$$

In determining which of the two methods to use in a particular case, a comparison may be carried out. Gibbons and Vance (1983), for example, carried out such a comparison for the two-parameter (scale/shape) Weibull distribution through a simulation study, and found that for larger progressively Type-I right censored samples, the traditional maximum likelihood method generally performed better than the modified method. Thomas and Wilson (1972) conducted a comparison of various linear estimators for the Weibull (and extreme value) distribution, which may have been one of the first steps towards also comparing various censoring schemes to determine the best one to employ. We discuss optimal censoring schemes in detail in Chapter 10. It is worthwhile to note that the location-scale extreme value random variable is equivalent to the natural logarithm of a two-parameter [scale(σ) and shape(θ)] Weibull random variable. The location parameter for the extreme value distribution is given by $\ln\left(\sigma^{1/\theta}\right)$ and the scale parameter by $1/\theta$.

7.3.4 The Uniform Distribution

We have already considered, in Chapter 6, best linear unbiased estimation of both the one- and two-parameter *uniform* distributions. In the one-parameter *uniform*$(0, \theta)$ case, the MLE of θ does not exist in an explicit form and has to be determined from the likelihood equation by a numerical method. The equation to be solved numerically for θ is given by (denoting the progressively censored sample by ${}_rY_{r+i:m:n},\ i = 1, 2, \cdots, m-r$)

$$\sum_{i=r+1}^{m} \frac{R_i\ {}_rY_{i:m:n}}{\theta - {}_rY_{i:m:n}} = m, \quad \theta \geq {}_rY_{m:m:n}; \tag{7.11}$$

see also Aggarwala and Balakrishnan (1998a).

Notice that at $\theta = {}_rY_{m:m:n}$, the left-hand side of (7.11) is infinite, while as θ goes to infinity, the left-hand side goes to 0. Furthermore, the left-hand side of (7.11) is monotonically decreasing in θ between ${}_rY_{m:m:n}$ and ∞. Thus, there must be a unique solution to this equation in $({}_rY_{m:m:n}, \infty)$.

In the two-parameter *uniform*$(\mu, \mu+\sigma)$ case, the two equations to be solved simultaneously are:

$$\begin{aligned}
\sum_{i=r+1}^{m} \frac{R_i}{\mu + \sigma - {}_rY_{i:m:n}} &= \frac{r}{{}_rY_{r+1:m:n} - \mu}, \quad \mu \leq {}_rY_{r+1:m:n}, \\
\sum_{i=r+1}^{m} \frac{R_i\ {}_rY_{i:m:n}}{\mu + \sigma - {}_rY_{i:m:n}} &= \frac{r\ \mu}{{}_rY_{r+1:m:n} - \mu} + m, \quad \mu + \sigma \geq {}_rY_{m:m:n},
\end{aligned} \tag{7.12}$$

where ${}_rY_{r+1:m:n},\ {}_rY_{r+2:m:n}, \cdots,\ {}_rY_{m:m:n}$ now denote the general progressively Type-II censored order statistics from the Uniform $(\mu, \mu+\sigma)$ distribution. However, in the special case when $r = 0$, so that we are left with a progressively Type-II right censored sample $Y_{i:m:n},\ i = 1, \cdots, m$, the MLE of μ is given by $\widehat{\mu} = {}_rY_{r+1:m:n} = Y_{1:m:n} = Y_{1:n}$, and the following equation must then be solved for σ:

$$\sum_{i=1}^{m} \frac{R_i \left(Y_{i:m:n} - Y_{1:m:n}\right)}{\sigma - \left(Y_{i:m:n} - Y_{1:m:n}\right)} = m, \quad \sigma \geq Y_{m:m:n} - Y_{1:m:n}, \tag{7.13}$$

which is of the same form as (7.11). Using similar arguments then, we conclude that there must be exactly one solution to this equation in the interval $\left(\left(Y_{m:m:n} - Y_{1:m:n}\right), \infty\right)$.

7.3.5 The Pareto Distribution

Let $W_{1:m:n}, W_{2:m:n}, \cdots, W_{m:m:n}$ denote a progressively Type-II right censored sample from the location-scale shifted $Pareto\,(\nu)$ distribution. The likelihood function to be maximized for estimators of μ and σ (which we will denote by $\widehat{\mu}$ and $\widehat{\sigma}$) are [see Aggarwala and Balakrishnan (1998c)]

$$L\left(\mu,\sigma\right) = \left(Constant\right)\,\nu^m \sigma^{\nu n} \prod_{i=1}^{m} \left(W_{i:m:n} - \mu\right)^{-\nu(R_i+1)-1}, \quad W_{1:m:n} > \mu + \sigma. \tag{7.14}$$

Upon inspection of this likelihood, it is evident that it is an increasing function of both μ and σ. Therefore, the maximum must lie on the line $W_{1:m:n} = \mu + \sigma$. Substituting for σ in the above likelihood function and differentiating with respect to μ, the resulting equation to be solved for the MLE of μ is given by, provided $m > 1$,

$$\sum_{i=1}^{m} \frac{\nu\left(R_i+1\right)+1}{W_{i:m:n} - \mu} = \frac{\nu n}{W_{1:m:n} - \mu}. \tag{7.15}$$

The maximum likelihood estimator of σ can then be found by setting $\widehat{\sigma} = W_{1:m:n} - \widehat{\mu}$.

Two parameter scale-shape estimation is also feasible for progressively censored order statistics from Pareto distributions. Let $W_{1:m:n}, W_{2:m:n}, \cdots, W_{m:m:n}$ denote a progressively Type-II right censored sample from the scaled $Pareto\,(\nu)$ distribution, that is, $W_{i:m:n} = \sigma X_{i:m:n}$ where the parameters ν and σ are both to be estimated from the sample. We will denote the corresponding maximum likelihood estimates by $\widetilde{\nu}$ and $\widetilde{\sigma}$. The likelihood function to be maximized is then given by

$$L(\nu,\sigma) = \left(Constant\right)\,\nu^m \sigma^{\nu n} \prod_{i=1}^{m} W_{i:m:n}^{-\nu(R_i+1)-1}, \quad W_{1:m:n} > \sigma. \tag{7.16}$$

It is clear that this likelihood function is an increasing function of σ, and therefore, $\widetilde{\sigma} = W_{1:m:n}$. Upon partial differentiation of the logarithm of the likelihood function with respect to ν, we find that, for $m > 1$,

$$\widetilde{\nu} = \frac{m}{\sum_{i=1}^{m}\left(R_i+1\right)\ln\left(W_{i:m:n}\right) - n\ln\left(\widetilde{\sigma}\right)}. \tag{7.17}$$

In fact, we are able to obtain the distribution of $\widetilde{\nu}$ as follows.

Applying the order-preserving transformation $U_i = \ln\left(W_{i:m:n}/\sigma\right), i = 1, \cdots, m$, we find that $U_1, U_2, ..., U_m$ is a progressively Type-II right censored sample from the exponential distribution with mean $1/\nu$. We can then, using the results of Balakrishnan and Sandhu (1996) presented in Section 2.2.2, write the denominator of $\widetilde{\nu}$ as the sum of $m-1$ independent generalized spacings, each possessing an exponential distribution with mean $1/\nu$. More specifically,

$$\sum_{i=1}^{m}\left(R_i+1\right)\ln\left(W_{i:m:n}\right) - n\ln\left(W_{1:m:n}\right) = \sum_{i=1}^{m}\left(R_i+1\right)U_i - nU_1 = \sum_{i=2}^{m} Z_i,$$

where $Z_1 = nU_1$ and $Z_i = (n - R_1 - \cdots R_{i-1} - i + 1)(U_i - U_{i-1})$ $(i = 2, \cdots, m)$ are the mutually independent generalized spacings mentioned above, each having an exponential distribution with mean $1/\nu$.

We, therefore, find that $2\nu m/\widetilde{\nu}$ has a chi-square distribution with $2(m-1)$ degrees of freedom. This fact can be used to develop confidence intervals or tests of hypothesis about ν. Furthermore, a consistent and unbiased estimator of ν is $\widetilde{\widetilde{\nu}} = \frac{m-2}{m}\widetilde{\nu}$, with $Var\left(\widetilde{\widetilde{\nu}}\right) = \frac{\nu^2}{m-3}$.

From (2.31), we also know that $\widetilde{\sigma}/\sigma$ has a *Pareto* (νn) distribution. It is also easily seen that $\widetilde{\sigma}$ and $\widetilde{\nu}$ are independent random variables, since $\widetilde{\sigma}$ is a function of Z_1 alone and $\widetilde{\nu}$ is a function of only $Z_2, Z_3, \cdots, Z_m$. We can then determine an unbiased and consistent estimator for σ as $\widetilde{\widetilde{\sigma}} = \left[1 - \frac{m}{(m-1)n\widetilde{\nu}}\right]\widetilde{\sigma}$ with $Var(\widetilde{\widetilde{\sigma}}) = \frac{\sigma^2 m}{\nu n(m-1)(\nu n-2)}$.

Remark 7.13 It has been noted that the estimator $\widetilde{\nu}$ (and so $\widetilde{\widetilde{\nu}}$) exists for $m > 1$. However, the first and second moments of the estimator exist only if $m > 3$.

Remark 7.14 Notice that the distributions of $\widetilde{\sigma}$ and $\widetilde{\nu}$ (and so $\widetilde{\widetilde{\sigma}}$ and $\widetilde{\widetilde{\nu}}$) depend only upon m and/or n. Therefore, the progressive censoring scheme $(R_1, R_2, \cdots, R_m)$ does not affect, for example, the variances of the scale–shape estimators. If a practitioner is interested in minimizing the sum of the two variances or the determinant of the variance-covariance matrix of the estimators for fixed values of m and n in order to obtain the most efficient estimators with respect to these measures, he or she is free to choose $(R_1, R_2, \cdots, R_m)$ in any convenient manner. This property of progressively censored order statistics is also observed for maximum likelihood and best linear unbiased estimators of parameters of *location–scale* exponential distributions. Notice that for the Pareto distribution, minimization of variances of location-scale estimators for either the maximum likelihood or best linear unbiased methods will involve the progressive censoring scheme. We will discuss optimal censoring schemes in more detail later in Chapter 10.

Remark 7.15 In the special case where no censoring takes place, we have, setting $R_1 = R_2 = \cdots = R_n = 0$ in (7.17),

$$\widetilde{\nu} = \frac{n}{\sum_{i=1}^{n} \ln\left(\frac{W_{i:n}}{\sigma}\right)} \quad \text{and} \quad \widetilde{\sigma} = W_{1:n}.$$

This is a well-known result for complete samples from Pareto distributions; see, for example, Johnson, Kotz and Balakrishnan (1994).

Example 7.3 Let us consider the data in Example 6.4. In this case, we have $n = 15$, the following progressive Type-II right censoring scheme: $R_1 = 5, R_2 =$

$0, R_3 = 2, R_4 = 0, R_5 = 3$ (so that $m = 5$), and the sample as 5.11073, 5.34932, 5.36434, 5.70137, 5.90067.

In Example 6.4, we computed the BLUEs of the location and scale parameters based on this sample as

$$\mu^* = 1.87680, \quad \sigma^* = 3.16207.$$

From (7.15), we find the MLEs of μ and σ as

$$\widehat{\mu} = 2.51416, \quad \widehat{\sigma} = 2.59658.$$

A sensitivity analysis similar to that in Example 6.4 [see also Aggarwala and Balakrishnan (1998c)] presented below shows that approximate knowledge of the shape parameter ν will not alter the estimates greatly:

ν	$\widehat{\mu}$	$\widehat{\sigma}$
2.6	2.89079	2.21995
2.8	2.70259	2.40815
3.0	2.51416	2.59658
3.2	2.32555	2.78518
3.4	2.13679	2.97394

If it is known that the location parameter is 0, we may consider two-parameter maximum likelihood estimation of the scale (σ) parameter and shape (ν) parameter. In this case, we obtain from (7.17) that

$$\widetilde{\sigma} = w_{1:5:15} = 5.11073 \quad \text{and} \quad \widetilde{\nu} = 5.71320.$$

As explained earlier, we are also able to correct the above MLEs for bias. The resulting unbiased estimates are

$$\widetilde{\widetilde{\sigma}} = 5.03619 \quad \text{and} \quad \widetilde{\widetilde{\nu}} = 3.42792$$

with corresponding standard errors

$$SE(\widetilde{\widetilde{\sigma}}) = 0.11170 \quad \text{and} \quad SE(\widetilde{\widetilde{\nu}}) = 2.42390.$$

In addition to point estimation, we may also consider interval estimation of ν. A 90% confidence interval for ν is obtained as

$$(1.5612, 8.8596).$$

7.3.6 The Laplace Distribution

For the Laplace distribution with cumulative distribution function as in (6.52) and probability density function given by

$$f(y) = \frac{1}{2\sigma} e^{-|y-\mu|/\sigma}, \quad -\infty < y < \infty, \quad -\infty < \mu < \infty, \ \sigma > 0, \tag{7.18}$$

the maximum likelihood estimation based on complete samples has been discussed by a number of authors; see, for example, Johnson, Kotz and Balakrishnan (1995). Balakrishnan and Cutler (1995) have discussed maximum likelihood estimation of parameters of the Laplace distribution based on conventionally Type-II censored samples. They considered both symmetric and one-sided (right) censoring. Childs and Balakrishnan (1996) utilized these results to develop conditional inference procedures based on conventionally Type-II right censored samples. Proceeding similarly, Childs and Balakrishnan (1997) have also derived the MLEs of the parameters μ and σ based on general conventionally Type-II censored samples.

In this section, we derive the MLEs of the location and scale parameters of a Laplace distribution based on progressively Type-II right censored samples. The results obtained are generalizations of those given in Balakrishnan and Cutler (1995), where it is shown that for conventionally Type-II right censored samples in which only the first m failure times are observed, the MLE of μ is simply the usual sample median based on the full sample, provided $m \geq \frac{n}{2}$. For $m < \frac{n}{2}$, the MLE of μ turns out to be a linear function of the observed order statistics. In both cases, the MLE of μ is a linear function of the observed order statistics. The results presented in this section for maximum likelihood estimation based on progressively Type-II right censored samples from the Laplace distribution reduce to those presented by Balakrishnan and Cutler (1995) for the special case when $R_1 = R_2 = \cdots = R_{m-1} = 0$ and $R_m = n - m$, in which case we are left with a conventionally Type-II right censored sample.

Denoting the progressively Type-II right censored sample of size m with censoring scheme $(R_1, R_2, \cdots, R_m)$ from a random sample of size n from (7.18) by $Y_{i:m:n}$, $i = 1, \cdots, m$, we can write the likelihood function (L) for a progressively Type-II right censored sample as

$$L(\mu, \sigma) = c \prod_{i=1}^{m} f(Y_{i:m:n}) \left[1 - F(Y_{i:m:n})\right]^{R_i},$$

where c is as given in (1.3), and $f(\cdot)$ and $F(\cdot)$ are as given in (7.18) and (6.52), respectively. We will first maximize with respect to μ.

Notice that, for values of $\mu \leq Y_{1:m:n}$, the likelihood function reduces to

$$L_0(\mu, \sigma) = \frac{1}{2^n \sigma^m} e^{-\sum_{i=1}^{m}(R_i+1)(Y_{i:m:n}-\mu)/\sigma}, \tag{7.19}$$

which is a monotonically increasing function of μ. Next, we consider values of $\mu > Y_{m:m:n}$. For these values, the likelihood function reduces to

$$L_m(\mu, \sigma) = \frac{1}{(2\sigma)^m} e^{\sum_{i=1}^m (Y_{i:m:n}-\mu)/\sigma} \prod_{i=1}^m \left[1 - \frac{1}{2} e^{(Y_{i:m:n}-\mu)/\sigma}\right]^{R_i}. \tag{7.20}$$

Upon taking the logarithm of L_m and differentiating with respect to μ, we obtain

$$\frac{\partial \ln L_m}{\partial \mu} = -\frac{m}{\sigma} + \sum_{i=1}^m \frac{R_i\, e^{(Y_{i:m:n}-\mu)/\sigma}}{2\sigma\left[1 - \frac{1}{2}e^{(Y_{i:m:n}-\mu)/\sigma}\right]}. \tag{7.21}$$

Now, if $R_1 = R_2 = \cdots = R_m = 0$, then $m = n$, so that the right-hand side of (7.21) is simply $-\frac{n}{\sigma}$ which is strictly less than 0. If some $R_i > 0,\ i = 1, 2, \cdots, m$, then

$$\frac{\partial \ln L_m}{\partial \sigma} < -\frac{m}{\sigma} + \frac{\sum_{i=1}^m R_i}{\sigma} \le 0 \ \text{ if } \ \sum_{i=1}^m R_i \le m, \ \text{ ie. } n - m \le m, \ \text{ ie. } m \ge \frac{n}{2},$$

so that L_m is monotonically decreasing for these values of m. Thus, if the observed number of failures $m \ge \frac{n}{2}$, the MLE of μ lies in the interval $[Y_{1:m:n},\ Y_{m:m:n}]$.

Consider now the values of μ such that $Y_{j:m:n} < \mu \le Y_{j+1:m:n}$ for $j \in \{1, 2, \cdots, m-1\}$. In this case, the likelihood function reduces to

$$\begin{aligned} L_j(\mu, \sigma) &= \frac{1}{(2\sigma)^m} e^{\sum_{i=1}^j (Y_{i:m:n}-\mu)/\sigma} \prod_{i=1}^j \left[1 - \frac{1}{2} e^{(Y_{i:m:n}-\mu)/\sigma}\right]^{R_i} \\ &\quad \times e^{-\sum_{i=j+1}^m (Y_{i:m:n}-\mu)/\sigma} \prod_{i=j+1}^m \left[\frac{1}{2} e^{-(Y_{i:m:n}-\mu)/\sigma}\right]^{R_i}. \end{aligned} \tag{7.22}$$

Note that the likelihood function is a continuous function in μ $(-\infty < \mu < \infty)$, so that $L_j(Y_{j+1:m:n}, \sigma) = L_{j+1}(Y_{j+1:m:n}, \sigma)$, $j = 0, 1, \cdots, m-1$. Upon taking the logarithm of L_j and differentiating with respect to μ, we obtain

$$\frac{\partial \ln L_j}{\partial \mu} = \frac{1}{\sigma}\left[\sum_{i=1}^j \frac{R_i\, e^{(Y_{i:m:n}-\mu)/\sigma}}{2 - e^{(Y_{i:m:n}-\mu)/\sigma}} + n - \sum_{i=1}^j R_i - 2j\right]. \tag{7.23}$$

Now if $R_i = 0$ $(i = 1, 2, \cdots, j)$, then the right-hand side of (7.23) becomes simply $\frac{n-2j}{\sigma}$, which is strictly negative, provided $j > \frac{n}{2}$. If some $R_i > 0,\ i = 1, 2, \cdots, j$, then the right-hand side of (7.23) is strictly less than $\frac{n-2j}{\sigma} \le 0$ if $j \ge \frac{n}{2}$. Thus, in general,

$$\frac{\partial L_j}{\partial \mu} < 0 \ \text{ provided } \ j > \frac{n}{2}$$

so that L_j is monotonically decreasing for these values of j. Thus, if the observed number of failures $m > \frac{n}{2}$, the maximum likelihood estimator of μ lies in the interval $\left[Y_{1:m:n},\ Y_{[n/2]+1:m:n}\right]$.

Upon further inspection of (7.23), it is evident that if $n-R_1-\cdots-R_j-2j>0$, then the right-hand side is strictly positive. This can only be possible if $j<\frac{n}{2}$, in which case $n-R_1-\cdots-Rj-2j>\frac{n}{2}-j-R_1-\cdots-R_j\geq 0$ if $R_1+\cdots+R_j+j\leq\frac{n}{2}$. Thus,

$$\frac{\partial L_j}{\partial\mu}>0 \text{ provided } j<\frac{n}{2} \text{ and } R_1+\cdots+R_j+j\leq\frac{n}{2},$$

so that L_j is monotonically increasing for these values of j.

At this point, we can formulate the following algorithm [due to Aggarwala and Balakrishnan (1999)] for obtaining MLEs of the location and scale parameters from a Laplace distribution when a progressively Type-II right censored sample of size m from a sample of size n is observed, with censoring scheme $(R_1, R_2, \cdots, R_m)$. We will denote the progressively Type-II right censored order statistics from this sample by $Y_{i:m:n}$, $i=1,2,\cdots,m$, and the corresponding observed values of the order statistics by y_i, $i=1,2,\cdots,m$.

1. Find the largest number $k<\frac{n}{2}$ such that $R_1+\cdots+R_k+k\leq\frac{n}{2}$, assuming $R_0=0$.

   (a) If $m<\frac{n}{2}$, the maximum likelihood estimates are those corresponding to

   $$\max\left\{\max_{y_{k+1}\leq\mu\leq y_{k+2},\ \sigma>0} L_{k+1}, \max_{y_{k+2}<\mu\leq y_{k+3},\ \sigma>0} L_{k+2}, \cdots, \max_{\mu>y_m,\ \sigma>0} L_m\right\}.$$

   (b) If $m>\frac{n}{2}$, the maximum likelihood estimates are those corresponding to

   $$\max\left\{\max_{y_{k+1}\leq\mu\leq y_{k+2},\ \sigma>0} L_{k+1}, \max_{y_{k+2}<\mu\leq y_{k+3},\ \sigma>0} L_{k+2}, \cdots, \max_{y_{n/2}<\mu\leq y_{(n/2)+1},\ \sigma>0} L_{n/2}\right\} \text{ if } n \text{ is even, and}$$

   $$\max\left\{\max_{y_{k+1}\leq\mu\leq y_{k+2},\ \sigma>0} L_{k+1}, \max_{y_{k+2}<\mu\leq y_{k+3},\ \sigma>0} L_{k+2}, \cdots, \max_{y_{(n-1)/2}<\mu\leq y_{(n+1)/2},\ \sigma>0} L_{(n-1)/2}\right\} \text{ if } n \text{ is odd.}$$

   For the case when n is odd, notice that we may have $k=\frac{n-1}{2}$. This simply means that the likelihood function is monotonically increasing for $\mu<Y_{(n+1)/2:m:n}$ and the likelihood function is monotonically decreasing for $\mu>Y_{(n+1)/2:m:n}$. Thus, the MLE of μ is $Y_{(n+1)/2:m:n}$, which we can use to solve for the MLE of σ. The resulting likelihood equation to be maximized with respect to σ is proportional to

$$\frac{1}{\sigma^m} e^{\sum_{i=1}^{(n-1)/2}(Y_{i:m:n}-Y_{(n+1)/2:m:n})/\sigma} \times \prod_{i=1}^{(n-1)/2}\left\{1-\frac{1}{2}e^{[Y_{i:m:n}-Y_{(n+1)/2}]/\sigma}\right\}^{R_i} \times e^{-\sum_{i=(n+3)/2}^{m}(R_i+1)[Y_{i:m:n}-Y_{(n+1)/2}]/\sigma}.$$

(This corresponds to maximizing either $L_{(n-1)/2}$ or $L_{(n+1)/2}$, due to the continuity of the likelihood function discussed earlier.)

(c) If $m = \frac{n}{2}$, the MLEs are those corresponding to

$$\max\left\{\max_{y_{k+1}\le\mu\le y_{k+2},\ \sigma>0} L_{k+1}, \max_{y_{k+2}<\mu\le y_{k+3},\ \sigma>0} L_{k+2}, \cdots, \max_{y_{(n/2)-1}<\mu\le y_{(n/2)},\ \sigma>0} L_{(n/2)-1}\right\}.$$

Notice that here it is possible to obtain $k = \frac{n}{2} - 1 = m - 1$. This means that for $\mu < Y_{m:m:n}$, the likelihood function is monotonically increasing and for $\mu > Y_{m:m:n}$, the likelihood function is monotonically decreasing. Therefore, the MLE of μ is $Y_{m:m:n}$. This can be used to solve for the MLE of σ. The resulting likelihood equation to be maximized with respect to σ is proportional to

$$\frac{1}{\sigma^m} e^{\sum_{i=1}^{m-1}(Y_{i:m:n}-Y_{m:m:n})/\sigma} \prod_{i=1}^{m-1}\left[1-\frac{1}{2}e^{(Y_{i:m:n}-Y_{m:m:n})/\sigma}\right]^{R_i}.$$

Remark 7.16 For the special case of conventional Type-II right censoring, where $R_1 = R_2 = \cdots = R_{m-1} = 0$ and $R_m = n - m$, this algorithm reduces to that given in Balakrishnan and Cutler (1995). For $m < \frac{n}{2}$, we have $k = m - 1$, and we just maximize L_m. For $m > \frac{n}{2}$, for n odd, we have $k = \frac{n-1}{2}$, and the MLE of μ is $Y_{(n+1)/2:m:n}$. For n even, we have $k = \frac{n}{2} - 1$, and we must maximize $L_{n/2}$. From (7.23), $\frac{\partial L_{n/2}}{\partial\mu}$ is obviously zero, and so the MLE of μ is any value in $\left[Y_{n/2:m:n},\ Y_{(n/2)+1:m:n}\right]$. Finally, for $m = \frac{n}{2}$, we have $k = \frac{n}{2} - 1$, so that the MLE of μ is $Y_{n/2:m:n}$. These estimates of μ may then be used to obtain MLEs of σ.

Example 7.4 Let us consider the data in Example 6.5. In this case, we have a progressively Type-II right censored sample of size $m = 10$ from a sample of size $n = 20$ from the Laplace distribution with $\mu = 25$ and $\sigma = 5$, with censoring scheme $\mathbf{R}$ = (2,0,0,2,0,0,0,2,0,4), and the progressively Type-II right censored sample observed is as follows:

$$19.21167876,\ 21.97364262,\ 23.41776818,\ 23.66253070,\ 23.80222832,\ 24.23017797,\ 25.62072188,\ 25.86990938,\ 26.47997028, 27.55344134.$$

From the algorithm presented above, we first find $k = 6$. Thus, we must find μ and σ which correspond to

$$\max\left\{\max_{y_7\le\mu\le y_8,\ \sigma>0} L_7,\ \max_{y_8<\mu\le y_9,\ \sigma>0} L_8,\ \max_{y_9<\mu\le y_{10},\ \sigma>0} L_9\right\}.$$

Using Maple V Release 3, the maximum value of the likelihood function is obtained when we maximize $L_8(\mu,\sigma)$ over the region specified above. The corresponding MLEs are $\widehat{\mu} = 26.31069$ and $\widehat{\sigma} = 2.67091$.

Recall that in Example 6.5 we determined the best linear unbiased estimates of μ and σ and their standard errors as

$$\mu^* = 26.26607,\ SE(\mu^*) = 0.72333, \qquad \sigma^* = 2.64071,\ SE(\sigma^*) = 0.28972.$$

These values agree well with the MLEs we have just obtained.

Remark 7.17 It should be noted here that to obtain standard errors of the MLEs, a simulational study had to be conducted. Furthermore, since the class of distributions under study does not possess "regularity" properties due to its lack of differentiability, it may not be appropriate to approximate the asymptotic variance-covariance matrix of the MLEs using the method of inverting the matrix of second derivatives.

7.3.7 Other Distributions (Log-Normal, Gamma, Burr)

Cohen (1976) and Cohen and Norgaard (1977) have discussed maximum likelihood estimation and modified maximum likelihood estimation under progressive censoring for the three-parameter log-normal distribution with probability density function

$$f(y) = \frac{1}{\sigma\sqrt{2\pi}\,(y-\mu)} e^{-[ln(y-\mu)-\theta]^2/2\sigma^2},\ y > \mu,$$

and the three-parameter gamma distribution with probability density function

$$f(y) = \frac{1}{\Gamma(\theta)\,\sigma^\theta}(y-\mu)^{\theta-1} e^{-(y-\mu)/\sigma},\ \ y > \mu,$$

respectively. The reader is referred to these papers for detailed discussion of the resulting equations to be solved numerically for the various parameters of interest. The initial steps followed are similar to those for the Weibull distribution which we have outlined in Section 7.3.3, and a few special cases are also considered. Note that the cumulative distribution function for these two cases can not be written in an explicit form, as in the case of the normal distribution.

Gajjar and Khatri (1968) considered MLE's under progressively Type-I right censoring for the two-parameter log-normal and logistic distributions when the location and scale parameters change at each stage of censoring.

An interesting paper on maximum likelihood estimation under progressively Type-I right censored samples is due to Wingo (1993). In this paper, Wingo has considered the Burr Type XII distribution with cumulative distribution function

$$F(y) = 1 - \left(y^{\theta_4} + 1\right)^{-\theta_3}, \quad y > 0, \ \theta_3 > 0, \ \theta_4 > 0.$$

Note that neither of the parameters is a location or a scale parameter. In maximizing the likelihood function, the MLE of θ_3 is given by

$$\widehat{\theta}_3 = \frac{k_2}{\sum_{i=1}^{k_2} \ln\left(y_i^{\theta_4} + 1\right) + \sum_{i=1}^{m} R_i \ln\left(T_i^{\theta_4} + 1\right)},$$

and the MLE of θ_4 is obtained by substituting the above MLE of θ_3 into the following equation and solving for θ_4:

$$\begin{aligned}\frac{\partial \ln L}{\partial \theta_4} &= 0 = \frac{k_2}{\theta_4} + \sum_{i=1}^{k_2} \ln(y_i) - (\theta_3 + 1) \sum_{i=1}^{k_2} \frac{y_i^{\theta_4} \ln(y_i)}{y_i^{\theta_4} + 1} \\ &\qquad - \theta_3 \sum_{i=1}^{m} \frac{R_i T_i^{\theta_4} \ln(T_i)}{T_i^{\theta_4} + 1}.\end{aligned}$$

Wingo (1993) has discussed the existence of the estimators, and also applied a progressive Type-I right censoring scheme to a clinical trial designed to assess the effectiveness of an antibiotic ointment in relieving pain. For $n = 30$ patients, the progressive censoring scheme was chosen as $T_1 = 0.25$, $T_2 = 0.50$ and $T_3 = 0.75$ (in hours). Furthermore, the scheme $R_1 = 5$, $R_2 = 1$ and $R_3 = 4$ was employed. The observed progressively Type-I right censored sample of size 20 was (notice that a "failure" here is actually the time until a patient feels relief from pain!)

$$\begin{gathered}0.529, 0.554, 0.566, 0.653, 0.665, 0.683, 0.698, 0.786, 0.788, 0.828,\\ 0.829, 0.866, 0.879, 0.881, 0.899, 0.917, 1.037, 1.050, 1.110, 1.138.\end{gathered}$$

Some interesting observations here are that no failure was observed before the first censoring time, and there were many observed failures after the final censoring time. That is, there was no pre-set time at which experimentation was terminated. The ML estimates for this data turned out to be $\widehat{\theta}_3 = 2.597$ and

$\widehat{\theta}_4 = 6.560$. The resulting maximum likelihood estimate for the mean time to relief is 0.817 hours, from the expression given in Wingo (1983),

$$E(Y) = \frac{\Gamma\left(\theta_3 - \frac{1}{\theta_4}\right)\Gamma\left(1 + \frac{1}{\theta_4}\right)}{\Gamma(\theta_3)},$$

where the parameters θ_3 and θ_4 are replaced by their corresponding MLEs. Wingo (1993) has also derived expressions for the asymptotic variance-covariance matrix of MLEs as the inverse of the Fisher information matrix. The approximations calculated from the observed data are $Var\left(\widehat{\theta}_3\right) = 0.017$, $Var\left(\widehat{\theta}_4\right) = 0.014$, and $Cov\left(\widehat{\theta}_3, \widehat{\theta}_4\right) = 0.002$.

8

Linear Prediction

8.1 Introduction

Until now, we have discussed many properties of progressively Type-II right censored order statistics and also the estimation of location and scale parameters of different distributions based on progressively censored samples.

In this chapter, after assuming that we have observed a progressively Type-II right censored sample from a life-testing experiment, we will discuss the prediction of times to failure of the last R_m units still surviving at the time of termination of the life-testing experiment.

To be specific, let n items be placed on a life-test. At the time of the first failure, let R_1 of the remaining $n-1$ surviving units be randomly withdrawn from further testing; at the time of the next failure, let R_2 of the remaining $n-2-R_1$ surviving units be randomly withdrawn from further testing, and so on; finally, at the time of the m-th failure, let the remaining $R_m = n-m-R_1-\cdots-R_{m-1}$ surviving units be all censored from the life-testing experiment. As before, let us use $Y_{1:m:n}^{(R_1,\cdots,R_m)}, Y_{2:m:n}^{(R_1,\cdots,R_m)}, \cdots, Y_{m:m:n}^{(R_1,\cdots,R_m)}$ to denote such a progressively Type-II right censored sample. Here, based on this progressively censored sample, we are interested in predicting the times to failure of the last R_m (≥ 1) units still surviving at the observation $Y_{m:m:n}^{(R_1,\cdots,R_m)}$.

8.2 The Exponential Case

Let us consider the case when the lifetimes of the n units placed on the life-test are distributed as one-parameter exponential with probability density function

$$f(y) = \frac{1}{\theta}\, e^{-y/\theta}, \qquad y \geq 0,\ \theta > 0. \tag{8.1}$$

In this case, as seen earlier in Chapter 2, we have

$$\begin{aligned}
Z_1 &= nY_{1:m:n}^{(R_1,\cdots,R_m)}/\theta, \\
Z_2 &= (n-1-R_1)\left(Y_{2:m:n}^{(R_1,\cdots,R_m)} - Y_{1:m:n}^{(R_1,\cdots,R_m)}\right)\Big/\theta, \\
Z_3 &= (n-2-R_1-R_2)\left(Y_{3:m:n}^{(R_1,\cdots,R_m)} - Y_{2:m:n}^{(R_1,\cdots,R_m)}\right)\Big/\theta, \\
\cdots &\quad \cdots \\
\cdots &\quad \cdots
\end{aligned}$$

$$
\begin{aligned}
Z_i &= (n-i+1-R_1-\cdots-R_{i-1})\left(Y_{i:m:n}^{(R_1,\cdots,R_m)}-Y_{i-1:m:n}^{(R_1,\cdots,R_m)}\right)\Big/\theta,\\
\cdots &\quad \cdots\\
\cdots &\quad \cdots\\
Z_m &= (n-m+1-R_1-\cdots-R_{m-1})\left(Y_{m:m:n}^{(R_1,\cdots,R_m)}-Y_{m-1:m:n}^{(R_1,\cdots,R_m)}\right)\Big/\theta
\end{aligned}
\tag{8.2}
$$

to be independent and identically distributed as standard exponential.

Suppose we are now interested in predicting the time to failure of the smallest of the last R_m items censored at the observation $Y_{m:m:n}^{(R_1,\cdots,R_m)}$. It is then important to realize that this variable we are interested in predicting may be regarded as $Y_{m+1:m+1:n}^{(R_1,\cdots,R_{m-1},0,R_m-1)}$ (see, for example, Theorem 2.3), with the already observed data of $Y_{1:m:n}^{(R_1,\cdots,R_m)},\cdots,Y_{m:m:n}^{(R_1,\cdots,R_m)}$ being simply regarded as $Y_{1:m+1:n}^{(R_1,\cdots,R_{m-1},0,R_m-1)},\cdots,\ Y_{m:m+1:n}^{(R_1,\cdots,R_{m-1},0,R_m-1)}$. In this case, the above stated result readily implies that

$$
\begin{aligned}
Z_1 &= nY_{1:m+1:n}^{(R_1,\cdots,R_{m-1},0,R_m-1)}/\theta = S_1/\theta,\\
Z_2 &= (n-1-R_1)\left(Y_{2:m+1:n}^{(R_1,\cdots,R_{m-1},0,R_m-1)}-Y_{1:m+1:n}^{(R_1,\cdots,R_{m-1},0,R_m-1)}\right)\Big/\theta\\
&= S_2/\theta,\\
Z_3 &= (n-2-R_1-R_2)\left(Y_{3:m+1:n}^{(R_1,\cdots,R_{m-1},0,R_m-1)}-Y_{2:m+1:n}^{(R_1,\cdots,R_{m-1},0,R_m-1)}\right)\Big/\theta\\
&= S_3/\theta,\\
\cdots &\quad \cdots\\
\cdots &\quad \cdots\\
Z_i &= (n-i+1-R_1-\cdots-R_{i-1})\\
&\quad\times\left(Y_{i:m+1:n}^{(R_1,\cdots,R_{m-1},0,R_m-1)}-Y_{i-1:m+1:n}^{(R_1,\cdots,R_{m-1},0,R_m-1)}\right)\Big/\theta - S_i/\theta,\\
\cdots &\quad \cdots\\
\cdots &\quad \cdots\\
Z_m &= (n-m+1-R_1-\cdots-R_{m-1})\\
&\quad\times\left(Y_{m:m+1:n}^{(R_1,\cdots,R_{m-1},0,R_m-1)}-Y_{m-1:m+1:n}^{(R_1,\cdots,R_{m-1},0,R_m-1)}\right)\Big/\theta = S_m/\theta,\\
Z_{m+1} &= (n-m-R_1-\cdots-R_{m-1})\\
&\quad\times\left(Y_{m+1:m+1:n}^{(R_1,\cdots,R_{m-1},0,R_m-1)}-Y_{m:m+1:n}^{(R_1,\cdots,R_{m-1},0,R_m-1)}\right)\Big/\theta = S_{m+1}/\theta
\end{aligned}
\tag{8.3}
$$

are independent and identically distributed as standard exponential.

For the purpose of predicting $Y_{m+1:m+1:n}^{(R_1,\cdots,R_{m-1},0,R_m-1)}$, let us now consider a linear predictor of the form

$$Y_{m+1:m+1:n}^{(R_1,\cdots,R_{m-1},0,R_m-1)*} = \sum_{i=1}^{m} c_i Y_{i:m+1:n}^{(R_1,\cdots,R_{m-1},0,R_m-1)} \tag{8.4}$$

or, equivalently, of the form

$$Y_{m+1:m+1:n}^{(R_1,\cdots,R_{m-1},0,R_m-1)*} = \sum_{i=1}^{m} c_i' S_i, \tag{8.5}$$

where the coefficients c_i (or $c_i{}'$) need to be determined appropriately. For example, if we require the predictor to be unbiased, then we have the condition [from (8.5)]

$$\theta \sum_{i=1}^{m} c_i{}' = \theta \mu_{m+1:m+1:n}^{(R_1,\cdots,R_{m-1},0,R_m-1)},$$

where $\mu_{m+1:m+1:n}^{(R_1,\cdots,R_{m-1},0,R_m-1)}$ denotes the expected value of that progressively Type-II right censored order statistic from the standard exponential distribution. The above condition immediately reduces to

$$\begin{aligned}\sum_{i=1}^{m} c_i{}' &= \mu_{m+1:m+1:n}^{(R_1,\cdots,R_{m-1},0,R_m-1)} \\ &= \frac{1}{n} + \frac{1}{n-1-R_1} + \frac{1}{n-2-R_1-R_2} + \cdots \\ &\quad + \frac{1}{n-m+1-R_1-\cdots-R_{m-1}} + \frac{1}{n-m-R_1-\cdots-R_{m-1}}.\end{aligned} \tag{8.6}$$

Then, in order to derive the best linear unbiased predictor (BLUP), all we need to do is to minimize

$$\text{Var}\left(Y_{m+1:m+1:n}^{(R_1,\cdots,R_{m-1},0,R_m-1)} - \sum_{i=1}^{m} c_i{}' S_i\right)$$

with respect to $c_i{}'$ under the unbiasedness condition in (8.6); in other words, we need to minimize

$$\begin{aligned}\text{Var}\bigg(&\frac{S_1}{n} + \frac{S_2}{n-1-R_1} + \cdots + \frac{S_m}{n-m+1-R_1-\cdots-R_{m-1}} \\ &+ \frac{S_{m+1}}{n-m-R_1-\cdots-R_{m-1}} - \sum_{i=1}^{m} c_i{}' S_i\bigg)\end{aligned}$$

subject to the unbiasedness condition in (8.6). This optimization can be achieved by the Lagrangian method by starting with the objective function

$$\begin{aligned}
&\left(\frac{1}{n}-c_1'\right)^2+\left(\frac{1}{n-1-R_1}-c_2'\right)^2 \\
&+\cdots+\left(\frac{1}{n-m+1-R_1-\cdots-R_{m-1}}-c_m'\right)^2 \\
&+\frac{1}{(n-m-R_1-\cdots-R_{m-1})^2}-2\lambda\left\{\sum_{i=1}^{m}c_i'-\frac{1}{n}-\frac{1}{n-1-R_1}\right. \\
&\left.-\cdots-\frac{1}{n-m+1-R_1-\cdots R_{m-1}}-\frac{1}{n-m-R_1-\cdots-R_{m-1}}\right\},
\end{aligned} \tag{8.7}$$

where λ is the Lagrangian multiplier. Taking partial derivatives of (8.7) with respect to $c_1', \cdots, c_m'$ and then equating them to zero, we obtain:

$$\begin{aligned}
c_1' &= \frac{1}{n}+\lambda, \\
c_2' &= \frac{1}{n-1-R_1}+\lambda, \\
\cdots & \quad \cdots \\
\cdots & \quad \cdots \\
c_m' &= \frac{1}{n-m+1-R_1-\cdots R_{m-1}}+\lambda.
\end{aligned} \tag{8.8}$$

Upon adding these equations and using the unbiasedness condition in (8.6), we simply obtain

$$\lambda = \frac{1}{m(n-m-R_1-\cdots-R_{m-1})}$$

which, when substituted in (8.8), yields the solution for the coefficients $c_1', \cdots, c_m'$ as

$$\begin{aligned}
c_1' &= \frac{1}{n}+\frac{1}{m(n-m-R_1-\cdots-R_{m-1})}, \\
c_2' &= \frac{1}{n-1-R_1}+\frac{1}{m(n-m-R_1-\cdots-R_{m-1})}, \\
\cdots & \quad \cdots \\
\cdots & \quad \cdots \\
c_m' &= \frac{1}{n-m+1-R_1-\cdots-R_{m-1}}+\frac{1}{m(n-m-R_1-\cdots-R_{m-1})}.
\end{aligned}$$

Using these expressions, we derive the coefficients $c_1, c_2, \cdots, c_m$ in (8.4) as

$$c_1 = nc_1' - (n-1-R_1)c_2'$$

$$
\begin{aligned}
&= \frac{R_1+1}{m(n-m-R_1-\cdots-R_{m-1})} \\
&= \frac{R_1+1}{mR_m}, \\
c_2 &= (n-1-R_1)c_2{}' - (n-2-R_1-R_2)c_3{}' \\
&= \frac{R_2+1}{m(n-m-R_1-\cdots-R_{m-1})} \\
&= \frac{R_2+1}{mR_m}, \\
\cdots & \quad \cdots \\
\cdots & \quad \cdots \\
c_{m-1} &= (n-m+2-R_1-\cdots-R_{m-2})c'_{m-1} \\
&\qquad -(n-m+1-R_1-\cdots-R_{m-1})c'_m \\
&= \frac{R_{m-1}+1}{m(n-m-R_1-\cdots-R_{m-1})} \\
&= \frac{R_{m-1}+1}{mR_m}, \\
c_m &= (n-m+1-R_1-\cdots-R_{m-1})c'_m \\
&= 1+\frac{n-m+1-R_1-\cdots-R_{m-1}}{m(n-m-R_1-\cdots-R_{m-1})} \\
&= 1+\frac{R_m+1}{mR_m}.
\end{aligned}
$$

Substituting for these coefficients, we obtain the BLUP of $Y_{m+1:m+1:n}^{(R_1,\cdots,R_{m-1},0,R_m-1)}$ from (8.4) to be

$$
\begin{aligned}
Y_{m+1:m+1:n}^{(R_1,\cdots,R_{m-1},0,R_m-1)*} &= \sum_{i=1}^{m} c_i Y_{i:m+1:n}^{(R_1,\cdots,R_{m-1},0,R_m-1)} \\
&= \sum_{i=1}^{m} c_i Y_{i:m:n}^{(R_1,\cdots,R_m)} \\
&= Y_{m:m:n}^{(R_1,\cdots,R_m)} + \frac{1}{mR_m}\sum_{i=1}^{m}(R_i+1)Y_{i:m:n}^{(R_1,\cdots,R_m)}.
\end{aligned}
\tag{8.9}
$$

Furthermore, the precision of this BLUP may be determined by its variance computed from (8.7) as

$$
\theta^2\left\{m\lambda^2 + \frac{1}{(n-m-R_1-\cdots-R_{m-1})^2} -2\lambda\left(m\lambda - \frac{1}{n-m-R_1-\cdots-R_{m-1}}\right)\right\}
$$

$$= \frac{(m+1)\theta^2}{m(n-m-R_1-\cdots-R_{m-1})^2} = \frac{(m+1)\theta^2}{mR_m^2}\,. \qquad (8.10)$$

Proceeding similarly, we can also develop the BLUP of the time to failure of the second smallest of the last R_m items censored at the observation $Y_{m:m:n}^{(R_1,\cdots,R_m)}$, viz., $Y_{m+2:m+2:n}^{(R_1,\cdots,R_{m-1},0,0,R_m-2)}$, and so on.

Remark 8.1 The approach taken here can be easily modified in order to develop the BLUPs for the case when the lifetimes of the n units placed on the life-test are distributed as two-parameter exponential with probability density function

$$f(y) = \frac{1}{\sigma}\, e^{-(y-\mu)/\sigma}, \qquad y \geq \mu,\ \sigma > 0.$$

In this case, once again the normalized spacings $S_1, S_2, \cdots, S_{m+1}$, with the only change that S_1 is defined as $S_1 = n(Y_{1:m+1:n}^{(R_1,\cdots,R_{m-1},0,R_m-1)} - \mu)$, are independent and identically distributed as exponential with scale parameter σ.

Remark 8.2 If we do not require the predictor to be unbiased, then without placing the unbiasedness condition we could minimize the mean square error

$$E\left(Y_{m+1:m+1:n}^{(R_1,\cdots,R_{m-1},0,R_m-1)} - \sum_{i=1}^{m} c_i Y_{i:m+1:n}^{(R_1,\cdots,R_{m-1},0,R_m-1)}\right)^2.$$

This can be done by following exactly the same steps as those used in the derivation of the BLUP. Such a linear predictor will be the best linear invariant predictor (BLIP) of $Y_{m+1:m+1:n}^{(R_1,\cdots,R_{m-1},0,R_m-1)}$. One can also proceed similarly to derive the BLIP of $Y_{m+2:m+2:n}^{(R_1,\cdots,R_{m-1},0,0,R_m-2)}$, and so on.

Remark 8.3 Though the approach adopted here works very well in the case of the exponential distribution (due to the fact that the normalized spacings are independently and identically distributed as exponential in this case), similar derivation of BLUPs and BLIPs are possible in the case of other distributions as well. The resulting linear predictors may, however, not be in such simple and explicit algebraic forms.

8.3 Case of General Distributions

8.3.1 Scale-Parameter Distributions

Let us first consider the case when $Y_{1:m:n}^{(R_1,\cdots,R_m)}, Y_{2:m:n}^{(R_1,\cdots,R_m)}, \cdots, Y_{m:m:n}^{(R_1,\cdots,R_m)}$ is a progressively Type-II right censored sample observed from a general scale-parameter distribution with its probability density function of the form $\frac{1}{\theta}\, f\left(\frac{y}{\theta}\right)$. As before, let us consider the prediction of the time to failure of the smallest of the last R_m units censored at the observation $Y_{m:m:n}^{(R_1,\cdots,R_m)}$. Equivalently stated, having observed $Y_{1:m+1:n}^{(R_1,\cdots,R_{m-1},0,R_m-1)}, \cdots, Y_{m:m+1:n}^{(R_1,\cdots,R_{m-1},0,R_m-1)}$, we wish to predict $Y_{m+1:m+1:n}^{(R_1,\cdots,R_{m-1},0,R_m-1)}$. For convenience, let us now denote

$$\begin{aligned}
\boldsymbol{Y} &= \left(Y_{1:m+1:n}^{(R_1,\cdots,R_{m-1},0,R_m-1)}, \cdots, Y_{m:m+1:n}^{(R_1,\cdots,R_{m-1},0,R_m-1)}\right)^T, \\
E(\boldsymbol{Y}) &= \theta\boldsymbol{\mu} = \theta\left(\mu_{1:m+1:n}, \cdots, \mu_{m:m+1:n}\right)^T, \\
\mathrm{Var}(\boldsymbol{Y}) &= \theta^2\boldsymbol{\Sigma} = \theta^2 \begin{pmatrix} \sigma_{1,1:m+1:n} & \cdots & \sigma_{1,m:m+1:n} \\ \cdots & \cdots & \cdots \\ \sigma_{m,1:m+1:n} & \cdots & \sigma_{m,m:m+1:n} \end{pmatrix},
\end{aligned}$$

$$\begin{aligned}
E\left(Y_{m+1:m+1:n}^{(R_1,\cdots,R_{m-1},0,R_m-1)}\right) &= \theta\mu_{m+1:m+1:n}, \\
\mathrm{Var}\left(Y_{m+1:m+1:n}^{(R_1,\cdots,R_{m-1},0,R_m-1)}\right) &= \theta^2\sigma_{m+1,m+1:m+1:n},
\end{aligned}$$

and

$$\mathrm{Cov}\left(Y_{m+1:m+1:n}^{(R_1,\cdots,R_{m-1},0,R_m-1)}, \boldsymbol{Y}\right) = \theta^2\boldsymbol{\omega} = \theta^2\left(\sigma_{m+1,1:m+1:n}, \cdots, \sigma_{m+1,m:m+1:n}\right)^T.$$

Note that we have suppressed the superscript denoting the censoring scheme in our definition of $\boldsymbol{\mu}$ and $\boldsymbol{\Sigma}$ as it remains the same throughout this discussion. As before, let us consider a general linear predictor of $Y_{m+1:m+1:n}^{(R_1,\cdots,R_{m-1},0,R_m-1)}$ as

$$Y_{m+1:m+1:n}^{(R_1,\cdots,R_{m-1},0,R_m-1)*} = \sum_{i=1}^{m} c_i Y_{i:m+1:n}^{(R_1,\cdots,R_{m-1},0,R_m-1)} = \boldsymbol{c}^T\boldsymbol{Y}, \tag{8.11}$$

where the coefficients $c_1, c_2, \cdots, c_m$ need to be determined appropriately. If we require the predictor in (8.11) to be unbiased, for example, we place the restriction

$$\boldsymbol{c}^T\boldsymbol{\mu} = \mu_{m+1:m+1:n}. \tag{8.12}$$

Then, in order to derive the BLUP of $Y_{m+1:m+1:n}^{(R_1,\cdots,R_{m-1},0,R_m-1)}$, all we need to do is to minimize

$$\mathrm{Var}\left(Y_{m+1:m+1:n}^{(R_1,\cdots,R_{m-1},0,R_m-1)} - \boldsymbol{c}^T\boldsymbol{Y}\right)$$

under the unbiasedness condition in (8.12); that is, we need to minimize

$$\sigma_{m+1,m+1:m+1:n} + \boldsymbol{c}^T\boldsymbol{\Sigma}\boldsymbol{c} - 2\boldsymbol{c}^T\boldsymbol{\omega}$$

subject to the unbiasedness condition in (8.12). Following the Lagrangian method, we need to minimize the objective function

$$\sigma_{m+1,m+1:m+1:n} + \boldsymbol{c}^T\boldsymbol{\Sigma}\boldsymbol{c} - 2\boldsymbol{c}^T\boldsymbol{\omega} - 2\lambda(\boldsymbol{c}^T\boldsymbol{\mu} - \mu_{m+1:m+1:n}), \tag{8.13}$$

where λ is the Lagrangian multiplier. Differentiating (8.13) with respect to $\boldsymbol{c}$, we obtain

$$2\boldsymbol{\Sigma}\boldsymbol{c} - 2\boldsymbol{\omega} - 2\lambda\boldsymbol{\mu} = 0$$

which gives the solution for $\boldsymbol{c}$ as

$$\boldsymbol{c} = \boldsymbol{\Sigma}^{-1}\boldsymbol{\omega} + \lambda\boldsymbol{\Sigma}^{-1}\boldsymbol{\mu}. \tag{8.14}$$

From (8.14), we get

$$\boldsymbol{c}^T\boldsymbol{\mu} = \boldsymbol{\omega}^T\boldsymbol{\Sigma}^{-1}\boldsymbol{\mu} + \lambda\boldsymbol{\mu}^T\boldsymbol{\Sigma}^{-1}\boldsymbol{\mu}$$

which, when used with the unbiasedness condition, simply yields the optimal choice of the Lagrangian multiplier λ to be

$$\lambda = \frac{\mu_{m+1:m+1:n} - \boldsymbol{\omega}^T\boldsymbol{\Sigma}^{-1}\boldsymbol{\mu}}{\boldsymbol{\mu}^T\boldsymbol{\Sigma}^{-1}\boldsymbol{\mu}}.$$

Upon substituting this expression of λ in (8.14), we obtain the optimal choice of the coefficient vector $\boldsymbol{c}$ as

$$\begin{aligned} \boldsymbol{c} &= \boldsymbol{\Sigma}^{-1}\boldsymbol{\omega} + \frac{\mu_{m+1:m+1:n}\boldsymbol{\Sigma}^{-1}\boldsymbol{\mu}}{\boldsymbol{\mu}^T\boldsymbol{\Sigma}^{-1}\boldsymbol{\mu}} - \frac{\boldsymbol{\omega}^T\boldsymbol{\Sigma}^{-1}\boldsymbol{\mu}\boldsymbol{\Sigma}^{-1}\boldsymbol{\mu}}{\boldsymbol{\mu}^T\boldsymbol{\Sigma}^{-1}\boldsymbol{\mu}} \\ &= \boldsymbol{\Sigma}^{-1}\boldsymbol{\omega} + \frac{(\mu_{m+1:m+1:n} - \boldsymbol{\omega}^T\boldsymbol{\Sigma}^{-1}\boldsymbol{\mu})\boldsymbol{\Sigma}^{-1}\boldsymbol{\mu}}{\boldsymbol{\mu}^T\boldsymbol{\Sigma}^{-1}\boldsymbol{\mu}}. \end{aligned} \tag{8.15}$$

Substituting this expression of $\boldsymbol{c}$ in (8.11), we obtain the BLUP of $Y_{m+1:m+1:n}^{(R_1,\cdots,R_{m-1},0,R_m-1)}$ as

$$\begin{aligned} Y_{m+1:m+1:n}^{(R_1,\cdots,R_{m-1},0,R_m-1)*} &= \boldsymbol{c}^T\boldsymbol{Y} \\ &= \boldsymbol{\omega}^T\boldsymbol{\Sigma}^{-1}\boldsymbol{Y} + \frac{(\mu_{m+1:m+1:n} - \boldsymbol{\omega}^T\boldsymbol{\Sigma}^{-1}\boldsymbol{\mu})\boldsymbol{\mu}^T\boldsymbol{\Sigma}^{-1}\boldsymbol{Y}}{\boldsymbol{\mu}^T\boldsymbol{\Sigma}^{-1}\boldsymbol{\mu}}. \end{aligned} \tag{8.16}$$

Further, the precision of this BLUP can be determined by its variance computed from (8.13) as

$$\theta^2\left\{\sigma_{m+1,m+1:m+1:n} - \boldsymbol{\omega}^T\boldsymbol{\Sigma}^{-1}\boldsymbol{\omega} + \frac{(\mu_{m+1:m+1:n} - \boldsymbol{\omega}^T\boldsymbol{\Sigma}^{-1}\boldsymbol{\mu})^2}{\boldsymbol{\mu}^T\boldsymbol{\Sigma}^{-1}\boldsymbol{\mu}}\right\}. \tag{8.17}$$

Remark 8.4 Following similar steps, we can derive the BLUP of the second smallest of the last R_m items censored at the observation $Y_{m:m:n}^{(R_1,\cdots,R_m)}$, viz., $Y_{m+2:m+2:n}^{(R_1,\cdots,R_{m-1},0,0,R_m-2)}$. The BLUP in this case will also be given by the expression in (8.16) with the following two changes:

$$\boldsymbol{\omega} = \left(\sigma_{m+2,1:m+2:n},\cdots,\sigma_{m+2,m:m+2:n}\right)^T$$

and $\mu_{m+2:m+2:n}$ replacing $\mu_{m+1:m+1:n}$. Of course, the progressive censoring scheme adopted here is given by $(R_1,\cdots,\ R_{m-1},0,0,R_m-2)$ and the means, variances and covariances of progressively censored order statistics (from the standard distribution) present in the above formulas all correspond to this censoring scheme. One can similarly derive the BLUP of the times to failure of other censored items (amongst the last R_m censored) as well.

Remark 8.5 Without requiring the predictor to be unbiased, if we instead minimize the mean square error

$$E\left(Y_{m+1:m+1:n}^{(R_1,\cdots,R_{m-1},0,R_m-1)} - \sum_{i=1}^{m} c_i Y_{i:m+1:n}^{(R_1,\cdots,R_{m-1},0,R_m-1)}\right)^2$$

by proceeding exactly as before, we will derive the BLIP of $Y_{m+1:m+1:n}^{(R_1,\cdots,R_{m-1},0,R_m-1)}$ for a general scale-parameter distribution.

Remark 8.6 For the scale-parameter exponential distribution in (8.1), upon making use of the explicit expressions of the means, variances and covariances of the progressively censored order statistics presented earlier in Chapter 2, and performing algebraic simplifications, we can show that the BLUP of $Y_{m+1:m+1:n}^{(R_1,\cdots,R_{m-1},0,R_m-1)}$ in (8.16) reduces to the simple explicit expression in (8.9). For all other scale-parameter distributions, the formula in (8.16) may be used to numerically determine the BLUP of $Y_{m+1:m+1:n}^{(R_1,\cdots,R_{m-1},0,R_m-1)}$.

8.3.2 Location-Scale Distributions

Next, let us consider the case when $Y_{1:m:n}^{(R_1,\cdots,R_m)}, Y_{2:m:n}^{(R_1,\cdots,R_m)}, \cdots, Y_{m:m:n}^{(R_1,\cdots,R_m)}$ is a progressively Type-II right censored sample observed from a general location-scale distribution with its probability density function of the form $\frac{1}{\sigma} f\left(\frac{y-\mu}{\sigma}\right)$. As before, let us consider the prediction of the time to failure of the smallest of the last R_m units censored at the observation $Y_{m:m:n}^{(R_1,\cdots,R_m)}$. As mentioned earlier, this is equivalent to the prediction of $Y_{m+1:m+1:n}^{(R_1,\cdots,R_{m-1},0,R_m-1)}$ based on $Y_{1:m+1:n}^{(R_1,\cdots,R_{m-1},0,R_m-1)}, \cdots, Y_{m:m+1:n}^{(R_1,\cdots,R_{m-1},0,R_m-1)}$.

Let us denote

$$\begin{aligned}
\boldsymbol{Y} &= \left(Y_{1:m+1:n}^{(R_1,\cdots,R_{m-1},0,R_m-1)}, \cdots, Y_{m:m+1:n}^{(R_1,\cdots,R_{m-1},0,R_m-1)}\right)^T, \\
E(\boldsymbol{Y}) &= \mu\mathbf{1} + \sigma\boldsymbol{\mu} = \left(\mu + \sigma\mu_{1:m+1:n}, \cdots, \mu + \sigma\mu_{m:m+1:n}\right)^T, \\
\mathrm{Var}(\boldsymbol{Y}) &= \sigma^2\boldsymbol{\Sigma} = \sigma^2 \begin{pmatrix} \sigma_{1,1:m+1:n} & \cdots & \sigma_{1,m:m+1:n} \\ \cdots & \cdots & \cdots \\ \sigma_{m,1:m+1:n} & \cdots & \sigma_{m,m:m+1:n} \end{pmatrix},
\end{aligned}$$

$$\begin{aligned}
E\left(Y_{m+1:m+1:n}^{(R_1,\cdots,R_{m-1},0,R_m-1)}\right) &= \mu + \sigma\mu_{m+1:m+1:n}, \\
\mathrm{Var}\left(Y_{m+1:m+1:n}^{(R_1,\cdots,R_{m-1},0,R_m-1)}\right) &= \sigma^2\sigma_{m+1,m+1:m+1:n},
\end{aligned}$$

and

$$\mathrm{Cov}\left(Y_{m+1:m+1:n}^{(R_1,\cdots,R_{m-1},0,R_m-1)}, \boldsymbol{Y}\right) = \sigma^2\boldsymbol{\omega} = \sigma^2\left(\sigma_{m+1,1:m+1:n}, \cdots, \sigma_{m+1,m:m+1:n}\right)^T.$$

Here, $\mu_{i:m+1:n}$ and $\sigma_{i,j:m+1:n}$ denote the means and covariances of the progressively Type-II right censored order statistics from the standard (with $\mu = 0$ and $\sigma = 1$) distribution. For reasons stated in Section 8.3.1, we are again suppressing the superscript denoting the censoring scheme.

Let us now consider a general linear predictor of $Y_{m+1:m+1:n}^{(R_1,\cdots,R_{m-1},0,R_m-1)}$ as

$$Y_{m+1:m+1:n}^{(R_1,\cdots,R_{m-1},0,R_m-1)*} = \sum_{i=1}^{m} c_i Y_{i:m+1:n}^{(R_1,\cdots,R_{m-1},0,R_m-1)} = \boldsymbol{c}^T \boldsymbol{Y}, \tag{8.18}$$

where the coefficients $c_1, c_2, \cdots, c_m$ need to be determined appropriately. If we require the predictor in (8.18) to be unbiased, then we require

$$\mu \boldsymbol{c}^T \mathbf{1} + \sigma \boldsymbol{c}^T \boldsymbol{\mu} = \mu + \sigma \mu_{m+1:m+1:n}$$

which yields the unbiasedness conditions as

$$\boldsymbol{c}^T \mathbf{1} = 1 \qquad \text{and} \qquad \boldsymbol{c}^T \boldsymbol{\mu} = \mu_{m+1:m+1:n}. \tag{8.19}$$

Now in order to derive the BLUP of $Y_{m+1:m+1:n}^{(R_1,\cdots,R_{m-1},0,R_m-1)}$, all we need to do is to minimize

$$\text{Var}\left(Y_{m+1:m+1:n}^{(R_1,\cdots,R_{m-1},0,R_m-1)} - \boldsymbol{c}^T \boldsymbol{Y}\right)$$

subject to the unbiasedness conditions in (8.19). Following the Lagrangian method, we need to minimize the objective function

$$\sigma_{m+1,m\,|\,1:m+1:n} + \boldsymbol{c}^T \boldsymbol{\Sigma} \boldsymbol{c} - 2\boldsymbol{c}^T \boldsymbol{\omega} - 2\lambda_1(\boldsymbol{c}^T \mathbf{1} - 1) - 2\lambda_2(\boldsymbol{c}^T \boldsymbol{\mu} - \mu_{m+1:m+1:n}), \tag{8.20}$$

where λ_1 and λ_2 are the Lagrangian multipliers. Differentiating (8.20) with respect to $\boldsymbol{c}$, we obtain

$$2\boldsymbol{\Sigma} \boldsymbol{c} - 2\boldsymbol{\omega} - 2\lambda_1 \mathbf{1} - 2\lambda_2 \boldsymbol{\mu} = 0$$

which gives the solution for $\boldsymbol{c}$ as

$$\boldsymbol{c} = \boldsymbol{\Sigma}^{-1} \boldsymbol{\omega} + \lambda_1 \boldsymbol{\Sigma}^{-1} \mathbf{1} + \lambda_2 \boldsymbol{\Sigma}^{-1} \boldsymbol{\mu}. \tag{8.21}$$

From (8.21), we get

$$\boldsymbol{c}^T \mathbf{1} = \boldsymbol{\omega}^T \boldsymbol{\Sigma}^{-1} \mathbf{1} + \lambda_1 \mathbf{1}^T \boldsymbol{\Sigma}^{-1} \mathbf{1} + \lambda_2 \boldsymbol{\mu}^T \boldsymbol{\Sigma}^{-1} \mathbf{1}$$

and

$$\boldsymbol{c}^T \boldsymbol{\mu} = \boldsymbol{\omega}^T \boldsymbol{\Sigma}^{-1} \boldsymbol{\mu} + \lambda_1 \boldsymbol{\mu}^T \boldsymbol{\Sigma}^{-1} \mathbf{1} + \lambda_2 \boldsymbol{\mu}^T \boldsymbol{\Sigma}^{-1} \boldsymbol{\mu}$$

which, when used with the unbiasedness conditions in (8.19), gives the equations

$$\left.\begin{array}{l} \lambda_1 \mathbf{1}^T \boldsymbol{\Sigma}^{-1} \mathbf{1} + \lambda_2 \boldsymbol{\mu}^T \boldsymbol{\Sigma}^{-1} \mathbf{1} = 1 - \boldsymbol{\omega}^T \boldsymbol{\Sigma}^{-1} \mathbf{1} \\ \lambda_1 \boldsymbol{\mu}^T \boldsymbol{\Sigma}^{-1} \mathbf{1} + \lambda_2 \boldsymbol{\mu}^T \boldsymbol{\Sigma}^{-1} \boldsymbol{\mu} = \mu_{m+1:m+1:n} - \boldsymbol{\omega}^T \boldsymbol{\Sigma}^{-1} \boldsymbol{\mu}. \end{array}\right\}$$

From these equations, we obtain the optimal choice of the Lagrangian multipliers λ_1 and λ_2 to be

$$\lambda_1 = \frac{\boldsymbol{\mu}^T\boldsymbol{\Sigma}^{-1}\boldsymbol{\mu} - \boldsymbol{\mu}^T\boldsymbol{\Sigma}^{-1}\boldsymbol{\mu}\boldsymbol{\omega}^T\boldsymbol{\Sigma}^{-1}\mathbf{1} - \mu_{m+1:m+1:n}\boldsymbol{\mu}^T\boldsymbol{\Sigma}^{-1}\mathbf{1} + \boldsymbol{\mu}^T\boldsymbol{\Sigma}^{-1}\mathbf{1}\boldsymbol{\omega}^T\boldsymbol{\Sigma}^{-1}\boldsymbol{\mu}}{\Delta} \tag{8.22}$$

and

$$\lambda_2 = \frac{-\boldsymbol{\mu}^T\boldsymbol{\Sigma}^{-1}\mathbf{1} + \boldsymbol{\mu}^T\boldsymbol{\Sigma}^{-1}\mathbf{1}\boldsymbol{\omega}^T\boldsymbol{\Sigma}^{-1}\mathbf{1} + \mu_{m+1:m+1:n}\mathbf{1}^T\boldsymbol{\Sigma}^{-1}\mathbf{1} - \mathbf{1}^T\boldsymbol{\Sigma}^{-1}\mathbf{1}\boldsymbol{\omega}^T\boldsymbol{\Sigma}^{-1}\boldsymbol{\mu}}{\Delta}, \tag{8.23}$$

where

$$\Delta = (\boldsymbol{\mu}^T\boldsymbol{\Sigma}^{-1}\boldsymbol{\mu})(\mathbf{1}^T\boldsymbol{\Sigma}^{-1}\mathbf{1}) - (\boldsymbol{\mu}^T\boldsymbol{\Sigma}^{-1}\mathbf{1})^2. \tag{8.24}$$

Upon substituting these expressions of λ_1 and λ_2 in (8.21), we obtain the optimal choice of the coefficient vector $\boldsymbol{c}$ as

$$\begin{aligned}
\boldsymbol{c} &= \boldsymbol{\Sigma}^{-1}\boldsymbol{\omega} + \frac{\boldsymbol{\mu}^T\boldsymbol{\Sigma}^{-1}\boldsymbol{\mu}\boldsymbol{\Sigma}^{-1}\mathbf{1}}{\Delta} - \frac{\boldsymbol{\mu}^T\boldsymbol{\Sigma}^{-1}\mathbf{1}\boldsymbol{\Sigma}^{-1}\boldsymbol{\mu}}{\Delta} \\
&\quad + \mu_{m+1:m+1:n}\left\{\frac{\mathbf{1}^T\boldsymbol{\Sigma}^{-1}\mathbf{1}\boldsymbol{\Sigma}^{-1}\boldsymbol{\mu}}{\Delta} - \frac{\boldsymbol{\mu}^T\boldsymbol{\Sigma}^{-1}\mathbf{1}\boldsymbol{\Sigma}^{-1}\mathbf{1}}{\Delta}\right\} \\
&\quad + \frac{\boldsymbol{\omega}^T\boldsymbol{\Sigma}^{-1}\boldsymbol{\mu}\boldsymbol{\mu}^T\boldsymbol{\Sigma}^{-1}\mathbf{1}\boldsymbol{\Sigma}^{-1}\mathbf{1}}{\Delta} - \frac{\boldsymbol{\omega}^T\boldsymbol{\Sigma}^{-1}\mathbf{1}\boldsymbol{\mu}^T\boldsymbol{\Sigma}^{-1}\boldsymbol{\mu}\boldsymbol{\Sigma}^{-1}\mathbf{1}}{\Delta} \\
&\quad + \frac{\boldsymbol{\omega}^T\boldsymbol{\Sigma}^{-1}\mathbf{1}\boldsymbol{\mu}^T\boldsymbol{\Sigma}^{-1}\mathbf{1}\boldsymbol{\Sigma}^{-1}\boldsymbol{\mu}}{\Delta} - \frac{\boldsymbol{\omega}^T\boldsymbol{\Sigma}^{-1}\boldsymbol{\mu}\mathbf{1}^T\boldsymbol{\Sigma}^{-1}\mathbf{1}\boldsymbol{\Sigma}^{-1}\boldsymbol{\mu}}{\Delta}.
\end{aligned} \tag{8.25}$$

Substituting this expression of $\boldsymbol{c}$ in (8.18), we obtain the BLUP of $Y_{m+1:m+1:n}^{(R_1,\cdots,R_{m-1},0,R_m-1)}$ as

$$\begin{aligned}
&Y_{m+1:m+1:n}^{(R_1,\cdots,R_{m-1},0,R_m-1)*} \\
&= \boldsymbol{c}^T\boldsymbol{Y} \\
&= \boldsymbol{\omega}^T\boldsymbol{\Sigma}^{-1}\boldsymbol{Y} + \frac{\boldsymbol{\mu}^T\boldsymbol{\Sigma}^{-1}\boldsymbol{\mu}\mathbf{1}^T\boldsymbol{\Sigma}^{-1}\boldsymbol{Y}}{\Delta} - \frac{\boldsymbol{\mu}^T\boldsymbol{\Sigma}^{-1}\mathbf{1}\boldsymbol{\mu}^T\boldsymbol{\Sigma}^{-1}\boldsymbol{Y}}{\Delta} \\
&\quad + \mu_{m+1:m+1:n}\left\{\frac{\mathbf{1}^T\boldsymbol{\Sigma}^{-1}\mathbf{1}\boldsymbol{\mu}^T\boldsymbol{\Sigma}^{-1}\boldsymbol{Y}}{\Delta} - \frac{\boldsymbol{\mu}^T\boldsymbol{\Sigma}^{-1}\mathbf{1}\mathbf{1}^T\boldsymbol{\Sigma}^{-1}\boldsymbol{Y}}{\Delta}\right\} \\
&\quad + \frac{\boldsymbol{\omega}^T\boldsymbol{\Sigma}^{-1}\mathbf{1}\boldsymbol{\mu}^T\boldsymbol{\Sigma}^{-1}\mathbf{1}\boldsymbol{\mu}^T\boldsymbol{\Sigma}^{-1}\boldsymbol{Y}}{\Delta} - \frac{\boldsymbol{\omega}^T\boldsymbol{\Sigma}^{-1}\mathbf{1}\boldsymbol{\mu}^T\boldsymbol{\Sigma}^{-1}\boldsymbol{\mu}\mathbf{1}^T\boldsymbol{\Sigma}^{-1}\boldsymbol{Y}}{\Delta} \\
&\quad + \frac{\boldsymbol{\omega}^T\boldsymbol{\Sigma}^{-1}\boldsymbol{\mu}\boldsymbol{\mu}^T\boldsymbol{\Sigma}^{-1}\mathbf{1}\mathbf{1}^T\boldsymbol{\Sigma}^{-1}\boldsymbol{Y}}{\Delta} - \frac{\boldsymbol{\omega}^T\boldsymbol{\Sigma}^{-1}\boldsymbol{\mu}\mathbf{1}^T\boldsymbol{\Sigma}^{-1}\mathbf{1}\boldsymbol{\mu}^T\boldsymbol{\Sigma}^{-1}\boldsymbol{Y}}{\Delta} \\
&= \mu^* + \mu_{m+1:m+1:n}\sigma^* + \boldsymbol{\omega}^T\boldsymbol{\Sigma}^{-1}(\boldsymbol{Y} - \mu^*\mathbf{1} - \sigma^*\boldsymbol{\mu})
\end{aligned} \tag{8.26}$$

as established earlier by Goldberger (1962); in the above formula, μ^* and σ^* denote the best linear unbiased estimators (BLUEs) of μ and σ based on $\boldsymbol{Y}$ (see Section 6.2 for details).

Further, the precision of this BLUP can be determined by its variance computed from (8.20) as

$$\sigma^2\{\sigma_{m+1,m+1:m+1:n} - \boldsymbol{\omega}^T\boldsymbol{\Sigma}^{-1}\boldsymbol{\omega} + \lambda_1^2\mathbf{1}^T\boldsymbol{\Sigma}^{-1}\mathbf{1} + \lambda_2^2\boldsymbol{\mu}^T\boldsymbol{\Sigma}^{-1}\boldsymbol{\mu} \\ +2\lambda_1\lambda_2\boldsymbol{\mu}^T\boldsymbol{\Sigma}^{-1}\mathbf{1}\}, \qquad (8.27)$$

where λ_1 and λ_2 are as given in (8.22) and (8.23), respectively.

Remark 8.7 As already mentioned in Remark 8.4, the formula in (8.26) may be used for the BLUP of the times to failure of other censored items (amongst the last R_m censored) after making a couple of minor changes in the notation.

Remark 8.8 For the location-scale exponential distribution, upon making use of the explicit expressions of the means, variances and covariances of the progressively censored order statistics presented earlier in Chapter 2, and performing algebraic simplifications, we can derive an explicit expression for the BLUP of $Y_{m+1:m+1:n}^{(R_1,\cdots,R_{m-1},0,R_m-1)}$. For other location-scale distributions, the formula in (8.26) may be used to numerically determine the BLUP of $Y_{m+1:m+1:n}^{(R_1,\cdots,R_{m-1},0,R_m-1)}$.

8.4 A Simple Approach Based on BLUEs

In order to explain this approach due to Doganaksoy and Balakrishnan (1997), let us first consider the scale-parameter exponential distribution in (8.1). In this case, as already seen in Chapter 6, based on the observed progressively Type-II right censored sample $Y_{1:m:n}^{(R_1,\cdots,R_m)}, Y_{2:m:n}^{(R_1,\cdots,R_m)}, \cdots, Y_{m:m:n}^{(R_1,\cdots,R_m)}$, the BLUE of θ is (see Section 6.1)

$$\theta^* = \frac{1}{m}\sum_{i=1}^{m}(R_i+1)Y_{i:m:n}^{(R_1,\cdots,R_m)}. \qquad (8.28)$$

Let us consider the prediction of the time to failure of the smallest of the last R_m items censored at the observation $Y_{m:m:n}^{(R_1,\cdots,R_m)}$. Then, as before, let us view this variable as $Y_{m+1:m+1:n}^{(R_1,\cdots,R_{m-1},0,R_m-1)}$ and the observed progressively Type-II right censored sample as $Y_{1:m+1:n}^{(R_1,\cdots,R_{m-1},0,R_m-1)}, \cdots, Y_{m:m+1:n}^{(R_1,\cdots,R_{m-1},0,R_m-1)}$. Based on this progressively Type-II right censored sample of size $m+1$, we have the BLUE of θ as (see Section 6.1)

$$\theta^* = \frac{1}{m+1}\left\{\sum_{i=1}^{m-1}(R_i+1)Y_{i:m+1:n}^{(R_1,\cdots,R_{m-1},0,R_m-1)} + Y_{m:m+1:n}^{(R_1,\cdots,R_{m-1},0,R_m-1)}\right.$$

$$+ R_m Y_{m+1:m+1:n}^{(R_1,\cdots,R_{m-1},0,R_m-1)}\Big\}$$

$$= \frac{1}{m+1}\Big\{\sum_{i=1}^{m-1}(R_i+1)Y_{i:m:n}^{(R_1,\cdots,R_m)} + Y_{m:m:n}^{(R_1,\cdots,R_m)} + R_m Y_{m+1:m+1:n}^{(R_1,\cdots,R_{m-1},0,R_m-1)}\Big\}. \tag{8.29}$$

Though the BLUE of θ in (8.28) can be computed from the observed sample and the BLUE of θ in (8.29) can not be computed, they may be equated and solved for $Y_{m+1:m+1:n}^{(R_1,\cdots,R_{m-1},0,R_m-1)}$ in order to derive its predictor. The equation

$$\frac{1}{m+1}\Big\{\sum_{i=1}^{m-1}(R_i+1)Y_{i:m:n}^{(R_1,\cdots,R_m)} + Y_{m:m:n}^{(R_1,\cdots,R_m)} + R_m Y_{m+1:m+1:n}^{(R_1,\cdots,R_{m-1},0,R_m-1)}\Big\} = \frac{1}{m}\sum_{i=1}^{m}(R_i+1)Y_{i:m:n}^{(R_1,\cdots,R_m)}$$

readily gives the solution for $Y_{m+1:m+1:n}^{(R_1,\cdots,R_{m-1},0,R_m-1)}$ as

$$Y_{m+1:m+1:n}^{(R_1,\cdots,R_{m-1},0,R_m-1)*} = Y_{m:m:n}^{(R_1,\cdots,R_m)} + \frac{1}{mR_m}\sum_{i=1}^{m}(R_i+1)Y_{i:m:n}^{(R_1,\cdots,R_m)}. \tag{8.30}$$

This is exactly the same as the BLUP of $Y_{m+1:m+1:n}^{(R_1,\cdots,R_{m-1},0,R_m-1)}$ presented earlier in (8.9).

Remark 8.9 As established by Doganaksoy and Balakrishnan (1997), this is a very general property of BLUEs and BLUPs. Specifically, Doganaksoy and Balakrishnan (1997) established that the BLUEs remain unchanged if the BLUPs of future failures are treated as observed values. Hence, the explicit expressions of BLUEs or the numerical determination of BLUEs of parameters as discussed in Chapter 6 may be used in this simple manner in order to determine the BLUPs of the times to failure of the last R_m items censored at the observation $Y_{m:m:n}^{(R_1,\cdots,R_m)}$.

From Section 6.1, based on the progressively Type-II right censored sample $Y_{1:m+2:n}^{(R_1,\cdots,R_{m-1},0,0,R_m-2)}, Y_{2:m+2:n}^{(R_1,\cdots,R_{m-1},0,0,R_m-2)}, \cdots, Y_{m+2:m+2:n}^{(R_1,\cdots,R_{m-1},0,0,R_m-2)}$ from a scale-parameter exponential distribution, we have the BLUE of θ as

$$\theta^* = \frac{1}{m+2}\Big\{\sum_{i=1}^{m+1}(R_i+1)Y_{i:m+2:n}^{(R_1,\cdots,R_{m-1},0,0,R_m-2)} + Y_{m:m+2:n}^{(R_1,\cdots,R_{m-1},0,0,R_m-2)} + Y_{m+1:m+2:n}^{(R_1,\cdots,R_{m-1},0,0,R_m-2)} + (R_m-1)Y_{m+2:m+2:n}^{(R_1,\cdots,R_{m-1},0,0,R_m-2)}\Big\}$$

$$= \frac{1}{m+2}\left\{\sum_{i=1}^{m-1}(R_i+1)Y_{i:m:n}^{(R_1,\cdots,R_m)} + Y_{m:m:n}^{(R_1,\cdots,R_m)} + Y_{m+1:m+1:n}^{(R_1,\cdots,R_{m-1},0,R_m-1)} + (R_m-1)Y_{m+2:m+2:n}^{(R_1,\cdots,R_{m-1},0,0,R_m-2)}\right\} \quad (8.31)$$

in which when we replace $Y_{m+1:m+1:n}^{(R_1,\cdots,R_{m-1},0,R_m-1)}$ by its BLUP in (8.30), we obtain

$$\theta^* = \frac{1}{m+2}\left\{\sum_{i=1}^{m-1}(R_i+1)Y_{i:m:n}^{(R_1,\cdots,R_m)} + 2X_{m:m:n}^{(R_1,\cdots,R_m)} + \frac{1}{mR_m}\sum_{i=1}^{m}(R_i+1)Y_{i:m:n}^{(R_1,\cdots,R_m)} + (R_m-1)Y_{m+2:m+2:n}^{(R_1,\cdots,R_{m-1},0,0,R_m-2)}\right\}. \quad (8.32)$$

Upon equating the expression of θ^* in (8.32) with that in (8.28) and solving for $Y_{m+2:m+2:n}^{(R_1,\cdots,R_{m-1},0,0,R_m-2)}$, we obtain

$$Y_{m+2:m+2:n}^{(R_1,\cdots,R_{m-1},0,0,R_m-2)*} = Y_{m:m:n}^{(R_1,\cdots,R_m)} + \left(\frac{1}{R_m}+\frac{1}{R_{m-1}}\right)\frac{1}{m}\sum_{i=1}^{m}(R_i+1)Y_{i:m:n}^{(R_1,\cdots,R_m)}, \quad (8.33)$$

which is the BLUP of $Y_{m+2:m+2:n}^{(R_1,\cdots,R_{m-1},0,0,R_m-2)}$ as determined by following the steps outlined in Section 8.2 for the one-parameter exponential distribution. Proceeding similarly, we can derive the BLUPs of the times to failure of the other items among the last R_m censored at the observation $Y_{m:m:n}^{(R_1,\cdots,R_m)}$.

Remark 8.10 In the case of other scale-parameter distributions, upon making use of the tables of coefficients for best linear unbiased estimators based on progressively Type-II right censored samples, we can use the above approach to numerically determine the BLUPs of the times to failure of the last R_m items censored at the observation $Y_{m:m:n}^{(R_1,\cdots,R_m)}$.

Next, let us consider the two-parameter exponential distribution as considered earlier in Chapter 6. In this case, based on the observed progressively Type-II right censored sample $Y_{1:m:n}^{(R_1,\cdots,R_m)}, Y_{2:m:n}^{(R_1,\cdots,R_m)}, \cdots, Y_{m:m:n}^{(R_1,\cdots,R_m)}$, the BLUE of σ is (see Section 6.2)

$$\sigma^* = \frac{1}{m-1}\sum_{i=2}^{m}(R_i+1)\left(Y_{i:m:n}^{(R_1,\cdots,R_m)} - Y_{1:m:n}^{(R_1,\cdots,R_m)}\right). \quad (8.34)$$

Let us consider the prediction of the time to failure of the smallest of the last R_m items censored at the observation $Y_{m:m:n}^{(R_1,\cdots,R_m)}$, viz., $Y_{m+1:m+1:n}^{(R_1,\cdots,R_{m-1},0,0,R_m-1)}$. Then based on the progressively Type-II right censored sample of size $m+1$, we have the BLUE of σ as (see Section 6.2)

$$\begin{aligned}
\sigma^* &= \frac{1}{m}\Bigg\{\sum_{i=2}^{m-1}(R_i+1)\left(Y_{i:m+1:n}^{(R_1,\cdots,R_{m-1},0,R_m-1)} - Y_{1:m+1:n}^{(R_1,\cdots,R_{m-1},0,R_m-1)}\right) \\
&\quad + \left(Y_{m:m+1:n}^{(R_1,\cdots,R_{m-1},0,R_m-1)} - Y_{1:m+1:n}^{(R_1,\cdots,R_{m-1},0,R_m-1)}\right) \\
&\quad + R_m\left(Y_{m+1:m+1:n}^{(R_1,\cdots,R_{m-1},0,R_m-1)} - Y_{1:m+1:n}^{(R_1,\cdots,R_{m-1},0,R_m-1)}\right)\Bigg\} \\
&= \frac{1}{m}\Bigg\{\sum_{i=2}^{m-1}(R_i+1)\left(Y_{i:m:n}^{(R_1,\cdots,R_m)} - Y_{1:m:n}^{(R_1,\cdots,R_m)}\right) \\
&\quad + \left(Y_{m:m:n}^{(R_1,\cdots,R_m)} - Y_{1:m:n}^{(R_1,\cdots,R_m)}\right) \\
&\quad + R_m\left(Y_{m+1:m+1:n}^{(R_1,\cdots,R_{m-1},0,R_m-1)} - Y_{1:m:n}^{(R_1,\cdots,R_m)}\right)\Bigg\}. \qquad (8.35)
\end{aligned}$$

Upon equating the BLUE of σ in (8.35) with that in (8.34) and solving for $Y_{m+1:m+1:n}^{(R_1,\cdots,R_{m-1},0,R_m-1)}$, we obtain its BLUP as

$$\begin{aligned}
&Y_{m+1:m+1:n}^{(R_1,\cdots,R_{m-1},0,R_m-1)*} \\
&\quad = Y_{m:m:n}^{(R_1,\cdots,R_m)} \\
&\quad\quad + \frac{1}{(m-1)R_m}\sum_{i=2}^{m}(R_i+1)\left(Y_{i:m:n}^{(R_1,\cdots,R_m)} - Y_{1:m:n}^{(R_1,\cdots,R_m)}\right). \qquad (8.36)
\end{aligned}$$

If we had instead equated the BLUEs of the location parameter μ based on the two progressively Type-II right censored samples, we would have obtained the equation

$$\begin{aligned}
&Y_{1:m:n}^{(R_1,\cdots,R_m)} - \frac{1}{n(m-1)}\sum_{i=2}^{m}(R_i+1)\left(Y_{i:m:n}^{(R_1,\cdots,R_m)} - Y_{1:m:n}^{(R_1,\cdots,R_m)}\right) \\
&= Y_{1:m:n}^{(R_1,\cdots,R_m)} - \frac{1}{nm}\Bigg\{\sum_{i=2}^{m-1}(R_i+1)\left(Y_{i:m:n}^{(R_1,\cdots,R_m)} - Y_{1:m:n}^{(R_1,\cdots,R_m)}\right) \\
&\quad + \left(Y_{m:m:n}^{(R_1,\cdots,R_m)} - Y_{1:m:n}^{(R_1,\cdots,R_m)}\right) \\
&\quad + R_m\left(Y_{m+1:m+1:n}^{(R_1,\cdots,R_{m-1},0,R_m-1)} - Y_{1:m:n}^{(R_1,\cdots,R_m)}\right)\Bigg\}
\end{aligned}$$

which would have resulted in exactly the same BLUP of $Y_{m+1:m+1:n}^{(R_1,\cdots,R_{m-1},0,R_m-1)}$ as the one in (8.36).

Using the same procedure, we can derive the BLUPs of the times to failure of the other items among the last R_m censored at the observation $Y_{m:m:n}^{(R_1,\cdots,R_m)}$.

Remark 8.11 In the case of other location-scale distributions, upon making use of the tables of coefficients for best linear unbiased estimators (of either the location parameter or the scale parameter) based on progressively Type-II right censored samples, we can use the above approach to numerically determine the BLUPs of the times to failure of the last R_m items censored at the observation $Y_{m:m:n}^{(R_1,\cdots,R_m)}$.

Remark 8.12 Though the BLUPs have all been developed here in the case of progressively Type-II right censored samples, it is clear that this approach can be adopted easily for the case of general progressively Type-II censored samples. For example, in the case of the exponential distributions, upon making use of the explicit expressions of the BLUEs based on general progressively Type-II censored samples presented earlier in Chapter 6, we can derive explicit expressions for the BLUPs of the times to failure of the last R_m items censored at the observation ${}_rY_{m:m:n}^{(R_{r+1},\cdots,R_m)}$.

8.5 First-Order Approximations to BLUPs

We can use the first-order approximations to the BLUEs presented earlier in Sections 6.1.5 and 6.2.7 in order to develop first-order approximations to BLUPs for the case of scale-parameter distributions as well as the case of location-scale distributions. For this task, we will once again utilize the approach of Doganaksoy and Balakrishnan (1997) as detailed already in Section 8.4.

Let us first consider the case of scale-parameter distributions. Let us assume that we have observed a progressively Type-II right censored sample $Y_{1:m:n}^{(R_1,\cdots,R_m)},\cdots,Y_{m:m:n}^{(R_1,\cdots,R_m)}$, and that we are interested in predicting the smallest lifetime among the last R_m items censored at the observation $Y_{m:m:n}^{(R_1,\cdots,R_m)}$. Then, as displayed earlier in Section 6.1.5, we can write the first-order approximation to the BLUE of the scale parameter θ as

$$\theta^* = \frac{1}{\Delta}\sum_{i=1}^{m} c_i Y_{i:m:n}^{(R_1,\cdots,R_m)}, \tag{8.37}$$

where

$$c_1 = \frac{F^{-1}(\pi_1)k_2}{(F^{-1^{(1)}}(\pi_1))^2 k_1(k_2b_1-k_1b_2)} - \frac{F^{-1}(\pi_2)}{F^{-1^{(1)}}(\pi_1)F^{-1^{(1)}}(\pi_2)(k_2b_1-k_1b_2)},$$

$$
\begin{aligned}
c_i &= \frac{-F^{-1}(\pi_{i-1})}{F^{-1^{(1)}}(\pi_{i-1})F^{-1^{(1)}}(\pi_i)(k_i b_{i-1} - k_{i-1} b_i)} \\
&\quad - \frac{F^{-1}(\pi_{i+1})}{F^{-1^{(1)}}(\pi_i)F^{-1^{(1)}}(\pi_{i+1})(k_{i+1} b_i - k_i b_{i+1})} \\
&\quad + \frac{F^{-1}(\pi_i)(k_{i+1} b_{i-1} - k_{i-1} b_{i+1})}{\{F^{-1^{(1)}}(\pi_i)\}^2 (k_i b_{i-1} - k_{i-1} b_i)(k_{i+1} b_i - k_i b_{i+1})}, \\
&\qquad\qquad i = 2, 3, \cdots, m-1,
\end{aligned}
$$

$$
\begin{aligned}
c_m &= \frac{-F^{-1}(\pi_{m-1})}{F^{-1^{(1)}}(\pi_{m-1})F^{-1^{(1)}}(\pi_m)(k_m b_{m-1} - k_{m-1} b_m)} \\
&\quad + \frac{F^{-1}(\pi_m) b_{m-1}}{\{F^{-1^{(1)}}(\pi_m)\}^2 b_m (k_m b_{m-1} - k_{m-1} b_m)}, \\
\Delta &= \frac{\{F^{-1}(\pi_1)\}^2 k_2}{\{F^{-1^{(1)}}(\pi_1)\}^2 k_1 (k_2 b_1 - k_1 b_2)} + \frac{\{F^{-1}(\pi_m)\}^2 b_{m-1}}{\{F^{-1^{(1)}}(\pi_m)\}^2 b_m (k_m b_{m-1} - k_{m-1} b_m)} \\
&\quad + \sum_{i=2}^{m-1} \frac{\{F^{-1}(\pi_i)\}^2 (k_{i+1} b_{i-1} - k_{i-1} b_{i+1})}{\{F^{-1^{(1)}}(\pi_i)\}^2 (k_i b_{i-1} - k_{i-1} b_i)(k_{i+1} b_i - k_i b_{i+1})} \\
&\quad - 2 \sum_{i=1}^{m-1} \frac{F^{-1}(\pi_i) F^{-1}(\pi_{i+1})}{F^{-1^{(1)}}(\pi_i) F^{-1^{(1)}}(\pi_{i+1})(k_{i+1} b_i - k_i b_{i+1})}, \\
k_i &= \prod_{k=1}^{i} \left\{ \frac{m-k+2+R_k+R_{k+1}+\cdots+R_m}{m-k+3+R_k+R_{k+1}+\cdots+R_m} \right\} \\
&\quad - \prod_{k=1}^{i} \left\{ \frac{m-k+1+R_k+R_{k+1}+\cdots+R_m}{m-k+2+R_k+R_{k+1}+\cdots+R_m} \right\}, \quad i = 1, 2, \cdots, m, \\
b_i &= \prod_{k=1}^{i} \left\{ \frac{m-k+1+R_k+R_{k+1}+\cdots+R_m}{m-k+2+R_k+R_{k+1}+\cdots+R_m} \right\}, \quad i = 1, 2, \cdots, m,
\end{aligned}
$$

and

$$\pi_i = 1 - b_i, \qquad i = 1, 2, \cdots, m.$$

Similarly, based on the progressively Type-II right censored sample $Y_{1:m+1:n}^{(R_1,\cdots,R_{m-1},0,R_m-1)}, \cdots, Y_{m+1:m+1:n}^{(R_1,\cdots,R_{m-1},0,R_m-1)}$, we can write the first-order approximation to the BLUE of the scale parameter θ as

$$
\begin{aligned}
\theta^* &= \frac{1}{\Delta^*} \sum_{i=1}^{m+1} c_i^* Y_{i:m+1:n}^{(R_1,\cdots,R_{m-1},0,R_m-1)} \\
&= \frac{1}{\Delta^*} \left\{ \sum_{i=1}^{m} c_i^* Y_{i:m:n}^{(R_1,\cdots,R_m)} + c_{m+1}^* Y_{m+1:m+1:n}^{(R_1,\cdots,R_{m-1},0,R_m-1)} \right\}, \qquad (8.38)
\end{aligned}
$$

where

$$c_1^* = \frac{F^{-1}(\pi_1^*)k_2^*}{(F^{-1^{(1)}}(\pi_1^*))^2 k_1^*(k_2^* b_1^* - k_1^* b_2^*)} - \frac{F^{-1}(\pi_2^*)}{F^{-1^{(1)}}(\pi_1^*)F^{-1^{(1)}}(\pi_2^*)(k_2^* b_1^* - k_1^* b_2^*)},$$

$$\begin{aligned} c_i^* &= \frac{-F^{-1}(\pi_{i-1}^*)}{F^{-1^{(1)}}(\pi_{i-1}^*)F^{-1^{(1)}}(\pi_i^*)(k_i^* b_{i-1}^* - k_{i-1}^* b_i^*)} \\ &\quad - \frac{F^{-1}(\pi_{i+1}^*)}{F^{-1^{(1)}}(\pi_i^*)F^{-1^{(1)}}(\pi_{i+1}^*)(k_{i+1}^* b_i^* - k_i^* b_{i+1}^*)} \\ &\quad + \frac{F^{-1}(\pi_i^*)(k_{i+1}^* b_{i-1}^* - k_{i-1}^* b_{i+1}^*)}{\{F^{-1^{(1)}}(\pi_i^*)\}^2(k_i^* b_{i-1}^* - k_{i-1}^* b_i^*)(k_{i+1}^* b_i^* - k_i^* b_{i+1}^*)}, \\ &\qquad i = 2, 3, \cdots, m, \end{aligned}$$

$$\begin{aligned} c_{m+1}^* &= \frac{-F^{-1}(\pi_m^*)}{F^{-1^{(1)}}(\pi_m^*)F^{-1^{(1)}}(\pi_{m+1}^*)(k_{m+1}^* b_m^* - k_m^* b_{m+1}^*)} \\ &\quad + \frac{F^{-1}(\pi_{m+1}^*) b_m^*}{\{F^{-1^{(1)}}(\pi_{m+1}^*)\}^2 b_{m+1}^*(k_{m+1}^* b_m^* - k_m^* b_{m+1}^*)}, \end{aligned}$$

$$\begin{aligned} \Delta^* &= \frac{\{F^{-1}(\pi_1^*)\}^2 k_2^*}{\{F^{-1^{(1)}}(\pi_1^*)\}^2 k_1^*(k_2^* b_1^* - k_1^* b_2^*)} \\ &\quad + \frac{\{F^{-1}(\pi_{m+1}^*)\}^2 b_m^*}{\{F^{-1^{(1)}}(\pi_{m+1}^*)\}^2 b_{m+1}^*(k_{m+1}^* b_m^* - k_m^* b_{m+1}^*)} \\ &\quad + \sum_{i=2}^{m} \frac{\{F^{-1}(\pi_i^*)\}^2(k_{i+1}^* b_{i-1}^* - k_{i-1}^* b_{i+1}^*)}{\{F^{-1^{(1)}}(\pi_i^*)\}^2(k_i^* b_{i-1}^* - k_{i-1}^* b_i^*)(k_{i+1}^* b_i^* - k_i^* b_{i+1}^*)} \\ &\quad - 2\sum_{i=2}^{m} \frac{F^{-1}(\pi_i^*)F^{-1}(\pi_{i+1}^*)}{F^{-1^{(1)}}(\pi_i^*)F^{-1^{(1)}}(\pi_{i+1}^*)(k_{i+1}^* b_i^* - k_i^* b_{i+1}^*)}, \end{aligned}$$

$$\begin{aligned} k_i^* &= \prod_{k=1}^{i} \left\{ \frac{m - k + 2 + R_k + R_{k+1} + \cdots + R_{m-1} + R_m}{m - k + 3 + R_k + R_{k+1} + \cdots + R_{m-1} + R_m} \right\} \\ &\quad - \prod_{k=1}^{i} \left\{ \frac{m - k + 1 + R_k + R_{k+1} + \cdots + R_{m-1} + R_m}{m - k + 2 + R_k + R_{k+1} + \cdots + R_{m-1} + R_m} \right\}, \\ &\qquad i = 1, 2, \cdots, m, \end{aligned}$$

$$\begin{aligned} k_{m+1}^* &= \prod_{k=1}^{m} \left\{ \frac{m - k + 2 + R_k + R_{k+1} + \cdots + R_{m-1} + R_m}{m - k + 3 + R_k + R_{k+1} + \cdots + R_{m-1} + R_m} \right\} \cdot \left(\frac{R_m + 1}{R_m + 2} \right) \\ &\quad - \prod_{k=1}^{m} \left\{ \frac{m - k + 1 + R_k + R_{k+1} + \cdots + R_{m-1} + R_m}{m - k + 2 + R_k + R_{k+1} + \cdots + R_{m-1} + R_m} \right\} \cdot \left(\frac{R_m}{R_m + 1} \right), \end{aligned}$$

$$b_i^* = \prod_{k=1}^{i} \left\{ \frac{m-k+1+R_k+R_{k+1}+\cdots+R_{m-1}+R_m}{m-k+2+R_k+R_{k+1}+\cdots+R_{m-1}+R_m} \right\}, \qquad i=1,2,\cdots,m,$$

$$b_{m+1}^* = \prod_{k=1}^{m} \left\{ \frac{m-k+1+R_k+R_{k+1}+\cdots+R_{m-1}+R_m}{m-k+2+R_k+R_{k+1}+\cdots+R_{m-1}+R_m} \right\} \cdot \left(\frac{R_m}{R_m+1} \right),$$

and

$$\pi_i^* = 1 - b_i^*, \qquad i = 1, 2, \cdots, m+1.$$

Upon equating the expressions of θ^* in (8.37) and (8.38) and solving for $Y_{m+1:m+1:n}^{(R_1,\cdots,R_{m-1},0,R_m-1)}$, we obtain the first-order approximation to its BLUP as

$$Y_{m+1:m+1:n}^{(R_1,\cdots,R_{m-1},0,R_m-1)*} = \frac{1}{c_{m+1}^*} \sum_{i=1}^{m} \left(\frac{\Delta^*}{\Delta} c_i - c_i^* \right) Y_{i:m:n}^{(R_1,\cdots,R_m)}; \tag{8.39}$$

see Balakrishnan and Rao (1997a).

Proceeding as we did earlier in Section 8.4, we can sequentially derive the first-order approximations to the BLUPs of the times to failure of the other items among the last R_m censored at the observation $Y_{m:m:n}^{(R_1,\cdots,R_m)}$.

Remark 8.13 For the location-scale distributions, by using the results on first-order approximations to the BLUEs presented in Section 6.2.7, we may similarly derive the first-order approximations to the BLUPs of the failure times of the last R_m items censored at the observation $Y_{m:m:n}^{(R_1,\cdots,R_m)}$. For this derivation, we may either equate the first-order approximations to the BLUEs of the location parameter μ or equate those of the BLUEs of the scale parameter σ as both approaches will lead exactly to the same first-order approximations to the BLUPs.

8.6 Prediction Intervals

So far, our discussion focused only on point prediction. Quite often in practical situations, however, one may be interested in obtaining prediction intervals for the future failures. This may be done by considering appropriate pivotal quantities and determining their percentage points either exactly or through simulations.

As an example, let us consider the scale-parameter exponential distribution in (8.1). As before, let us assume that we have observed the progressively Type-II right censored sample $Y_{1:m:n}^{(R_1,\cdots,R_m)}, \cdots, Y_{m:m:n}^{(R_1,\cdots,R_m)}$ and that we are now interested in obtaining a prediction interval for $Y_{m+1:m+1:n}^{(R_1,\cdots,R_{m-1},0,R_m-1)}$. For this purpose, we may consider the pivotal quantity

$$P_1 = \frac{Y_{m+1:m+1:n}^{(R_1,\cdots,R_{m-1},0,R_m-1)} - Y_{m+1:m+1:n}^{(R_1,\cdots,R_{m-1},0,R_m-1)*}}{\theta^*}, \tag{8.40}$$

where θ^* is the BLUE of θ in (8.28) and $Y_{m+1:m+1:n}^{(R_1,\cdots,R_{m-1},0,R_m-1)*}$ is the BLUP of $Y_{m+1:m+1:n}^{(R_1,\cdots,R_{m-1},0,R_m-1)}$ in (8.30). This is equivalent to considering the pivotal quantity

$$P_2 = \frac{Y_{m+1:m+1:n}^{(R_1,\cdots,R_{m-1},0,R_m-1)} - Y_{m:m:n}^{(R_1,\cdots,R_m)}}{\theta^*} \tag{8.41}$$

since, as shown earlier in Section 8.4 [Eq. (8.30)],

$$Y_{m+1:m+1:n}^{(R_1,\cdots,R_{m-1},0,R_m-1)*} = Y_{m:m:n}^{(R_1,\cdots,R_m)} + \frac{1}{R_m}\,\theta^*.$$

In this case,

$$\begin{aligned}
Z_1 &= nY_{1:m:n}^{(R_1,\cdots,R_m)}/\theta \\
&= nY_{1:m+1:n}^{(R_1,\cdots,R_{m-1},0,R_m-1)}/\theta, \\
Z_2 &= (n-1-R_1)\left(Y_{2:m:n}^{(R_1,\cdots,R_m)} - Y_{1:m:n}^{(R_1,\cdots,R_m)}\right)\Big/\theta \\
&= (n-1-R_1)\left(Y_{2:m+1:n}^{(R_1,\cdots,R_{m-1},0,R_m-1)} - Y_{1:m+1:n}^{(R_1,\cdots,R_{m-1},0,R_m-1)}\right)\Big/\theta, \\
\cdots & \quad \cdots \\
\cdots & \quad \cdots \\
Z_m &= (n-m+1-R_1-\cdots-R_{m-1})\left(Y_{m:m:n}^{(R_1,\cdots,R_m)} - Y_{m-1:m:n}^{(R_1,\cdots,R_m)}\right)\Big/\theta \\
&= (n-m+1-R_1-\cdots-R_{m-1}) \\
&\quad \times \left(Y_{m:m+1:n}^{(R_1,\cdots,R_{m-1},0,R_m-1)} - Y_{m-1:m+1:n}^{(R_1,\cdots,R_{m-1},0,R_m-1)}\right)\Big/\theta, \\
Z_{m+1} &= (n-m-R_1-\cdots-R_{m-1})\left(Y_{m+1:m+1:n}^{(R_1,\cdots,R_{m-1},0,R_m-1)} - Y_{m:m:n}^{(R_1,\cdots,R_m)}\right)\Big/\theta \\
&= (n-m-R_1-\cdots-R_{m-1}) \\
&\quad \times \left(Y_{m+1:m+1:n}^{(R_1,\cdots,R_{m-1},0,R_m-1)} - Y_{m:m+1:n}^{(R_1,\cdots,R_{m-1},0,R_m-1)}\right)\Big/\theta
\end{aligned}$$

are all independent and identically distributed as standard exponential. Consequently, we have

$$2m\,\frac{\theta^*}{\theta} = \sum_{i=1}^{m} 2Z_i$$

and

$$2(n-m-R_1-\cdots-R_{m-1})\left(\frac{Y_{m+1:m+1:n}^{(R_1,\cdots,R_{m-1},0,R_m-1)}-Y_{m:m:n}^{(R_1,\cdots,R_m)}}{\theta}\right)=2Z_{m+1}$$

to be independently distributed as chi-square with $2m$ degrees of freedom and chi-square with 2 degrees of freedom, respectively. As a result, we have $(n-m-R_1-\cdots-R_{m-1})P_2 = R_mP_2$, where P_2 is as in (8.41), to be distributed as F with $(2,2m)$ degrees of freedom. Then, using the percentage points of $F_{2,2m}$ distribution, we can easily construct exact prediction intervals for $Y_{m+1:m+1:n}^{(R_1,\cdots,R_{m-1},0,R_m-1)}$.

Proceeding similarly, we can also construct exact prediction intervals for the times to failure of the other items among the last R_m censored at the observation $Y_{m:m:n}^{(R_1,\cdots,R_m)}$.

Remark 8.14 For the two-parameter exponential distribution, we can proceed similarly starting with the pivotal quantity given in (8.40) and develop exact prediction intervals.

Remark 8.15 In the case of other distributions, we may simulate the percentage points of the pivotal quantity P_2 in (8.41) which may then be used to construct the necessary prediction intervals. Note that the BLUE of θ is needed (either in an explicit form or in a tabular form) as well for this purpose. Note that although the pivotal quantity P_1 in (8.40) can be used for the same purpose, there is no guarantee that the prediction interval so obtained will be larger than $Y_{m:m:n}^{(R_1,\cdots,R_m)}$.

8.7 Illustrative Examples

In this section, we will present several examples to illustrate the predictions discussed in the preceding sections of this chapter.

Example 8.1 Let us consider the progressively Type-II censored sample [obtained from Nelson's (1982, p. 228, Table 6.1) data on times to breakdown of an insulating fluid in an accelerated test at 34 kilovolts] used earlier in Section 6.1.5:

i	1	2	3	4	5	6	7	8
$y_{i:8:19}$	0.19	0.78	0.96	1.31	2.78	4.85	6.50	7.35
R_i	0	0	3	0	3	0	0	5

Let us now assume a scale-parameter exponential distribution in (8.1) as the time-to-breakdown distribution. From Chapter 6 (Section 6.1), we then determine the BLUE of θ to be

$$\begin{aligned}\theta^* &= \frac{1}{m}\sum_{i=1}^{m}(R_i+1)y_{i:m:n}^{(R_1,\cdots,R_m)}\\ &= \frac{1}{8}\{0.19+0.78+(4\times 0.96)+1.31+(4\times 2.78)+4.85+6.50\\ &\qquad +(6\times 7.35)\}\\ &= 9.08625\end{aligned}$$

and its standard error to be

$$\mathrm{SE}(\theta^*) = \frac{\theta^*}{\sqrt{m}} = \frac{9.08625}{\sqrt{8}} = 3.21247.$$

From (8.9), we obtain the BLUP of the smallest of the last 5 items censored at $y_{8:8:19} = 7.35$ to be

$$y_{9:9:19}^* = y_{8:8:19} + \frac{\theta^*}{R_m} = 7.35 + \frac{9.08625}{5} = 9.16725$$

and its standard error from (8.10) to be

$$\mathrm{SE}(y_{9:9:19}^*) = \frac{\theta^*}{R_m}\sqrt{\frac{m+1}{m}} = \frac{3(9.08625)}{5\sqrt{8}} = 1.92748.$$

Alternatively, based on the progressively Type-II right censored sample $y_{1:8:19},\cdots,y_{8:8:19},y_{9:9:19}$, we have the BLUE of θ (from Section 6.1) to be

$$\begin{aligned}\theta^* &= \frac{1}{9}\{0.19+0.78+(4\times 0.96)+1.31+(4\times 2.78)+4.85\\ &\qquad +6.50+7.35+(5\times y_{9:9:19})\}\\ &= 3.99333+0.55556\,y_{9:9:19}.\end{aligned}$$

Upon equating this with $\theta^* = 9.08625$ and solving, we obtain the BLUP of $y_{9:9:19}$ to be

$$y_{9:9:19}^* = \frac{9.08625-3.99333}{0.55556} = 9.167$$

which is exactly the same as obtained before.

Next, based on the progressively Type-II right censored sample $y_{1:8:19},\cdots,$ $y_{8:8:19},y_{9:9:19}^*,y_{10:10:19}$, we have the BLUE of θ to be

$$\begin{aligned}\theta^* &= \frac{1}{10}\{0.19+0.78+(4\times 0.96)+1.31+(4\times 2.78)+4.85\\ &\qquad +6.50+7.35+9.167+(4\times y_{10:10:19})\}\\ &= 4.5107+0.4\,y_{10:10:19}.\end{aligned}$$

Upon equating this with $\theta^* = 9.08625$ and solving, we obtain the BLUP of $y_{10:10:19}$ to be

$$y^*_{10:10:19} = \frac{9.08625 - 4.5107}{0.4} = 11.439;$$

that is, the BLUP of the second smallest of the last 5 items censored at $y_{8:8:19} = 7.35$ is found to be 11.439.

Similarly, based on the progressively Type-II right censored sample $y_{1:8:19}, \cdots, y_{8:8:19}, y^*_{9:9:19}, y^*_{10:10:19}, y_{11:11:19}$, we have the BLUE of θ to be

$$\begin{aligned}\theta^* &= \frac{1}{11}\{0.19 + 0.78 + (4 \times 0.96) + 1.31 + (4 \times 2.78) + 4.85 \\ &\quad + 6.50 + 7.35 + 9.167 + 11.439 + (3 \times y_{11:11:19})\} \\ &= 5.14055 + 0.27273\, y_{11:11:19}.\end{aligned}$$

Upon equating this with $\theta^* = 9.08625$ and solving, we obtain the BLUP of $y_{11:11:19}$ to be

$$y^*_{11:11:19} = \frac{9.08625 - 5.14055}{0.27273} = 14.467;$$

that is, the BLUP of the third smallest of the last 5 items censored at $y_{8:8:19} = 7.35$ is found to be 14.467.

Next, based on the progressively Type-II right censored sample $y_{1:8:19}, \cdots, y_{8:8:19}, y^*_{9:9:19}, y^*_{10:10:19}, y^*_{11:11:19}, y_{12:12:19}$, we have the BLUE of θ to be

$$\begin{aligned}\theta^* &= \frac{1}{12}\{0.19 + 0.78 + (4 \times 0.96) + 1.31 + (4 \times 2.78) + 4.85 \\ &\quad + 6.50 + 7.35 + 9.167 + 11.439 + 14.467 + (2 \times y_{12:12:19})\} \\ &= 5.91775 + 0.16667\, y_{12:12:19}.\end{aligned}$$

Upon equating this with $\theta^* = 9.08625$ and solving, we obtain the BLUP of $y_{12:12:19}$ to be

$$y^*_{12:12:19} = \frac{9.08625 - 5.91775}{0.16667} = 19.011;$$

that is, the BLUP of the fourth smallest of the last 5 items censored at $y_{8:8:19} = 7.35$ is found to be 19.011.

Finally, based on the progressively Type-II right censored sample $y_{1:8:19}, \cdots, y_{8:8:19}, y^*_{9:9:19}, y^*_{10:10:19}, y^*_{11:11:19}, y^*_{12:12:19}, y_{13:13:19}$, we have the BLUE of θ to be

$$\begin{aligned}\theta^* &= \frac{1}{13}\{0.19 + 0.78 + (4 \times 0.96) + 1.31 + (4 \times 2.78) + 4.85 \\ &\quad + 6.50 + 7.35 + 9.167 + 11.439 + 14.467 + 19.011 + y_{13:13:19}\} \\ &= 6.92492 + 0.07692\ y_{13:13:19}.\end{aligned}$$

Upon equating this with $\theta^* = 9.08625$ and solving, we obtain the BLUP of $y_{13:13:19}$ to be

$$y^*_{13:13:19} = \frac{9.08625 - 6.92492}{0.07692} = 28.098;$$

that is, the BLUP of the largest of the last 5 items censored at $y_{8:8:19} = 7.35$ is found to be 28.098.

Example 8.2 Let us use the data in Example 8.1 in order to illustrate the first-order approximation to the BLUPs. In this case, since $X = Y/\theta$ has a standard exponential distribution, we immediately have

$$F(x) = 1 - e^{-x}, \ F^{-1}(u) = -\ln(1-u) \ \text{ and } \ F^{-1^{(1)}}(u) = \frac{1}{1-u} \, .$$

Then, as seen already in Section 6.1.5, based on the progressively Type-II right censored sample $y_{1:8:19}, \cdots, y_{8:8:19}$, we have the first-order approximation to the BLUE of θ as [see Eq. (8.37)]

$$\begin{aligned}
\theta^* &= \frac{1}{\Delta} \sum_{i=1}^{8} c_i y_{i:8:19} \\
&= (0.12305 \times 0.19) + (0.12305 \times 0.78) + (0.47255 \times 0.96) \\
&\quad + (0.12458 \times 1.31) + (0.47153 \times 2.78) + (0.12798 \times 4.85) \\
&\quad + (0.12808 \times 6.50) + (0.82641 \times 7.35) \\
&= 9.57.
\end{aligned}$$

Similarly, based on the progressively Type-II right censored sample $y_{1:8:19}, \cdots,$ $y_{8:8:19}, y_{9:9:19}$, we have the first-order approximation to the BLUE of θ as [see Eq. (8.38)]

$$\begin{aligned}
\theta^* &= \frac{1}{\Delta^*} \left\{ \sum_{i=1}^{8} c_i^* y_{i:8:19} + c_9^* y_{9:9:19} \right\} \\
&= (0.10915 \times 0.19) + (0.10916 \times 0.78) + (0.41919 \times 0.96) \\
&\quad + (0.11051 \times 1.31) + (0.41829 \times 2.78) + (0.11353 \times 4.85) \\
&\quad + (0.11362 \times 6.50) + (0.11376 \times 7.35) + (0.61933 \times x_{9:9:19}) \\
&= 3.9412 + (0.61933 \times y_{9:9:19}).
\end{aligned}$$

Upon equating this with $\theta^* = 9.57$ and solving, we obtain the first-order approximation to the BLUP of $y_{9:9:19}$ as

$$y_{9:9:19}^* = \frac{9.57 - 3.9412}{0.61933} = 9.0885.$$

It is of interest to note here that this first-order approximation to the BLUP of $y_{9:9:19}$ is quite close to the exact value of 9.167 determined earlier.

Proceeding along the same lines, we can develop similar first-order approximations to the BLUPs of the times to failure of the other items among the last 5 items censored at $y_{8:8:19} = 7.35$.

Example 8.3 Consider the log-times to breakdown of the insulating fluid from the data presented in Example 8.1:

i	1	2	3	4	5	6	7	8
$y_{i:8:19}$	−1.6608	−0.2485	−0.0409	0.2700	1.0224	1.5789	1.8718	1.9947
R_i	0	0	3	0	3	0	0	5

Let us now assume an extreme value distribution with density function

$$f(y;\mu,\sigma) = \frac{1}{\sigma}\, e^{(y-\mu)/\sigma} e^{-e^{(y-\mu)/\sigma}}, \qquad -\infty < y < \infty,$$

for the distribution of log-times to breakdown. In this case, since $X = (Y-\mu)/\sigma$ has a standard extreme value distribution with probability density function

$$f(x) = e^{x}\, e^{-e^{x}}, \qquad -\infty < x < \infty,$$

we immediately have

$$F(x) = 1 - e^{-e^{x}},\ F^{-1}(u) = \ln(-\ln(1-u)) \text{ and } F^{-1^{(1)}}(u) = \frac{-1}{(1-u)\ln(1-u)}\,.$$

Then, as seen already in Section 6.2.7, based on the progressively Type-II right censored sample $y_{1:8:19}, \cdots, y_{8:8:19}$, we have the first-order approximations to the BLUEs of μ and σ as

$$\begin{aligned}\mu^* &= (-0.09888 \times -1.6608) + (-0.06737 \times -0.2485) + (-0.00296 \times -0.0409)\\ &\quad + (-0.04081 \times 0.2700) + (0.12238 \times 1.0224) + (0.00760 \times 1.5789)\\ &\quad + (0.04516 \times 1.8718) + (1.03488 \times 1.9947)\\ &= 2.456\end{aligned}$$

and

$$\begin{aligned}\sigma^* &= (-0.15392 \times -1.6608) + (-0.11755 \times -0.2485) + (-0.11670 \times -0.0409)\\ &\quad + (-0.10285 \times 0.2700) + (-0.03942 \times 1.0224) + (-0.07023 \times 1.5789)\\ &\quad + (0.04037 \times 1.8718) + (0.64104 \times 1.9947)\\ &= 1.31377,\end{aligned}$$

respectively. Similarly, based on the progressively Type-II right censored sample $y_{1:8:19}, \cdots, y_{8:8:19}, y_{9:9:19}$, we have the first-order approximations to the BLUE of μ as

$$\begin{aligned}\mu^* &= (-0.06042 \times -1.6608) + (-0.03942 \times -0.2485) + (0.01365 \times -0.0409)\\ &\quad + (-0.01952 \times 0.2700) + (0.10649 \times 1.0224) + (0.01596 \times 1.5789)\\ &\quad + (0.04286 \times 1.8718) + (0.07576 \times 1.9947) + (0.86464 \times y_{9:9:19})\\ &= 0.46973 + (0.86464 \times y_{9:9:19}).\end{aligned}$$

Upon equating this with $\mu^* = 2.456$ and solving, we obtain the first-order approximation to the BLUP of $y_{9:9:19}$ as

$$y^*_{9:9:19} = \frac{2.456 - 0.46973}{0.86464} = 2.2972.$$

It is important to note here that we would have come up with the same prediction had we used the equality of σ^* instead. To see this, let us consider the first-order approximation to the BLUE of σ based on $y_{1:8:19}, \cdots, y_{8:8:19}, y_{9:9:19}$ given by

$$\begin{aligned}\sigma^* &= (-0.12772 \times -1.6608) + (-0.09850 \times -0.2485) + (-0.10539 \times -0.0409)\\ &\quad + (-0.08834 \times 0.2700) + (-0.05025 \times 1.0224) + (-0.06454 \times 1.5789)\\ &\quad + (-0.04194 \times 1.8718) + (-0.01236 \times 1.9947) + (0.58903 \times y_{9:9:19})\\ &= -0.039382 + (0.58903 \times y_{9:9:19}).\end{aligned}$$

Upon equating this with $\sigma^* = 1.31377$ and solving, we obtain the first-order approximation to the BLUP of $y_{9:9:19}$ as

$$y^*_{9:9:19} = \frac{1.31377 + 0.039382}{0.58903} = 2.2972$$

which is exactly the same as the predicted value obtained earlier through the equality of μ^*.

Finally, we observe that the predicted value of the log-time of breakdown obtained from the exponential model (which, incidentally, is not the BLUP) turns out to be

$$\ln(9.0885) = 2.2070$$

which is quite close to the value determined above with the extreme value model.

Example 8.4 Let us consider the progressively Type-II right censored data from the *Pareto*($\nu = 3$) distribution (with $\mu = 0$ and $\sigma = 5$), presented earlier in Example 6.4. In this case, with n=15, $m = 5$, $\boldsymbol{R} = (5,0,2,0,3)$, and the data as

$$5.11073, 5.34932, 5.36434, 5.70137, 5.90067,$$

we determined the BLUEs of μ and σ as (see Example 6.4)

$$\mu^* = 1.87680 \quad \text{and} \quad \sigma^* = 3.16207.$$

Then, the best linear unbiased prediction for the failure following $Y^{(5,0,2,0,3)}_{5:5:15}$ may be determined simply by equating μ^* (or σ^*) based on the sample of size $m = 5$ to μ^* (or σ^*) based on the sample of size 6 with progressive censoring scheme (5,0,2,0,0,2), whose coefficients may be computed from (6.46) [or (6.47)], and then solving for $Y^{(5,0,2,0,0,2)}_{6:6:15}$. This quantity can in turn be used to predict $Y_{7:7:15}^{(5,0,2,0,0,0,1)}$ in a similar way upon using the predictor just obtained for $Y_{6:6:15}^{(5,0,2,0,0,2)}$, and so on.

Using the above progressively Type-II right censored sample, we can compute in this manner the best linear unbiased predictors for the three failure times (following the last observation 5.90067) as

$$Y_{6:6:15}^{*(5,0,2,0,0,2)} = 6.40365, \; Y_{7:7:15}^{*(5,0,2,0,0,0,1)} = 7.30902, \; Y_{8:8:15}^{*(5,0,2,0,0,0,0,0)} = 10.02513.$$

9

Conditional Inference

9.1 Introduction

So far, all the inferential methods we have discussed are unconditional in nature. In this chapter, we will demonstrate how exact confidence intervals or prediction intervals may be obtained using the conditional method. Conditional inference, first proposed by Fisher (1934), has been successfully applied by Lawless (1973, 1978, 1982) to develop inference based on complete as well as conventionally Type-II right censored samples. As a matter of fact, Lawless (1982, p. 199) indicated the use of conditional inference based on progressively Type-II right censored data; but a full length account of this topic has been provided by Viveros and Balakrishnan (1994) which naturally forms a basis for much of the discussion in this chapter.

Though considerable attention is usually paid to point estimation, in many practical situations an experimenter may also require uncertainty information such as those conveyed by confidence intervals, tolerance intervals or prediction intervals for the life characteristics of interest. The primary motivation in constructing confidence intervals is that they provide a range of plausible values for the life parameter of interest based on the observed progressively censored data at a required level of confidence. The practical relevance of confidence intervals and their construction has been well demonstrated by Hahn and Meeker (1991).

In Section 9.2, we discuss the derivation of conditional confidence intervals for location and scale parameters. In Section 9.3, we present conditional confidence intervals for quantiles and reliability and also conditional prediction intervals for the smallest lifetime in a future sample. In Section 9.4, we develop the necessary computational formulae for confidence intervals and prediction intervals for the case when the progressively censored data is assumed to have come from an extreme value distribution. In Section 9.5, we present the results for the case of the two-parameter exponential distribution. In Section 9.6, we present examples to illustrate all the methods of inference developed in the preceding sections. Finally, in Section 9.7, we discuss conditional inference for the two-parameter Pareto distribution.

9.2 Inference for Location and Scale Parameters

Let Y denote the lifetime of a unit placed on a life-test. Let us assume that Y has a distribution belonging to the location-scale family with probability density function $f(y; \mu, \sigma)$ and cumulative distribution function $F(y; \mu, \sigma)$ of the form

$$f(y;\mu,\sigma)=\frac{1}{\sigma}\,g\left(\frac{y-\mu}{\sigma}\right)\ \text{and}\ F(y;\mu,\sigma)=G\left(\frac{y-\mu}{\sigma}\right), \tag{9.1}$$

respectively. Here, μ is the location parameter and σ is the scale parameter. In (9.1), $g(\cdot)$ is the probability density function of the standard member of that location-scale family of distributions (with $\mu=0$ and $\sigma=1$) and $G(\cdot)$ is the corresponding cumulative distribution function.

Let $y_{1:m:n}^{(R_1,\cdots,R_m)},\cdots,y_{m:m:n}^{(R_1,\cdots,R_m)}$ denote a progressively Type-II right censored sample of size m obtained from n items placed on a life-test and adopting the progressive censoring scheme $(R_1,\cdots,R_m)$. Their joint density function, as seen earlier in Chapter 2, is given by

$$\begin{aligned} &f(y_{1:m:n},\cdots,y_{m:m:n};\mu,\sigma)\\ &\quad=\frac{C}{\sigma^m}\prod_{i=1}^{m} g\left(\frac{y_{i:m:n}-\mu}{\sigma}\right)\left\{1-G\left(\frac{y_{i:m:n}-\mu}{\sigma}\right)\right\}^{R_i}, \end{aligned} \tag{9.2}$$

where

$$C=n(n-R_1-1)\cdots(n-R_1-\cdots-R_{m-1}-m+1). \tag{9.3}$$

From (9.2), we observe that the joint density function preserves the location-scale structure which is evident from the fact that the standardized variables $\frac{Y_{1:m:n}-\mu}{\sigma},\cdots,\frac{Y_{m:m:n}-\mu}{\sigma}$ have a joint distribution functionally independent of μ and σ. Hence, the method of conditional inference for location and scale parameters apply in this situation.

Let $(\hat\mu,\hat\sigma)$ denote the MLE of (μ,σ) obtained by maximizing the likelihood function of (μ,σ) in (9.2). Then,

$$Z_1=\frac{\hat\mu-\mu}{\hat\sigma}\qquad\text{and}\qquad Z_2=\frac{\hat\sigma}{\sigma} \tag{9.4}$$

are pivotal quantities as their joint density function does not involve either μ or σ. With

$$\boldsymbol{A}=(A_{1:m:n},A_{2:m:n},\cdots,A_{m:m:n}), \tag{9.5}$$

where $A_{i:m:n}=(Y_{i:m:n}-\hat\mu)/\hat\sigma$, being an ancillary statistic, we may develop inference for μ and σ based on the joint distribution of (Z_1,Z_2) conditional on the observed value $\boldsymbol{a}$ of $\boldsymbol{A}$. Since the ancillary statistic in (9.5) also satisfies the maximum likelihood equations, only $m-2$ of the $A_{i:m:n}$'s will be functionally independent. Furthermore, the MLEs $(\hat\mu,\hat\sigma)$ may be replaced by any pair of equivariant estimators such as the best linear unbiased estimators or the best linear invariant estimators as long as the inferences are made conditionally on $\boldsymbol{A}$ in (9.5).

Since

$$\frac{y_{i:m:n}-\mu}{\sigma} = \frac{(y_{i:m:n}-\hat{\mu})+(\hat{\mu}-\mu)}{\hat{\sigma}}\cdot\frac{\hat{\sigma}}{\sigma} = a_{i:m:n}z_2+z_1z_2, \tag{9.6}$$

conditionally on the observed value of the ancillary statistic $\boldsymbol{a}$ in (9.5), we obtain the joint density function of (Z_1, Z_2) from (9.2) as

$$f(z_1, z_2 \mid \boldsymbol{a}) = K(\boldsymbol{a})z_2^{m-1}\prod_{i=1}^{m} g\left(a_{i:m:n}z_2+z_1z_2\right)\left\{1-G\left(a_{i:m:n}z_2+z_1z_2\right)\right\}^{R_i}, \tag{9.7}$$

where $K(\boldsymbol{a})$ is the normalizing constant.

The marginal conditional density functions of Z_1 and Z_2 given $\boldsymbol{a}$, namely, $f(z_1 \mid \boldsymbol{a})$, and $f(z_2 \mid \boldsymbol{a})$, can be obtained from (9.7) by integrating out the other variable. These marginal conditional density functions can then be used to develop conditional inference for the parameters μ and σ, respectively. As shall be seen later in Section 9.5, the necessary integration in (9.7) can be carried out explicitly in the case of the exponential distribution. In many other cases, including the extreme value distribution considered in Section 9.4, the required integration in (9.7) has to be carried out using numerical methods.

9.3 Inference for Quantiles and Reliability and Prediction Intervals

An experimenter, in addition to deriving inference about location and scale parameters, may also be interested in making inference about some other interesting aspects of the assumed lifetime distribution such as the quantiles, the reliability at some mission time, or predicting the lifetimes of units to be used in the future. The conditional approach explained in the last section can be extended easily in order to derive confidence intervals for the quantiles and the reliability and prediction intervals for the lifetimes of future units. These developments are closely related to those made by Lawless (1978).

9.3.1 Inference for Quantiles

For a specified p $(0 < p < 1)$, the *p-th quantile* is the time point y_p such that

$$\Pr(Y \le y_p) = F(y_p; \mu, \sigma) = p, \tag{9.8}$$

where $F(\cdot)$ is as given in (9.1). Then, for any location-scale family of distributions, it is easy to see that

$$y_p = \mu + \sigma\, G^{-1}(p) \tag{9.9}$$

and, consequently, the MLE of y_p is given by

$$\hat{y}_p = \hat{\mu} + \hat{\sigma}\, G^{-1}(p). \tag{9.10}$$

Inference on y_p can then be developed based on

$$Z_p = \frac{y_p - \hat{\mu}}{\hat{\sigma}} \tag{9.11}$$

which indeed is a pivotal quantity since $Z_p = \frac{G^{-1}(p)}{Z_2} - Z_1$, where Z_1 and Z_2 are as defined earlier in (9.4). By change of variables, we simply obtain from (9.7) the joint conditional density function of (Z_p, Z_2) as

$$\begin{aligned} f(z_p, z_2 \mid \boldsymbol{a}) &= K(\boldsymbol{a}) z_2^{m-1} \prod_{i=1}^{m} g\left(a_{i:m:n} z_2 - z_p z_2 + G^{-1}(p)\right) \\ &\quad \times \left\{1 - G\left(a_{i:m:n} z_2 - z_p z_2 + G^{-1}(p)\right)\right\}^{R_i}, \end{aligned} \tag{9.12}$$

where $K(\boldsymbol{a})$ is the normalizing constant. The marginal conditional density function of Z_p given $\boldsymbol{a}$, viz., $f(z_p \mid \boldsymbol{a})$, can then be obtained from (9.12) by integrating out z_2. This marginal conditional density function of Z_p can then be used to develop conditional confidence intervals for the p-th quantile y_p.

9.3.2 Inference for Reliability

For a specified mission time y_0, let

$$R(y_0) = \Pr(Y > y_0) = 1 - F(y_0; \mu, \sigma) = 1 - G\left(\frac{y_0 - \mu}{\sigma}\right) \tag{9.13}$$

denote the reliability at y_0. Let us now consider the determination of a lower α confidence limit for $R(y_0)$. If $\ell(\boldsymbol{Y}; p)$ denotes the lower α confidence bound for the quantile y_{1-p}, viz.,

$$\Pr\left(\ell(\boldsymbol{Y}; p) \le y_{1-p}\right) = \alpha \tag{9.14}$$

where $\boldsymbol{Y}$ denotes the progressively Type-II right censored sample at hand, then that choice of p such that $\ell(\boldsymbol{Y}; p) = y_0$ will yield a lower α confidence limit for the reliability $R(y_0)$; in other words, $p = p(\boldsymbol{Y})$ is such that

$$\Pr\left(R(y_0) \ge p(\boldsymbol{Y})\right) = \alpha. \tag{9.15}$$

9.3.3 Prediction Intervals for Future Failures

A prediction problem that is often of interest to an experimenter is in obtaining prediction bounds for the lifetimes of units from a future sample, say, the smallest lifetime $Y_{1:k}$ or the largest lifetime $Y_{k:k}$ from a future sample of k

units. Observe that this prediction problem is different from the one discussed earlier in Chapter 8. The prediction problem considered here is often referred to as *two-sample prediction* while that considered in Chapter 8 as *one-sample prediction.*

Since $Y_{1:k}$ has its density function to be

$$\begin{aligned} f_{1:k}(y) &= k\{1 - F(y;\mu,\sigma)\}^{k-1} f(y;\mu,\sigma) \\ &= \frac{k}{\sigma}\left\{1 - G\left(\frac{y-\mu}{\sigma}\right)\right\}^{k-1} g\left(\frac{y-\mu}{\sigma}\right), \end{aligned} \tag{9.16}$$

and independently of the joint conditional density function of (Z_1, Z_2) in (9.7) based on the first sample, we readily have the joint conditional density function of $\left(Z_1, Z_2, Z_3 = \frac{Y_{1:k}-\hat{\mu}}{\hat{\sigma}}\right)$ as

$$\begin{aligned} & f(z_1, z_2, z_3 \mid \boldsymbol{a}) \\ &= K^*(\boldsymbol{a}) k z_2^m \left[\prod_{i=1}^{m} g(a_{i:m:n} z_2 + z_1 z_2)\left\{1 - G(a_{i:m:n} z_2 + z_1 z_2)\right\}^{R_i}\right] \\ &\quad \times \{1 - G(z_3 z_2 + z_1 z_2)\}^{k-1} g(z_3 z_2 + z_1 z_2), \end{aligned} \tag{9.17}$$

where $K^*(\boldsymbol{a})$ is the normalizing constant. The marginal conditional density function of $Z_3 = \frac{Y_{1:k}-\hat{\mu}}{\hat{\sigma}}$ can be obtained from (9.17) by integrating out z_1 and z_2. This marginal conditional density function of Z_3 can then be used to develop conditional prediction intervals for the smallest lifetime $Y_{1:k}$ from a future sample of k units.

We can proceed similarly to develop conditional prediction intervals for $Y_{k:k}$. Since $Y_{k:k}$ has its density function to be

$$\begin{aligned} f_{k:k}(y) &= k\{F(y;\mu,\sigma)\}^{k-1} f(y;\mu,\sigma) \\ &= \frac{k}{\sigma}\left\{G\left(\frac{y-\mu}{\sigma}\right)\right\}^{k-1} g\left(\frac{y-\mu}{\sigma}\right), \end{aligned} \tag{9.18}$$

and independently of the joint conditional density function of (Z_1, Z_2) in (9.7) based on the first sample, we immediately find the joint conditional density function of $\left(Z_1, Z_2, Z_4 = \frac{Y_{k:k}-\hat{\mu}}{\hat{\sigma}}\right)$ as

$$\begin{aligned} & f(z_1, z_2, z_4 \mid \boldsymbol{a}) \\ &= K^{**}(\boldsymbol{a}) k z_2^m \left[\prod_{i=1}^{m} g(a_{i:m:n} z_2 + z_1 z_2)\left\{1 - G(a_{i:m:n} z_2 + z_1 z_2)\right\}^{R_i}\right] \\ &\quad \times \{G(z_4 z_2 + z_1 z_2)\}^{k-1} g(z_4 z_2 + z_1 z_2), \end{aligned} \tag{9.19}$$

where $K^{**}(\boldsymbol{a})$ is the normalizing constant. The marginal conditional density function of $Z_4 = \frac{Y_{k:k}-\hat{\mu}}{\hat{\sigma}}$ can be obtained from (9.19) by integrating out z_1 and z_2. This marginal conditional density function of Z_4 can then be used to develop conditional prediction intervals for the largest lifetime $Y_{k:k}$ from a future sample of k units.

9.4 Results for Extreme Value Distribution

Let us consider the extreme value distribution with probability density function

$$f(y;\mu,\sigma) = \frac{1}{\sigma}\, e^{(y-\mu)/\sigma}\, e^{-e^{(y-\mu)/\sigma}}, \quad -\infty < y < \infty, \tag{9.20}$$

and cumulative distribution function

$$F(y;\mu,\sigma) = 1 - e^{-e^{(y-\mu)/\sigma}}, \qquad -\infty < y < \infty. \tag{9.21}$$

Clearly, this conforms to the location-scale family in (9.1) with

$$g(x) = e^{x}\, e^{-e^{x}} \qquad \text{and} \qquad G(x) = 1 - e^{-e^{x}}. \tag{9.22}$$

Assuming this model for the logarithm of lifetimes is equivalent to assuming a two-parameter Weibull distribution for the original lifetimes. For a detailed discussion on various properties and applications of these two distributions, one may refer to Johnson, Kotz and Balakrishnan (1994, 1995).

Upon substituting the expressions in (9.22) into (9.7), we obtain the joint conditional density function of Z_1 and Z_2 in this case to be

$$\begin{aligned} f(z_1, z_2 \mid \boldsymbol{a}) &= K(\boldsymbol{a}) z_2^{m-1} e^{m z_1 z_2 + s_a z_2}\, e^{-\sum_{i=1}^{m}(R_i+1)e^{a_{i:m:n} z_2 + z_1 z_2}}, \\ & \qquad -\infty < z_1 < \infty,\ 0 < z_2 < \infty, \end{aligned} \tag{9.23}$$

where $s_a = \sum_{i=1}^{m} a_{i:m:n}$ and $K(\boldsymbol{a})$ is the normalizing constant. If we now wish to construct conditional confidence intervals for μ, σ, y_p and $R(y_0)$ and prediction intervals for $Y_{1:k}$, then all we need to do is to determine the appropriate percentage points of the pivotal quantities Z_1, Z_2, Z_p and Z_3. Convenient computational formulas for the conditional cumulative distribution functions

$$F_j(z \mid \boldsymbol{a}) = \Pr(Z_j \le z \mid \boldsymbol{a}) \quad \text{for } j = 1, 2, p, 3 \tag{9.24}$$

have been presented by Viveros and Balakrishnan (1994).

Each of the marginal conditional cumulative distribution functions of Z_1, Z_2 and Z_p, given the ancillary statistic $\boldsymbol{a}$, can be written as

$$F_j(z \mid \boldsymbol{a}) = \Pr(Z_j \le z \mid \boldsymbol{a}) = \frac{H_j(z;\boldsymbol{a})}{H_j(\infty;\boldsymbol{a})}, \tag{9.25}$$

where $H_j(z;\boldsymbol{a})$ depends on the specific pivotal quantity. For example, in the case of Z_1, we obtain

$$H_1(z;\boldsymbol{a}) = \int_0^{\infty} \frac{z_2^{m-1} e^{s_a z_2}\Gamma\left(\sum_{i=1}^{m} e^{a_{i:m:n} z_2 + z z_2}; m\right)}{\left\{\sum_{i=1}^{m}(R_i+1)e^{a_{i:m:n} z_2}\right\}^{m}}\, dz_2, \tag{9.26}$$

where $\Gamma(t;m)$ is the incomplete gamma ratio defined by

$$\Gamma(t;m) = \int_0^{t} \frac{1}{\Gamma(m)}\, e^{-w} w^{m-1} dw. \tag{9.27}$$

Since z_1 can be integrated out in (9.7), $H_2(z;\boldsymbol{a})$ becomes

$$H_2(z;\boldsymbol{a}) = \int_0^z \frac{z_2^{m-2} e^{s_a z_2}}{\{\sum_{i=1}^m (R_i+1)e^{a_{i:m:n} z_2}\}^m}\, dz_2. \tag{9.28}$$

For any specified p $(0 < p < 1)$, $H_p(z;\boldsymbol{a})$ can be written as

$$H_p(z;\boldsymbol{a}) = \int_0^\infty \frac{z_2^{m-2} e^{s_a z_2} \left\{1 - \Gamma\left(\sum_{i=1}^m (R_i+1)e^{a_{i:m:n} z_2 - z z_2 + G^{-1}(p)}; m\right)\right\}}{\{\sum_{i=1}^m (R_i+1)e^{a_{i:m:n} z_2 + G^{-1}(p)}\}^m}\, dz_2, \tag{9.29}$$

where $\Gamma(t;m)$ is the incomplete gamma ratio as defined in (9.27).

For the purpose of deriving conditional prediction intervals for $Y_{1:k}$, we can write the conditional cumulative distribution function of Z_3 as

$$F_{Z_3}(z \mid \boldsymbol{a}) = \Pr(Z_3 \le z \mid \boldsymbol{a}) = 1 - \frac{H_3(z;\boldsymbol{a})}{H_3(-\infty;\boldsymbol{a})}\,, \tag{9.30}$$

where

$$H_3(z;\boldsymbol{a}) = \int_0^\infty \frac{z_2^{m-2} e^{s_a z_2}}{\{k e^{z z_2} + \sum_{i=1}^m (R_i+1)e^{a_{i:m:n} z_2}\}^m}\, dz_2. \tag{9.31}$$

Now in order to develop all the conditional inference described above, we first need to numerically compute the MLEs $\hat{\mu}$ and $\hat{\sigma}$ based on the observed progressively Type-II right censored sample. This problem has already been discussed in Chapter 7. The use of Newton's method may be adequate for this task. Similarly, the use of Simpson's rule may be adequate for the numerical integrations required in (9.26), (9.28), (9.29) and (9.31). Since the integrands tend to be very small, they need to be scaled first before performing the integration. This scaling will not affect the results of (9.25) and (9.30). The scaling that Viveros and Balakrishnan (1994) found convenient was to work with the integrand of (9.26) divided by its value at $(z, z_2) = (0, 1)$, that of (9.28) by its value at $z_2 = 1$, and similarly for (9.29) and (9.31).

Remark 9.1 Should one wish to avoid the numerical computations involved in developing all the conditional inferences described above, one may employ some approximations that require the cross derivatives of the joint likelihood function by proceeding along the lines of DiCiccio (1987) and Viveros and Sprott (1987).

9.5 Results for Exponential Distribution

Let us assume that the lifetimes follow a two-parameter exponential distribution with probability density function

$$f(y;\mu,\sigma) = \frac{1}{\sigma}\, e^{-(y-\mu)/\sigma}, \qquad y \ge \mu, \tag{9.32}$$

and cumulative distribution function

$$F(y;\mu,\sigma) = 1 - e^{-(y-\mu)/\sigma}, \qquad y \geq \mu. \tag{9.33}$$

Clearly, this conforms to the location-scale family in (9.1) with

$$g(x) = e^{-x} \quad \text{and} \quad G(x) = 1 - e^{-x},\ x \geq 0. \tag{9.34}$$

The location parameter μ is referred to as the threshold parameter and is interpreted as the minimum lifetime for the items since all units survive this length of time. For a detailed discussion on various properties and applications of the exponential distribution, one may refer to Balakrishnan and Basu (1995).

Since a complete sufficient statistic for (μ, σ) exists in this case based on progressively Type-II right censored samples, in view of Basu's theorem [see Rao (1973)], the pivotal quantities Z_1, Z_2, Z_p and Z_3 are all statistically independent of the ancillary statistic $\boldsymbol{A}$. As a result, conditioning on $\boldsymbol{A}$ makes no difference in this instance.

Based on the observed progressively Type-II right censored sample, we have the MLEs of μ and σ as (see Chapter 7)

$$\hat{\mu} = Y_{1:m:n} \quad \text{and} \quad \hat{\sigma} = \frac{1}{m}\sum_{i=2}^{m}(R_i + 1)(Y_{i:m:n} - Y_{1:m:n}). \tag{9.35}$$

Based on the properties of progressively Type-II censored order statistics from the exponential distribution seen earlier in Chapter 2, it is known that $\frac{n(\hat{\mu}-\mu)}{\sigma}$ and $\frac{m\hat{\sigma}}{\sigma}$ are independently distributed as exponential and gamma with $m-1$ as shape parameter, respectively. Therefore,

$$\frac{n(m-1)}{m} Z_1 \overset{d}{=} F_{2,2m-2} \quad \text{and} \quad 2mZ_2 \overset{d}{=} \chi^2_{2m-2}, \tag{9.36}$$

where χ^2_{2m-2} denotes the central chi-square distribution with $2m-2$ degrees of freedom and $F_{2,2m-2}$ denotes the central F-distribution with $(2, 2m-2)$ degrees of freedom. Using the percentage points of F and chi-square distributions, we may construct exact confidence intervals for μ and σ, respectively.

Further, we observe in this case that

$$Z_p = \frac{y_p - \hat{\mu}}{\hat{\sigma}} = \frac{\mu + \sigma G^{-1}(p) - \hat{\mu}}{\hat{\sigma}} = -\frac{m}{V_2}\left(\frac{V_1}{n} + 2\ln(1-p)\right) \tag{9.37}$$

and

$$\begin{aligned} Z_3 = \frac{Y_{1:k} - \hat{\mu}}{\hat{\sigma}} &= \frac{(Y_{1:k} - \mu) - (\hat{\mu} - \mu)}{\hat{\sigma}} \\ &= \frac{m}{V_2}\left(\frac{V_3}{k} - \frac{V_1}{n}\right), \end{aligned} \tag{9.38}$$

where V_1, V_2 and V_3 are independently distributed as chi-square with $2, 2m-2$ and 2 degrees of freedom, respectively. Using these expressions, we may construct exact confidence intervals for the quantiles and the reliability and the prediction intervals for the minimum lifetime in a future sample of k units.

If we were to assume a scale-parameter exponential distribution instead, we have the maximum likelihood estimator of θ in this case as (see Chapter 7)

$$\hat{\theta} = \frac{1}{m}\sum_{i=1}^{m}(R_i+1)Y_{i:m:n}. \tag{9.39}$$

This is also a complete sufficient statistic for θ; further, we have seen earlier that

$$2m\,\frac{\hat{\theta}}{\theta} = 2mZ \stackrel{d}{=} \chi^2_{2m} \tag{9.40}$$

using which exact confidence intervals for θ can be easily constructed. Since the p-th quantile is $y_p = -\theta\ln(1-p)$ and the reliability at mission time y_0 is $R(y_0) = \Pr(Y \geq y_0) = e^{-y_0/\theta}$, MLEs and exact confidence intervals for y_p and $R(y_0)$ can be obtained from those of θ by direct transformation. Finally, exact prediction intervals for $Y_{1:k}$ (the smallest lifetime in a future sample of k items) may be obtained from the fact that

$$\frac{k\,Y_{1:k}}{\hat{\theta}} \stackrel{d}{=} F_{2,2m}. \tag{9.41}$$

9.6 Illustrative Examples

Example 9.1 Let us again consider the following progressively Type-II right censored data giving the log-times to breakdown of an insulating fluid tested at 34 kilovolts by Nelson (1982, Table 6.1):

i	1	2	3	4	5	6	7	8
$Y_{i:m:n}$	−1.6608	−0.2485	−0.0409	0.2700	1.0224	1.5789	1.8718	1.9947
R_i	0	0	3	0	3	0	0	5

In this case, we have $n = 19$, $m = 8$, $R_1 = 0$, $R_2 = 0$, $R_3 = 3$, $R_4 = 0$, $R_5 = 3$, $R_6 = 0$, $R_7 = 0$ and $R_8 = 5$. Let us now assume that the above progressively censored data has come from the extreme value distribution in (9.20). Through numerical maximization of the likelihood function, we find the MLEs of μ and σ as

$$\hat{\mu} = 2.222 \qquad \text{and} \qquad \hat{\sigma} = 1.026.$$

Using numerical integration methods, we next find

$$F_1(-1.76 \mid \boldsymbol{a}) = 0.025 \qquad \text{and} \qquad F_1(0.574 \mid \boldsymbol{a}) = 0.975.$$

Then an exact 95% conditional confidence interval for the location parameter μ is obtained from $-1.76 \le \frac{\hat{\mu}-\mu}{\hat{\sigma}} \le 0.574$ as $1.63 \le \mu \le 4.03$.

Next, we find numerically that

$$F_2(0.472 \mid \boldsymbol{a}) = 0.05 \qquad \text{and} \qquad F_2(1.418 \mid \boldsymbol{a}) = 0.95$$

using which an exact 90% conditional confidence interval for the scale parameter σ is obtained from $0.472 \le \frac{\hat{\sigma}}{\sigma} \le 1.418$ as $0.72 \le \sigma \le 2.18$.

Let us now consider the estimation of the 0.1-th quantile $y_{0.1}$. Since $G^{-1}(p) = \ln(-\ln(1-p))$ in this case, we immediately get the MLE of $y_{0.1}$ as

$$\hat{y}_{0.1} = \hat{\mu} + \hat{\sigma}\, G^{-1}(0.1) = -0.087.$$

We next find numerically that

$$\Pr(Z_{0.1} \le -4.64 \mid \boldsymbol{a}) = 0.025 \quad \text{and} \quad \Pr(Z_{0.1} \le -1.345 \mid \boldsymbol{a}) = 0.975$$

using which an exact 95% conditional confidence interval for the 0.1-th quantile $y_{0.1}$ is obtained from $-4.64 \le \frac{y_{0.1}-\hat{\mu}}{\hat{\sigma}} \le -1.345$ as $-2.54 \le y_{0.1} \le 0.84$.

Next, the estimated reliability at mission log-time $y_0 = \ln 2 = 0.693$ is

$$\hat{R}(y_0) = \hat{R}(0.693) = e^{-e^{(0.693-\hat{\mu})/\hat{\sigma}}} = 0.798.$$

We then numerically find that when $p = 0.637$ the 95% lower confidence bound for y_{1-p} is

$$\ell(\boldsymbol{y}; p) = 0.693 = y_0.$$

As a result, $p = 0.637$ becomes a 95% lower confidence limit for the reliability at mission log-time $y_0 = \ln 2 = 0.693$.

Suppose we are interested in predicting the smallest log-time to breakdown in a future sample of $k = 5$ insulating fluid specimens. For this purpose, we numerically determine that

$$\Pr(Z_3 \le -5.96 \mid \boldsymbol{a}) = 0.05 \quad \text{and} \quad \Pr(Z_3 \le -0.26 \mid \boldsymbol{a}) = 0.95$$

using which an exact 90% prediction interval for $Y_{1:5}$ is obtained from $-5.96 \le \frac{Y_{1:5}-\hat{\mu}}{\hat{\sigma}} \le -0.26$ as $-3.90 \le Y_{1:5} \le 1.96$.

Remark 9.2 The exact 90% conditional confidence interval for σ determined above to be (0.72, 2.18) is surprisingly close to 90% unconditional confidence interval for σ obtained to be (0.80, 1.80) by Nelson (1982, p. 230) using the complete data of $n = 19$ observations.

Remark 9.3 From the exact 90% conditional confidence interval for σ determined above to be (0.72, 2.18), we can make a transformation in order to obtain a 90% confidence interval for the shape parameter $c = \frac{1}{\sigma}$ of the associated Weibull distribution as $0.46 \le c \le 1.38$. It is of interest to observe that the value $c = 1$ lies well inside this interval thus providing evidence that an exponential distribution may be assumed for the progressively censored data in the original time scale.

Remark 9.4 The mean lifetime of products is a parameter that is usually of great interest to a reliability engineer with the natural scale of this parameter being the original time scale. Under the assumption of the Weibull distribution for the time to failure, we have the mean time to failure as

$$\lambda = E(e^Y) = e^{\mu}\,\Gamma(1+\sigma),$$

where $\Gamma(\cdot)$ denotes the complete gamma function. Though we readily obtain the MLE of λ as

$$\hat{\lambda} = e^{\hat{\mu}}\,\Gamma(1+\hat{\sigma}) = 9.331,$$

the lack of a pivotal quantity for λ under the Weibull model makes it impossible to derive exact conditional or unconditional confidence intervals for the mean lifetime.

Example 9.2 Let us consider the insulating fluid data considered already in Example 9.1, but this time in the original time scale of the data rather than in the logarithmic scale. Let us now assume a scale-parameter exponential distribution for this progressively censored data.

We then have the MLE of the mean lifetime θ as

$$\hat{\theta} = \frac{1}{m}\sum_{i=1}^{m} y_{i:m:n} = \frac{72.687}{8} = 9.1.$$

The MLE of the reliability at mission time $y_0 = 2$ is

$$\hat{R}(2) = e^{-2/\hat{\theta}} = 0.80.$$

A lower α confidence bound for $R(y_0)$ can easily be seen to be

$$e^{-y_0\chi^2_{1-\alpha,2m}/(2m\hat{\theta})},$$

where $\chi^2_{1-\alpha,2m}$ denotes the upper $1-\alpha$ percentage point of the central chi-square distribution with $2m$ degrees of freedom. Since we find from the chi-square tables that $\chi^2_{0.05,16} = 26.3$, we readily have the 95% lower confidence limit for $R(2)$ to be 0.70.

Suppose we are interested in constructing a 90% prediction interval for the smallest time to breakdown in a future sample of $k = 5$ insulating fluid specimens. Since we find from $F_{2,16}$ tables that $\Pr(0.052 \le \frac{5Y_{1:5}}{\hat{\theta}} \le 3.634) = 0.9$, we obtain the 90% prediction interval for $Y_{1:5}$ as (0.1, 6.6).

Remark 9.5 A comparison of the numerical results obtained in Example 9.1 under the extreme value distribution with those obtained in Example 9.2 under the exponential distribution reveals that the two sets of estimates are quite close. However, the intervals obtained under the extreme value distribution are seen to be wider than the corresponding intervals obtained under the exponential distribution. As aptly pointed out by Viveros and Balakrishnan (1994), this may be attributed to a slight loss in precision due to the uncertainty in the extra parameter involved in the extreme value distribution.

9.7 Results for Pareto Distribution

Let us now assume that the lifetimes follow a two-parameter Pareto distribution with probability density function

$$f(y;\mu,\sigma)=v\sigma^{v}(y-\mu)^{-(v+1)},\quad y\geq\mu+\sigma,\ v>0,\sigma>0, \tag{9.42}$$

and cumulative distribution function

$$F(y;\mu,\sigma)=1-\left(\frac{y-\mu}{\sigma}\right)^{-v},\quad y\geq\mu+\sigma,\ v>0,\ \sigma>0. \tag{9.43}$$

This conforms to the location-scale family in (9.1) with

$$g(x)=vx^{-(v+1)}\ \text{and}\ G(x)=1-x^{-v},\ x\geq 1. \tag{9.44}$$

For a detailed discussion on various aspects of the Pareto distribution, one may refer to Arnold (1983) and Chapter 20 of Johnson, Kotz and Balakrishnan (1994).

Let $y_{1:m:n}^{(R_1,\cdots,R_m)},\cdots,y_{m:m:n}^{(R_1,\cdots,R_m)}$ denote a progressively Type-II right censored sample of size m obtained from n items placed on a life-test and adopting a progressive censoring scheme $(R_1,\cdots,R_m)$. In this case, as seen earlier in Section 6.2.4, the BLUEs of μ and σ are

$$\mu^{*}=-\frac{1}{\Delta}\sum_{j=1}^{m}\sum_{i=1}^{m}\sum_{\ell=1}^{m}\sum_{k=1}^{m}\mu_i c^{i,k}(\mu_\ell-\mu_k)c^{\ell,j}y_{j:m:n}^{(R_1,\cdots,R_m)} \tag{9.45}$$

and

$$\sigma^{*}=\frac{1}{\Delta}\sum_{j=1}^{m}\sum_{i=1}^{m}\sum_{\ell=1}^{m}\sum_{k=1}^{m}c^{i,k}(\mu_\ell-\mu_k)c^{\ell,j}y_{j:m:n}^{(R_1,\cdots,R_m)}, \tag{9.46}$$

where

$$\Delta=\left(\sum_{i=1}^{m}\sum_{j=1}^{m}c^{i,j}\right)\left(\sum_{i=1}^{m}\sum_{j=1}^{m}\mu_i\mu_j c^{i,j}\right)-\left(\sum_{i=1}^{m}\sum_{j=1}^{m}\mu_i c^{i,j}\right)^{2},$$

$$c^{i,j}=\begin{cases}\dfrac{\gamma_i\gamma_{i+1}-\beta_i^2\beta_{i+1}^2}{\left(\prod_{k=1}^{i}\gamma_k\right)(\gamma_i-\beta_i^2)(\gamma_{i+1}-\beta_{i+1}^2)}, & i=j=1,2,\cdots,m-1,\\[2ex] \dfrac{1}{\left(\prod_{k=1}^{m-1}\gamma_k\right)(\gamma_m-\beta_m^2)}, & i=j=m,\\[2ex] \dfrac{\beta_{i+1}}{\left(\prod_{k=1}^{i}\gamma_k\right)(\beta_{i+1}^2-\gamma_{i+1})}, & j=i+1,\ i=1,2,\cdots,m-1,\\[2ex] 0, & \text{otherwise,}\end{cases}$$

$$\begin{aligned}\alpha_1 &= vn\ \text{and}\ \alpha_i=v(n-R_1-\cdots-R_{i-1}-i+1)\ \text{for}\ i=1,2\cdots,m,\\ \beta_i &= \frac{\alpha_i}{\alpha_i-1}\ \text{for}\ i=1,\cdots,m,\ \alpha_i>1,\\ \gamma_i &= \frac{\alpha_i}{\alpha_i-2}\ \text{for}\ i=1,\cdots,m,\ \alpha_i>2,\end{aligned}$$

and

$$\mu_i = \prod_{k=1}^{i} \beta_k \quad \text{for} \quad i = 1, 2, \cdots, m.$$

Let $Z_1 = \frac{\mu^* - \mu}{\sigma^*}$ and $Z_2 = \frac{\sigma^*}{\sigma}$ be the pivotal quantities and $\boldsymbol{A} = (A_{1:m:n}, \cdots, A_{m:m:n})$, where $A_{i:m:n} = \frac{Y_{i:m:n} - \mu^*}{\sigma^*}$, be an ancillary statistic. Then, upon substituting the expressions in (9.44) into (9.7), we obtain the joint conditional density function of Z_1 and Z_2 in this case to be [Aggarwala and Childs (1999)]

$$\begin{aligned} f(z_1, z_2 \mid \boldsymbol{a}) &= K(\boldsymbol{a}) \frac{z_1^m}{z_2^{vn+1}} \prod_{i=1}^{m} \frac{1}{(a_{i:m:n} + z_1)^{v(R_i+1)+1}}, \\ & \quad a_{1:m:n} + z_1 \geq 0, \ (a_{1:m:n} + z_1) z_2 \geq 1, \end{aligned} \tag{9.47}$$

where $K(\boldsymbol{a})$ is the normalizing constant. To obtain the marginal conditional density function of Z_1, given $\boldsymbol{a} = (a_{1:m:n}, \cdots, a_{m:m:n})$, we integrate (9.47) with respect to Z_2 to obtain

$$\begin{aligned} f(z_1 \mid \boldsymbol{a}) &= K(\boldsymbol{a}) \frac{v^{m-1} (a_{1:m:n} + z_1)^{vn}}{n} \prod_{i=1}^{m} \frac{1}{(a_{i:m:n} + z_1)^{v(R_i+1)+1}}, \\ & \quad z_1 \geq -a_{1:m:n}. \end{aligned} \tag{9.48}$$

Once the ancillary statistic $\boldsymbol{a} = (a_{1:m:n}, \cdots, a_{m:m:n})$ has been observed, conditional $100(1-\alpha)\%$ confidence intervals for the parameter μ can be constructed by first integrating (9.48) from $-a_{1:m:n}$ to ∞ to find the value of the constant $K(\boldsymbol{a})$, then finding two constants d and e such that

$$\Pr(Z_1 \geq e \mid \boldsymbol{a}) = \int_e^{\infty} f(z_1 \mid \boldsymbol{a}) dz_1 = \alpha/2 \tag{9.49}$$

and

$$\Pr(Z_1 \leq d \mid \boldsymbol{a}) = \int_{-u_1}^{d} f(z_1 \mid \boldsymbol{a}) dz_1 = \alpha/2. \tag{9.50}$$

The conditional $100(1-\alpha)\%$ confidence interval for μ is then $(\mu^* - e\sigma^*, \mu^* - d\sigma^*)$.

Conditional $100(1 - \alpha)\%$ confidence intervals for the parameter σ can be constucted from the joint density function in (9.47) to find two constants d and e such that

$$\Pr(Z_2 \geq e \mid \boldsymbol{a}) = \int_e^{\infty} \int_{\frac{1}{z_2} - a_{1:m:n}}^{\infty} f(z_1, z_2 \mid \boldsymbol{a}) dz_1 dz_2 = \alpha/2 \tag{9.51}$$

and

$$\Pr(Z_1 \leq d \mid \boldsymbol{a}) = \int_0^{d} \int_{\frac{1}{z_2} - a_{1:m:n}}^{\infty} f(z_1, z_2 \mid \boldsymbol{a}) dz_1 dz_2 = \alpha/2. \tag{9.52}$$

The conditional $100(1-\alpha)\%$ confidence interval for σ is then $\left(\frac{\sigma^*}{e}, \frac{\sigma^*}{d}\right)$.

To find a lower γ probability conditional tolerance interval for proportion β of the form $(\mu^* - q\sigma^*, \infty)$, we need to find q such that

$$\Pr\left(\int_{\mu^*-q\sigma^*}^{\infty} f(y;\mu,\sigma)dy \geq \beta \mid \boldsymbol{a}\right)$$
$$= \Pr\left(Z_2 \leq \frac{1}{(Z_1-q)\beta^{1/v}} \mid \boldsymbol{a}\right) = \gamma. \quad (9.53)$$

Equation (9.53) can be written as

$$\int_{-a_{1:m:n}}^{\infty} \int_{\frac{1}{a_{1:m:n}+z_1}}^{max\left\{\frac{1}{a_{1:m:n}+z_1}, \frac{1}{(z_1-q)\beta^{1/v}}\right\}} f(z_1, z_2 \mid \boldsymbol{a}) dz_2 dz_1 = \gamma$$

which yields

$$\int_{-a_{1:m:n}}^{\infty} K(\boldsymbol{a}) v^{m-1} \frac{(a_{1:m:n}+z_1)^{vn} - \left[min\{a_{1:m:n}+z_1, (z_1-q)\beta^{1/v}\}\right]^{vn}}{n}$$
$$\times \prod_{i=1}^{m} \frac{1}{(a_{i:m:n}+z_1)^{v(R_i+1)+1}} dz_1 = \gamma. \quad (9.54)$$

Once the ancillary statistic $\boldsymbol{a}$ has been observed, we may find the value of q by solving (9.54).

Similarly, to find an upper γ probability conditional tolerance interval for proportion β of the form $(-\infty, \mu^* - q\sigma^*)$, we need to solve for q from

$$\int_{-a_{1:m:n}}^{\infty} K(\boldsymbol{a}) v^{m-1} \frac{\left[min\{(z_1-q)(1-\beta)^{1/v}, a_{1:m:n}+z_1\}\right]^{vn}}{n}$$
$$\times \prod_{i=1}^{m} \frac{1}{(a_{i:m:n}+z_1)^{v(R_i+1)+1}} dz_1 = \gamma. \quad (9.55)$$

Example 9.3 Let us consider the data presented earlier in Example 6.4. In this case, we computed the BLUEs of μ and σ as $\mu^* = 1.87680$ and $\sigma^* = 3.16207$. In order to obtain a conditional 90% confidence interval for μ, we first integrate (9.48) from $-a_{1:m:n} = -1.0227$ to ∞ to obtain $K(\boldsymbol{a}) = 13.7598$. Then, solving for e from (9.49), we find $e = 2.2215$ and similarly $d = -0.5233$ from (9.50). So, the conditional 90% confidence interval for μ is

$$(\mu^* - e\sigma^*,\ \mu^* - d\sigma^*) = (-5.148,\ 3.531).$$

Next, solving (9.51) and (9.52), we get $e = 2.049$ and $d = 0.3155$. Then, the conditional 90% confidence interval for σ is

$$\left(\frac{\sigma^*}{e}, \frac{\sigma^*}{d}\right) = (1.543,\ 10.021).$$

To find a conditional 90% lower tolerance interval for proportion 0.8, we take $\beta = 0.8$ and $\gamma = 0.9$ and find from (9.54) $q = -1.043$. Then, a lower 0.9 probability tolerance interval for proportion 0.8 is $(\mu^* - q\sigma^*, \infty) = (5.175,\ \infty)$. Similarly, finding $q = -2.725$ from (9.55), we obtain an upper 0.9 probability tolerance interval for proportion 0.8 as $\{y \le \mu^* - q\sigma^* = 10.493\}$.

Proceeding in the same manner, Aggarwala and Childs (1999) varied the value of the shape parameter v and determined the following conditional 90% confidence intervals for μ and σ:

v	90% CI for μ	90% CI for σ
2.6	$(-3.791, 3.771)$	(1.304, 8.674)
2.8	$(-4.471, 3.652)$	(1.423, 9.350)
3.0	$(-5.148, 3.531)$	(1.543, 10.021)
3.2	$(-5.821, 3.411)$	(1.663, 10.688)
3.4	$(-6.490, 3.291)$	(1.784, 11.349)

Remark 9.6 It is important to mention here that by adopting the structural inference approach of Fraser (1968), Balakrishnan and Maswadah (1999a,b) have developed structural inference for the parameters of Fréchet-type extreme value and Rayleigh distributions, respectively, based on progressively Type-II right censored samples. They have also derived shortest confidence intervals for the parameters, the hazard rate and the reliability of these two distributions.

10

Optimal Censoring Schemes

10.1 Introduction

Up to this point, we have considered mathematical properties and problems of inference for progressively censored samples when a particular censoring scheme is to be employed. In reading this far, perhaps you yourself have asked the question, "How does a practitioner decide what the censoring scheme should be?" Is the decision made strictly on the basis of convenience, or can we choose a scheme which makes the most sense in some more statistical or mathematical setting? The question of choosing *optimal* values of $R_1, R_2, \cdots, R_m$ when considering a progressive Type-II right censoring scheme is certainly an important one to consider from a practical point of view, and as it turns out, it also gives rise to a number of interesting mathematical problems, in the areas of optimization, numerical analysis, simulation and programming, among others. We consider progressively Type-II right censored samples for the most part, since Type-II censored samples are far more tractable and interesting to consider from the point of linear inference and other mathematical properties, as we have already seen, and right censored samples will arise most frequently in life-testing applications, where it is possible and generally sensible to observe and monitor failures from the onset of experimentation at time $t = 0$.

The first question to consider is our definition of optimality. That is, what criterion will be used to determine whether one scheme is better than another? This is not a straightforward question! Consider a track athlete. Can we say one athlete is "better" than another if she or he are able to run faster, or run for a longer amount of time, or run for a longer distance, or throw farther, or jump higher, or lift more weight above their head, or all of the above? We realize at this point that the problem is not well enough defined to form a satisfactory answer. It depends upon the event or sport we are talking about. We could say with confidence that one track athlete is better than another if she or he could do *all* track events better; however, as we know, that is generally not the case. Even in the decathlon event, there has only been one time in history that one athlete has outperformed the field in every single event. The winner of a decathlon is the person who can obtain the highest value of some function of the ten "scores" corresponding to ten chosen events. If the function changes, we can no longer be certain that the same athlete would have won. The same can be said for swimmers, gymnasts, ice skaters, or team-sports athletes, or (in our case) progressive censoring schemes. We now see that we must narrow down the problem and define the "event" or function of many "events" when it comes to progressive censoring. In mathematical terms, we must specify an *objective function* to be optimized before we can determine the optimal progressive censoring scheme to be employed, and when the winning scheme is declared, it is with respect to that well-defined objective function, and perhaps, but not necessarily with respect to other objective functions as well.

So, let us think of a few different ways in which we can define optimality. Let us consider a practitioner who is interested in designing a warranty. This individual will be interested in the estimation of the time after which, say 95% of the product will survive, that is, the manufacturer of the product is willing to repair 5% per cent of all items produced. In this case, we are interested in estimating the fifth percentile of the product's failure-time distribution. This being a single (univariate) estimator, we could define the best scheme as the scheme which minimizes the mean squared error of the estimator. That is, the objective function (to be minimized in this case) would be

$$MSE\left(\widetilde{\xi}_{0.05}\right)$$

where $\widetilde{\xi}_{0.05}$ represents the estimator for the fifth percentile of the product's lifetime distribution, the parameter of interest. A similar objective function may be defined for any case where a single parameter is to be estimated. If a practitioner is interested in setting tolerance limits for a control chart, she or he may choose an objective function based on the variance of some pivotal quantity for $\xi_{0.05}$ which may or may not directly involve $\widetilde{\xi}_{0.05}$, in order to obtain a narrow confidence interval. One would hope the two optimal schemes corresponding to the two different optimality criterion would be similar in some sense. Again, this must be determined on a case-by-case basis.

We have discussed two-parameter (for example, location-scale) estimation extensively in Chapters 6 and 7. In the case where more than one parameter is to be estimated, there is a variance-covariance structure of the estimators which one may wish to consider. In the case of two-parameter estimation, optimality may be defined in terms of the trace or determinant of the variance-covariance matrix of the estimators, which makes sense particularly if the estimators are unbiased. In the case of biased estimators, we may want to replace the variance terms in our objective functions by the corresponding mean squared errors.

More complicated objective functions can also be considered, such as linear combinations or the inclusion in some other way of any number of factors of interest, perhaps including some of those mentioned thus far. For example, suppose the cost of experimentation is of concern, and will be proportional to the time on test of the experiment. Then one may wish to add in terms, containing

$$E\left(Y_{m:m:n}^{(R_1,R_2,\cdots,R_m)}\right) \quad \text{or} \quad P\left(Y_{m:m:n}^{(R_1,R_2,\cdots,R_m)} > T_U\right),$$

where $Y_{m:m:n}^{(R_1,R_2,\cdots,R_m)}$ is the last observed failure time in the experiment and T_U is a time by which we would like to end experimentation with high probability.

Similarly, one may want to include in the objective function terms involving the variance of $Y_{m:m:n}^{(R_1,R_2,\cdots,R_m)}$, or may have an interest in recycling items removed after the first failure, say, and therefore add a term such as

$$\text{constant} \times R_1,$$

where the constant may be related to the amount of utility each removed live item contains. It is clear that there are an infinite number of possibilities, and the practitioner is able to customize the objective function to be optimized according to her or his priorities. One point which must be considered before getting too fancy with the objective function is the ability to at least calculate it (perhaps with some computer algorithm) for any censoring scheme, if not to be able to express it as some closed-form function of $R_1, R_2, \cdots, R_m$. In the examples which we will present in this chapter, we will use results from Chapter 6 on best linear unbiased estimation and define objective functions based on the variances and where applicable, the covariance of BLUEs. In particular, for all of the distributions which we will consider, we will define both the trace and the determinant of the variance-covariance matrix of BLUEs for two-parameter location/scale distributions where both parameters are unknown, as objective functions of interest. Recall that we have discussed various ways of calculating these values, and as such, the actual value of the objective function may be determined, either in closed form or through a computer algorithm, for any scheme and for a number of distributions, including all of those discussed in Chapter 6. With reference to our discussion in Chapter 5 on obtaining moments of progressively Type-II right censored order statistics with only the knowledge of moments of usual order statistics from a sample of size n (see Section 5.4), we will be able to compute these objective functions for any location-scale distribution where moments of the usual order statistics are available. This will be useful for many distributions commonly encountered in reliability literature, and will be used in many of the calculations and table entries presented in this chapter.

The next question to consider, then, is one of actually finding the optimal censoring scheme once the objective function has been satisfactorily defined based on n units available to put on test. Suppose that a practitioner would like to observe m complete failure times, and must determine the best scheme to use in this case. (Notice that m and n are chosen in advance here, which depend on availability of units and experimental facilities. If one or both of these are also to be determined, one may use the tables presented in this chapter to decide upon values of m and n which are feasible given an agreeable value of the objective function, or construct similar tables if the chosen objective function is different from those presented.) In this finite sample situation, one may list all possible choices of censoring schemes and corresponding values of objective functions, and determine the best value (i.e., the value which optimizes the objective function) or a region of satisfactory values from this list, and either pick the best scheme or one which gives a value very close to the best but may be practically more convenient, much as one determines a best regression from a list of C_p statistics for all possible regressions computed using SAS. This will be the method employed for most distributions considered here, although there may be cases where a simple closed-form expression of the objective function is available, and optimization is somewhat more mathematically elegant.

Remark 10.1 All of the computation for this chapter was performed by programming in the Fortran language.

10.2 The Exponential Distribution

For the exponential distribution, it has been shown in Chapter 6 that the variance (for the one-parameter case) and the trace and determinant of the variance-covariance matrix (for the two-parameter case) of the BLUEs are independent of the censoring scheme employed. [We can see this from the expressions in (6.5), (6.31), (6.32), and (6.33).] Hence, there is no optimal censoring in terms of the objective functions we are interested in — that is, no one scheme is better than another under these objective functions; if we are concerned with completing the experiment quickly, we may employ a conventional Type-II right censoring scheme and simply observe the first m failures, whereas if we are concerned with saving experimental units for subsequent testing, we may employ a progressive Type-II right censoring scheme with $R_1 = n - m$ and all subsequent R_i's $= 0$. As mentioned earlier, if we also decide that we would like to have the possibility of observing some extreme lifetimes, but at the same time, with high probability, we do not want the experiment to run too long, then we can choose a censoring scheme to reflect this. (Essentially, by doing this, we are defining another relatively simple objective function to be optimized, or, equivalently, adding terms to the objective function of interest.)

For other distributions, particularly those used to model lifetime distributions, however, it is quite often the case that just the calculation of the variances and covariance of BLUEs for *all* possible censoring schemes, although straightforward with the methods we have presented, is time-consuming. For some of these distributions, we will present results for optimal censoring for some practical choices of n and m. Furthermore, we will see that when compared to the special case of conventional censoring schemes frequently employed by practitioners (for the same values of n and m), the increase in efficiencies when employing the optimal progressive Type-II right censoring scheme is often very substantial. For the objective functions which we will be considering, a sensible definition of the efficiency of Scheme A with respect to Scheme B is

$$\frac{Objective\ function(Scheme\ A)}{Objective\ function(Scheme\ B)};$$

hence, if one is interested in determining a region containing a number of satisfactory schemes, she or he can define that region in terms of efficiencies. (For example, an experimenter may be satisfied with choosing any scheme which is at least 98% as efficient as the optimal scheme, and choose the scheme out of these satisfactory ones which is most practical or convenient. Again, by doing this, we are building practicality into the objective function without actually complicating the calculations, but by using a "two-stage" approach to optimization.) It should be clear that for any fixed values of m and n, the efficiency of the optimal scheme with respect the conventional scheme will never be less than one (Why?); that is, there will never be a loss in efficiency over the conventional censoring scheme if the optimal scheme is employed. It should also be clear that in the case of the exponential distribution, if the objective function is chosen to be the variance of the one-parameter BLUE or the trace or determinant of the variance-covariance matrix of the two-parameter BLUEs, all efficiencies will be 1. (Why?)

10.3 The Normal Distribution

Tables of single and product moments of usual order statistics for standard normal, log-normal and extreme value distributions are available for various sample sizes; in fact, for these distributions, the moments of *usual* order statistics are obtained numerically, as they do not have simple closed form expressions. Using the results in Section 5.4 for obtaining moments of progressively Type-II right censored order statistics, and expressions for BLUEs and their variances and covariance for two-parameter distributions given in (6.27) and (6.28), we will be able to determine the coefficients of the observed progressively Type-II right censored order statistics for the best linear unbiased estimators, as well as their variances and covariance for any progressive Type-II right censoring scheme, and therefore for all progressive Type-II right censoring schemes.

The standard normal distribution, with density function given by

$$f(x) = \frac{1}{\sqrt{2\pi}} e^{-x^2/2}, \ -\infty < x < \infty, \tag{10.1}$$

is occasionally used in reliability studies to model lifetime distributions [see Bain and Engelhardt (1991)] perhaps due to the vast amount of knowledge that has accumulated regarding this distribution. It is also common to analyze transformed data under the normality assumption.

Consider a progressively Type-II right censored sample of size m from a sample of size n from the two-parameter normal, $N(\mu, \sigma^2)$, distribution. We will denote this sample by $Y_{i:m:n}^{(R_1,R_2,\cdots,R_m)}$, $1 \le i \le m$. Here, we can write $Y_{i:m:n}^{(R_1,R_2,\cdots,R_m)} = \mu + \sigma X_{i:m:n}^{(R_1,R_2,\cdots,R_m)}$, where $X_{i:m:n}^{(R_1,R_2,\cdots,R_m)}$ represents a progressively Type-II right censored sample with the same censoring scheme from the standard normal distribution in (10.1).

Tables of means, variances and covariances of usual order statistics from the standard normal distribution are tabulated up to sample size $n = 50$ in Tietjen, Kahaner and Beckman (1977). We can use these values to obtain moments of progressively Type-II right censored order statistics and then the coefficients for the BLUEs of μ and σ, as well as the variances and covariance of the BLUEs for every possible sampling scheme for selected values of m and n. We will then be able to identify the optimal censoring scheme for each pair of selected values of m and n when progressively Type-II right censored samples are from two-parameter normal distributions by identifying the censoring scheme corresponding to the optimal (which in our case is the minimum) value of the objective function. The two objective functions we will consider are: (1) the trace, and (2) the determinant of the variance-covariance matrix of BLUEs.

10.3.1 Discussion of Results

Tables 10.1–10.3 are presented in this chapter for the normal distribution. Tables 10.1 and 10.2 give, for selected values of m and n, efficiencies of the optimal progressive Type-II right censoring schemes with respect to conventional Type-II right censoring schemes, along with the corresponding optimal values of the objective function when the objective function is the trace and the determinant of the variance-covariance matrix of location/scale BLUEs, respectively. Remember that, by our definition of efficiency, high values of efficiency are desirable. Table 10.3 gives coefficients of the (ordered) progressively Type-II right censored order statistics for BLUEs of the location and scale parameters for various censoring schemes considered, up to $m = 6$, along with the variances and covariance of the BLUEs. This table also indicates which schemes are the optimal and which are least optimal for these same values of m up to 6. Since the only other values of m which we have run results for (with $n = 15$ excepted; the reasons for this are discussed next) are those for m very close to (or equal to) n, which will be uncommon from a practical point of view, we have excluded these schemes due to the large amount of space that would be required to present them.

The selected values of m and n when $m < \frac{n}{2}$ are such that the total number of censoring schemes to be considered is not larger than 11686, the number of possible schemes when $n = 20$ and $m = 6$. This is because the total number of schemes to consider becomes very large, even with today's computing facilities, certainly too large to do an exhaustive investigation for all the values of n of interest. For $m \geq \frac{n}{2}$, a few selected values were also considered, although generally in practice only a small fraction of the n complete lifetimes will be observed. Also, for larger values of m, as we will see when discussing the various tables, the efficiency gain from employing optimal progressive censoring versus conventional censoring decreases considerably; essentially any scheme will give a very high efficiency, since so many of the n failures will be observed.

The observations we will make here will be mainly concerned with trace-efficiency, although a similar analysis based on determinant-efficiency may be made using the corresponding tables.

Consider first Table 10.1. Efficiencies for the trace-optimal schemes (that is, the scheme for which minimum trace of the variance-covariance matrix of BLUEs is attained for some values of m and n) with respect to the conventional Type-II right censoring schemes are shown on the first line of each box, and the actual traces corresponding to these optimal schemes are shown on the second line. From the table, as one would expect, as m increases (with n being held constant), the gain in efficiency over the conventional scheme decreases; in other words, the smaller the proportion of failure times an experimenter would like to observe, the more precision can be gained by employing optimal progressive Type-II right censoring over conventional Type-II right censoring. Also, as n increases (with m being constant), the gain in trace-efficiency increases. For small values of m, we can see an efficiency with respect to conventional censoring of as much as 360% for the different cases tabulated.

Looking at the trace-optimal schemes denoted with a $^+$ in Table 10.3, it is quite evident that for small values of m, a scheme by which all but $m-1$ surviving units are randomly removed after the first observed failure (ie. $R_1 = n - m$) will result in the best precision of BLUEs. For $n = 15$, in which case a complete analysis was possible (but not shown, as explained earlier), the optimal censoring scheme starts to shift at about $m = 7$, with censoring occurring at points other than immediately following the first failure. However, the gain in efficiency for optimal schemes for $m \geq 7,\ n = 15$ diminishes quite rapidly.

An interesting question that arises here is the following: Suppose an experimenter wants to observe m failures. Is it wise to begin with a large number (n) of experimental units, in terms of precision? Intuitively, one would immediately say yes, since more experimental units, even if they are censored, results in more knowledge and should therefore lower variances of estimators. However, looking at the first table, it seems that for $m = 2$ and $m = 3$, the values of the optimal traces actually increase with n. (Notice that this is not so in the second table for determinants.) In searching for an explanation, one might consider the fact that with a larger number of units, there is also more uncertainty as to which units remain on test after censoring, or one may attribute small differences to round-off error and approximations used in calculation of the moments of usual order statistics. However, for $m \geq 4$, our intuition prevails: for fixed values of m, precision increases with n, and more drastically for larger m. Keep in mind that we have used the standard normal distribution for these tables; for populations with large variance, the gain in precision by starting with a larger value of n will be quite substantial, since we would multiply these values by the factor σ^2. Furthermore, if "time on test" is of concern to an experimenter, quite often, depending upon the censoring schemes employed, a large value of n will result in a smaller expected time on test for a fixed value of m.

Another question which may be raised by the experimenter is that if the optimal scheme is one in which $n-m$ units are removed after the first observed failures, how much precision is lost (or even possibly gained) if we begin with m units in the first place, and observe all m to fail? In some cases and for some distributions, it is quite possible that the loss is not substantial, and it is possible to get by without putting in a large number of units on initial test, whereas in other cases the information gained from the knowledge that $n-m$ units survived beyond the first failure (but nothing else is known about those units) could be quite helpful. Tables for comparing these values are available in Tietjen, Kahaner and Beckman (1977) for full samples for the case of the normal distribution.

Table 10.3 also shows least precise censoring schemes with respect to the trace (denoted $^-$) and determinant (denoted $^/$) of the variance-covariance matrix of BLUEs. It is very interesting to note that for most of the entries displayed corresponding to worst schemes (using either trace or determinant criterion), the commonly used conventional Type-II right censoring scheme is actually the least efficient of all progressive Type-II right censoring schemes!

Remark 10.2 While obtaining the results for $n = 50$, large values of m were not considered due to computational difficulties. The computations involved in Thomas and Wilson's (1972) result, combined with the manner in which the moments of usual order statistics were obtained, give rise to a number of problems in numerical analysis which have yet to be investigated in detail, including accuracy and error.

10.4 The Extreme Value Distribution

The standard extreme value distribution has density function [see (6.53)]

$$f(x) = e^x e^{-e^x}, \quad -\infty < x < \infty.$$

This distribution and its counterpart, which we will discuss in the next section, has been used extensively in modelling failure times of items; see, for example, Nelson (1982), Viveros and Balakrishnan (1994). The extreme value random variable may also be seen as a transformed Weibull random variable as we have already discussed in Chapter 7, and will therefore be an important distribution to consider. Furthermore, many asymptotic results for extreme order statistics from arbitrary distributions involve the extreme value distribution; see, for example, Arnold, Balakrishnan and Nagaraja (1992).

Tables of means, variances and covariances of order statistics for the standard extreme value distribution given above have been presented by Balakrishnan and Chan (1992) for sample sizes up to $n = 30$. We have used these values to obtain optimal censoring schemes for selected values of m and n when progressively Type-II right censored samples are from two-parameter (location-scale) extreme value distributions. We have obtained the optimal censoring schemes in much the same way as we did for the normal distribution.

10.4.1 Discussion of Results

Tables 10.4–10.6 presented in this chapter parallel those discussed for the normal distribution for selected values of m and n. We will consider first Table 10.4 giving optimal trace-efficiencies. Again, we see that for small values of m the improvement in precision of estimates is remarkable, with efficiencies as high as 575% for $m = 2$ and $n = 30$. Here, even for values of $\frac{m}{n}$ as high as $\frac{2}{3}$ (which we can only observe for $n = 15$), the gain in precision is still 21%.

Again, if one poses the question of whether or not a larger value of n will be beneficial if m is fixed, we find that for $m = 2$ or 3, the values of trace are actually increasing with n, indicating a decrease in precision, whereas for $m \geq 4$, the values decrease, indicating an increase in precision. From Table 10.5, we find again that this observation does not hold if one is considering determinant-efficiency.

The actual schemes corresponding to optimal censoring with respect to trace efficiency are shown in Table 10.6. The optimal schemes looks similar to those for the normal distribution, although the extreme value distribution has very different physical properties than the normal distribution. For small values of m, the optimal scheme is to randomly remove all but $m - 1$ live units from experimentation immediately following the first failure. This table also shows that, for all cases considered, the least efficient censoring scheme with respect to the trace of the variance-covariance matrix of BLUEs is the conventional Type-II right censoring scheme.

Again, a similar analysis based on determinants can be carried out. It is difficult to determine whether one should use the trace criterion or the determinant criterion in deciding upon a censoring pattern. However, a look at the table reveals that in most cases, both the worst and best schemes coincide regardless of the criterion chosen. Furthermore, when the schemes do differ, the actual removal pattern is similar, and if one calculates the alternate criterion (for example, if one calculates the trace resulting from the determinant-optimal scheme), the value is usually (actually, in all cases we have considered) very close to that of the optimal scheme for that criterion (note that we can only say this is true for the two criterion we have considered). Furthermore, if an experimenter decides to list, say the top 10 candidates or all schemes in the top 5% of the trace criterion, the optimal scheme for the determinant criterion will almost always be well within this "acceptable" range of values, and vice versa. This applies also to the other distributions that we have considered.

10.5 The Extreme Value (II) Distribution

Suppose W has the extreme value distribution discussed in the preceding section. If we make the transformation $X = -W$, then X is said to have the extreme value distribution of Type II or the extreme value (II) distribution with density function

$$f(x) = e^{-x} e^{-e^{-x}}, \quad -\infty < x < \infty.$$

This distribution is again seen in results involving extreme order statistics from many distributions. Its increasing failure rate and positive coefficient of skewness makes it a plausible candidate for modelling lifetime distributions of various items. Moments for usual order statistics from this distribution can be easily obtained from the moments of usual order statistics for the extreme value distribution. It will be interesting to see if results obtained for optimal censoring for the extreme value (II) distribution will be somehow "opposite" to those obtained for the extreme value distribution.

10.5.1 Discussion of Results

Tables 10.7–10.9 present results for the extreme value (II) distribution. Considering the trace-efficiencies in Table 10.7, it is evident that for this distribution, although the gains in precision over the conventional censoring scheme are still respectable for small values of n, they are no where near those for the extreme value distribution. Furthermore, from Table 10.9, for larger values of m, the optimal censoring scheme is seen to be the conventional Type-II right censoring scheme. However, for small values of m, the optimal schemes tend to have censoring at the beginning and end of the experimentation. We again find that for fixed values of $m > 3$, the value of the trace decreases with the sample size n employed.

We find here that the *least* efficient censoring schemes with respect to the trace for values of $m > 5$ is the scheme in which all but $m - 1$ units are removed from experimentation immediately following the first observed failure. Although this is in contrast to the extreme value distribution discussed in the previous section, for small values of m the least efficient scheme is still seen in many cases to be the conventional scheme.

Table 10.9 also enables us to compare the efficiency of the worst and the best (and also vice versa) censoring schemes; for example, for $m = 4$ and $n = 15$, the least precise censoring scheme $(0, 11, 0, 0)$ is 87% as precise as the most trace-efficient censoring scheme $(7, 0, 0, 4)$ and 90% as precise as the conventional censoring scheme, whereas for $m = 4$ and $n = 25$, the least precise censoring scheme $(0, 0, 0, 21)$ is only 78% as precise as the most precise censoring scheme $(17, 0, 0, 4)$.

10.6 The Log-Normal Distribution

The standard log-normal distribution has density function

$$f(x) = \frac{1}{x\sqrt{2\pi}} e^{-(\ln x)^2/2}, \quad 0 < x < \infty.$$

This distribution has been used widely in modelling real data, including life-testing data. [see Aitchison and Brown (1957)]. Single and product moments of usual order statistics from this distribution have been tabulated by Gupta, McDonald and Galarneau (1974) for sample sizes up to $n = 20$. These will be used here, again using Thomas and Wilson's (1972) method to obtain moments required to obtain coefficients for BLUEs, as well as their variances and covariance, to obtain optimal progressive censoring patterns when data are from the log-normal distribution.

10.6.1 Discussion of Results

Tables 10.10–10.12 summarize the results obtained for the log-normal distribution, and follow the same sequence as tables for the normal, extreme value, and extreme value (II) distributions. From Table 10.12, it is evident that for the log-normal distribution, the optimal progressive Type-II right censoring pattern (for both trace *and* determinant criterion) is the conventional Type-II right censoring scheme for the vast majority of cases, and in those cases where it is not, the efficiency of the optimal censoring pattern with respect to the conventional censoring pattern is very slightly favourable. Furthermore, without exception, the least precise censoring scheme for the cases considered is that in which all censoring is carried out immediately following the first observed failure. Notice that this is very different from the results for the normal and extreme value distributions, where this scheme was quite often the best scheme. Prior knowledge of the type of distribution to use in modelling the lifetimes of items, apart from unknown parameters, is therefore an important issue here.

It is interesting to note that for the log-normal distribution, a larger value of n for fixed m results in a decrease in the optimal value of the trace, even for $m = 2$ and $m = 3$. The increase in precision is quite significant.

Again, if one is interested in considering the trace-efficiency of the least precise versus the most precise censoring schemes, Table 10.12 will be of help. For example, for $m = 5$ and $n = 15$, the least precise censoring scheme $(10, 0, 0, 0, 0)$ is about 58% as precise as the most precise censoring scheme $(0, 0, 0, 0, 10)$. Thus, if an experimenter is concerned with saving units for further testing, this factor should be taken into account before embarking on the progressive censoring scheme $(10, 0, 0, 0, 0)$. As discussed earlier, the experimenter may want to alter the objective function to take into account the value of units removed early on. A similar analysis based on determinants can again be carried out using Table 10.11.

Remark 10.3 The tables of covariances of usual order statistics from the log-normal distribution given in Gupta, McDonald and Galarneau (1974) for $n = 20$ which were used in this section were given to only three-decimal accuracy, possibly with round-off error, and as such, there were problems in confirming that the variance-covariance matrices of progressively Type-II right censored order statistics from the log-normal distribution for large values of m were positive definite. Once again, this gives rise to interesting problems in numerical analysis. For this reason, only small values of m were considered for $n = 20$. Furthermore, although the results for $n = 20$ may not be as precise as others presented in this chapter, they should point us in the right direction for future study.

It is worthwhile to reiterate that if the objective function to optimize is altered, the optimal censoring criterion may also be different from those presented in the tables here, and that these tables are useful when the objective function is given by either the trace or the determinant of the variance-covariance matrix of the BLUEs, where there are two parameters to be estimated: a location and a scale. If, for example, only a location parameter is to be estimated, one would not be interested in the variance of the scale estimator and may choose the variance of the location estimator as the objective function. They would very likely find the optimal censoring pattern to be different from the case when both parameters were of interest, and the tables given here will not be very useful in that case, since a complete listing of all possible schemes is not available.

10.7 Tables

In this section, we present Tables 10.1–10.12 which deal with optimal censoring for the normal, extreme value, extreme value (II) and log-normal distributions, respectively.

Table 10.1: Efficiencies and traces for trace-optimal censoring schemes: Normal distribution (Line 1: Efficiency; Line 2: Trace)

$n \setminus m$	2	3	4	5	6	7	8	9	10	11	12	$n-2$	$n-1$	n
15	2.2757	1.8746	1.5827	1.3849	1.2509	1.1656	1.1104	1.0711	1.0454	1.0277	1.0159	1.0082	1.0033	1.0000
	1.0163	0.5015	0.3455	0.2700	0.2248	0.1934	0.1703	0.1532	0.1398	0.1292	0.1207	0.1137	0.1080	0.1033
20	2.5622	2.1533	1.8305	1.5987	1.4312							1.0051	1.0022	1.0000
	1.0371	0.5031	0.3423	0.2649	0.2189							0.0823	0.0794	0.0768
25	2.7976	2.3881	2.0464	1.7920								1.0036	1.0016	1.0000
	1.0513	0.5048	0.3409	0.2621								0.0645	0.0627	0.0612
30	2.9981	2.5909	2.2363										1.0013	1.0000
	1.0614	0.5063	0.3401										0.0519	0.0508
40	3.3288	2.9289	2.5583										1.0009	1.0000
	1.0748	0.5085	0.3393										0.0385	0.0380
50	3.6010	3.2088												
	1.0819	0.5094												

Table 10.2: Efficiencies and determinants for determinant-optimal censoring schemes: Normal distribution (Line 1: Efficiency; Line 2: Determinant)

$n \setminus m$	2	3	4	5	6	7	8	9	10	11	12	$n-2$	$n-1$	n
15	1.9029	1.6024	1.4241	1.3494	1.2951	1.2442	1.1994	1.1638	1.1331	1.1040	1.0765	1.0505	1.0254	1.0000
	0.0706	0.0306	0.0187	0.0127	0.0094	0.0074	0.0060	0.0051	0.0043	0.0038	0.0033	0.0030	0.0027	0.0024
20	2.0448	1.7070	1.5054	1.4431	1.3847							1.0382	1.0195	1.0000
	0.0599	0.0260	0.0159	0.0107	0.0109							0.0015	0.0014	0.0013
25	2.1585	1.7889	1.5676	1.5176								1.0310	1.0159	1.0000
	0.0530	0.0230	0.0141	0.0093								0.0009	0.0009	0.0008
30	2.2539	1.8564	1.6284										1.0134	1.0000
	0.0481	0.0210	0.0128										0.0006	0.0006
40	2.4089	1.9641	1.7339										1.0103	1.0000
	0.0416	0.0181	0.0109										0.0003	0.0003
50	2.5358	2.0515												
	0.0372	0.0163												

Table 10.3: Coefficients, variances, and covariance of BLUEs for selected schemes for normal distribution

n	m	Scheme	Coefficients (μ^*)	Coefficients (σ^*)	$\frac{Var(\mu^*)}{\sigma^2}$	$\frac{Var(\sigma^*)}{\sigma^2}$	$\frac{Cov(\mu^*,\sigma^*)}{\sigma^2}$
15	2	$0,13^{-/}$	-2.5574,3.5574	-2.049,2.049	1.5404	0.7723	1.0273
15	2	$13,0^{+*}$	0.0667,0.9333	-0.5377,0.5377	0.7213	0.2950	0.3771
15	3	$0,0,12^{-/}$	-1.0242,-0.4676,2.4918	-1.0601,-0.5477,1.6077	0.5713	0.3689	0.4020
15	3	$12,0,0^{+*}$	0.1656,0.2361,0.5983	-0.4338,0.0267,0.4071	0.3472	0.1543	0.1514
15	4	$0,0,0,11^{-/}$	-0.5462,-0.2448,-0.1148,1.9058	-0.7201,-0.3892,-0.2458,1.3552	0.3101	0.2368	0.2163
15	4	$11,0,0,0^{+*}$	0.1840,0.1401,0.2504,0.4255	-0.3787,-0.0554,0.1028,0.3313	0.2354	0.1101	0.0849
15	5	$0,0,0,0,10^{-/}$	-0.3217,-0.1364,-0.0560,0.0043,1.5097	-0.5459,-0.3050,-0.2002,-0.1211,1.1722	0.2022	0.1718	0.1325
15	5	$10,0,0,0,0^{+}$	0.1821,0.1101,0.1625,0.2215,0.3231	-0.3390,-0.0842,0.0196,0.1194,0.2842	0.1818	0.0882	0.0542
15	5	$0,10,0,0,0^{*}$	-0.1170,0.4216,0.1271,0.2109,0.3574	-0.3240.-0.0928,-0.0072,0.1125,0.3115	0.1816	0.1007	0.0747
15	6	$0,0,0,0,0,9^{-/}$	-0.1950,-0.0732,-0.0203,0.0196,0.0531, 1.2157	-0.4390,-0.2518,-0.1700,-0.1082,-0.0562, 1.0252	0.1480	0.1332	0.0868
15	6	$8,1,0,0,0,0^{+}$	0.1342,0.1410,0.1207,0.1534,0.1909, 0.2597	-0.3056,-0.0951,-0.0226,0.0466,0.1211, 0.2555	0.1485	0.0763	0.0392
15	6	$0,9,0,0,0,0^{*}$	-0.0528,0.3402,0.0996,0.1412,0.1910, 0.2808	-0.2715,-0.1221,-0.0370,0.0388,0.1222, 0.2695	0.1462	0.0812	0.0497
20	2	$0,18^{-/}$	-3.0609,4.0609	-2.1745,2.1745	1.8710	0.7863	1.1613
20	2	$18,0^{+*}$	0.0500,0.9500	-0.5087,0.5087	0.7668	0.2703	0.3839
20	3	$0,0,17^{-/}$	-1.2872,-0.6610,2.9482	-1.1244,-0.6212,1.7456	0.7056	0.3778	0.4713
20	3	$17,0,0^{+*}$	0.1523,0.2296,0.6181	-0.4153,0.0303,0.3850	0.3628	0.1403	0.1578

(contd.)

Table 10.3: Coefficients, variances, and covariance of BLUEs for selected schemes for normal distribution (contd.)

n	m	Scheme	Coefficients (μ^*)	Coefficients (σ^*)	$\frac{Var(\mu^*)}{\sigma^2}$	$\frac{Var(\sigma^*)}{\sigma^2}$	$\frac{Cov(\mu^*,\sigma^*)}{\sigma^2}$
20	4	$0,0,0,16^{-/}$	-0.7284,-0.3766,-0.2298,2.3348	-0.7645,-0.4380,-0.3014,1.5039	0.3828	0.2438	0.2634
20	4	$16,0,0,0^{+*}$	0.1748,0.1287,0.2529,0.4436	-0.3666,-0.0471,0.1019,0.3118	0.2424	0.0999	0.0911
20	5	$0,0,0,0,15^{-/}$	-0.4632,-0.2378,-0.1433,-0.0738,1.9180	-0.5804,-0.3417,-0.2414,-0.1673,1.3308	0.2456	0.1778	0.1682
20	5	$15,0,0,0,0^{+}$	0.1769,0.0970,0.1589,0.2271,0.3401	-0.3316,-0.0744,0.0235,0.1160,0.2664	0.1850	0.0799	0.0600
20	5	$0,15,0,0,0^{*}$	-0.1671,0.4537,0.1149,0.2148,0.3837	-0.3079,-0.0924,-0.0063,0.1095,0.2971	0.1908	0.0929	0.0841
20	6	$0,0,0,0,0,14^{-/}$	-0.3117,-0.1565,-0.0914,-0.0433,-0.0037, 1.6066	-0.4679,-0.2813,-0.2028,-0.1447,-0.0967, 1.1934	0.1747	0.1386	0.1155
20	6	$14,0,0,0,0,0^{+}$	0.1714,0.0837,0.1184,0.1564,0.1978, 0.2723	-0.3033,-0.0862,-0.0130,0.0502,0.1167, 0.2357	0.1512	0.0677	0.0425
20	6	$0,14,0,0,0,0^{*}$	-0.0961,0.3711,0.0856,0.1373,0.1979, 0.3042	-0.2560,-0.1235,-0.0339,0.0391,0.1186, 0.2557	0.1504	0.0745	0.0579
25	2	$0,23^{-/}$	-3.4564,4.4564	-2.2675,2.2675	2.1451	0.7960	1.2621
25	2	$23,0^{+*}$	0.0400,0.9600	-0.4885,0.4885	0.7984	0.2529	0.3859
25	3	$0,0,22^{-/}$	-1.4931,-0.8173,3.3104	-1.1719,-0.6754,1.8474	0.8218	0.3839	0.5237
25	3	$22,0,0^{+*}$	0.1435,0.2261,0.6304	-0.4023,0.0330,0.3693	0.3741	0.1307	0.1608
25	4	$0,0,0,21^{-/}$	-0.8707,-0.4828,-0.3245.2.6780	-0.7970,-0.4740,-0.3419,1.6129	0.4489	0.2486	0.2991
25	4	$21,0,0,0^{+*}$	0.1682,0.1218,0.2553,0.4547	-0.3579,-0.0412,0.1012,0.2979	0.2480	0.0928	0.0943
25	5	$0,0,0,0,20^{-/}$	-0.5733,-0.3192,-0.2151,-0.1393,2.2469	-0.6056,-0.3687,-0.2714,-0.2006,1.4463	0.2878	0.1819	0.1954
25	5	$20,0,0,0,0^{+}$	0.1727,0.0888,0.1573,0.2307,0.3505	-0.3261,-0.0675,0.0263,0.1135,0.2539	0.1879	0.0742	0.0633
25	5	$0,20,0,0,0^{*}$	-0.1991,0.4730,0.1071,0.2181,0.4008	-0.2954,-0.0934,-0.0053,0.1077,0.2865	0.1988	0.0872	0.0895

(contd.)

Table 10.3: Coefficients, variances, and covariance of BLUEs for selected schemes for normal distribution (contd.)

n	m	Scheme	Coefficients (μ^*)	Coefficients (σ^*)	$\frac{Var(\mu^*)}{\sigma^2}$	$\frac{Var(\sigma^*)}{\sigma^2}$	$\frac{Cov(\mu^*,\sigma^*)}{\sigma^2}$
30	2	$0,28^{-/}$	-3.7822,4.7822	-2.3410,2.3410	2.3790	0.8032	1.3425
30	2	$28,0^{+*}$	0.0333,0.9667	-0.4732,0.4732	0.8216	0.2398	0.3859
30	2	$0,0,27^{-/}$	-0.1663,-0.9485,3.6107	-1.2094,-0.7182,1.9275	0.9234	0.3884	0.5655
30	3	$27,0,0^{+*}$	0.1371,0.2241,0.6388	-0.3925,0.0351,0.3574	0.3828	0.1235	0.1623
30	3	$0,0,0,26^{-/}$	-0.9873,-0.5718,-0.4050,2.9640	-0.8226,-0.5023,-0.3735,1.6984	0.5084	0.2521	0.3277
30	4	$26,0,0,0^{+}$	0.1632,0.1172,0.2574,0.4622	-0.3513,-0.0367,0.1006,0.2874	0.2525	0.0876	0.0962
30	4	$0,26,0,0^{*}$	-0.3773,0.6446,0.1928,0.5400	-0.3815,-0.0181,0.0616,0.3380	0.2971	0.1156	0.1469
40	2	$0,38^{-/}$	-4.3004,5.3004	-2.4530,2.4530	2.7642	0.8137	1.4660
40	2	$38,0^{+*}$	0.0250,0.9750	-0.4512,0.4512	0.8538	0.2210	0.3836
40	3	$0,0,37^{-/}$	-1.9307,-1.1611,4.0918	-1.2663,-0.7829,2.0492	1.0944	0.3949	0.6298
40	3	$37,0,0^{+*}$	0.1283,0.2222,0.6495	-0.3783,0.0382,0.3401	0.3952	0.1133	0.1632
40	4	$0,0,0,36^{-/}$	-1.1719,-0.7158,-0.5369,3.4245	-0.8613,-0.5453,-0.4212,1.8277	0.6109	0.2572	0.3717
40	4	$36,0,0,0^{+}$	0.1558,0.1116,0.2610,0.4717	-0.3415,-0.0303,0.0998,0.2720	0.2592	0.0802	0.0980
40	4	$0,36,0,0^{*}$	-0.4156,0.6649,0.1904,0.5603	-0.3643,-0.0210,0.0604,0.3250	0.3157	0.1075	0.1517
50	2	$0,48^{-/}$	-4.7054,5.7054	-2.5368,2.5368	3.0749	0.8211	1.5590
50	2	$48,0^{+*}$	0.0200,0.9800	-04357,0.4357	0.8743	0.2076	0.3799
50	3	$0,0,47^{-/}$	-2.1399,-1.3302,4.4701	-1.3087,-0.8312,2.1399	1.2350	0.3995	0.6782
50	3	$47,0,0^{+*}$	0.1224,0.2212,0.6564	-0.3680,0.0400,0.3280	0.4033	0.1061	0.1628

Table 10.4: Efficiencies and traces for trace-optimal censoring schemes: Extreme value distribution (Line 1: Efficiency; Line 2: Trace)

$n \setminus m$	2	3	4	5	6	7	8	9	10	11	12	$n-2$	$n-1$	n
15	3.9670	3.3962	2.7037	2.1875	1.8255	1.5872	1.4220	1.2972	1.2093	1.1425	1.0847	1.0553	1.0260	1.0000
	1.6588	0.6836	0.4438	0.3380	0.2776	0.2357	0.2053	0.1835	0.1662	0.1529	0.1420	0.1333	0.1261	0.1201
20	4.6589	4.1973	3.4382	2.8225	2.3645							1.0409	1.0202	1.0000
	1.7188	0.6886	0.4402	0.3319	0.2707							0.0958	0.0922	0.0891
25	5.2419	4.8953	4.0989	3.4122								1.0330	1.0168	1.0000
	1.7607	0.6924	0.4381	0.3282								0.0748	0.0727	0.0707
30	5.7497	5.5159	4.6976										1.0145	1.0000
	1.7915	0.6951	0.4368										0.0600	0.0587

Table 10.5: Efficiencies and determinants for determinant-optimal censoring schemes: Extreme value distribution (Line 1: Efficiency; Line 2: Determinant)

$n \setminus m$	2	3	4	5	6	7	8	9	10	11	12	$n-2$	$n-1$	n
15	2.9835	2.6736	2.3606	2.1174	1.9924	1.8555	1.7242	1.6030	1.4916	1.3927	1.2756	1.1997	1.1041	1.0000
	0.2089	0.0701	0.0373	0.0238	0.0162	0.0120	0.0094	0.0076	0.0063	0.0053	0.0046	0.0040	0.0035	0.0031
20	3.3404	3.0255	2.6794	2.4434	2.3216							1.1603	1.0842	1.0000
	0.1882	0.0628	0.0335	0.0212	0.0144							0.0020	0.0019	0.0017
25	3.6362	3.3149	2.9380	2.7145								1.1348	1.0711	1.0000
	0.1738	0.0578	0.0309	0.0193								0.0012	0.0011	0.0011
30	3.8911	3.5629	3.1571										1.0619	1.0000
	0.1630	0.0540	0.0289										0.0008	0.0007

Table 10.6: Coefficients, variances, and covariance of BLUEs for selected schemes for extreme value distribution

n	m	Scheme	Coefficients (μ^*)	Coefficients (σ^*)	$\frac{Var(\mu^*)}{\sigma^2}$	$\frac{Var(\sigma^*)}{\sigma^2}$	$\frac{Cov(\mu^*,\sigma^*)}{\sigma^2}$
15	2	$0,13^{-/}$	-2.1745,3.1745	-0.9663,0.9663	5.6144	0.9661	2.1911
15	2	$13,0^{+*}$	-0.1323,1.1323	-0.3447,0.3447	1.3716	0.2872	0.4302
15	3	$0,0,12^{-/}$	-0.8285,-0.7651,2.5936	-0.4799,-0.4573,0.9372	1.8474	0.4743	0.8299
15	3	$12,0,0^{+*}$	0.0300,0.0978,0.8792	-0.2507,-0.1109,0.3617	0.5211	0.1625	0.1208
15	4	$0,0,0,11^{-/}$	-0.4341,-0.4035,-0.3457,2.1833	-0.3166,-0.3076,-0.2798,0.9040	0.8898	0.3101	0.4334
15	4	$11,0,0,0^{+*}$	0.0555,0.0542,0.1960,0.6944	-0.2044,-0.1199,-0.0353,0.3596	0.3204	0.1234	0.0473
15	5	$0,0,0,0,10^{-/}$	-0.2576,-0.2398,-0.2054,-0.1590,1.8618	-0.2343,-0.2312,-0.2145,-0.1880,0.8680	0.5115	0.2278	0.2570
15	5	$10,0,0,0,0^{+}$	0.0623,0.0462,0.1060,0.2228,0.5627	-0.1727,-0.1124,-0.0759,0.0086,0.3524	0.2351	0.1029	0.0182
15	5	$0,10,0,0,0^{*}$	-0.0484,0.1683,0.0607,0.1858,0.6335	-0.1073,-0.1730,-0.0773,0.0018,0.3557	0.2592	0.1026	0.0524
15	6	$0,0,0,0,0,9^{-/}$	-0.1618,-0.1499,-0.1272,-0.0971.-0.0598, 1.5958	-0.1845,-0.1845,-0.1738,-0.1558,-0.1312, 0.8297	0.3284	0.1783	0.1618
15	6	$9,0,0,0,0,0^{+}$	0.0610,0.0431,0.0754,0.1316,0.2218, 0.4671	-0.1486,-0.1033,-0.0836,-0.0447,0.0364, 0.3438	0.1881	0.0895	0.0039
15	6	$0,9,0,0,0,0^{*}$	-0.0210,0.1413,0.0477,0.1022,0.2067, 0.5231	-0.0908,-0.1592,-0.0800,-0.0452,0.0338, 0.3414	0.1956	0.0863	0.0250
20	2	$0,18^{-/}$	-2.4828,3.4828	-0.9748,0.9748	7.0332	0.9747	2.4953
20	2	$18,0^{+*}$	-0.1330,1.1330	-0.3171,0.3171	1.4638	0.2550	0.4302
20	3	$0,0,17^{-/}$	-0.9844,-0.9322,2.9167	-0.4851,-0.4682,0.9532	2.4096	0.4808	0.9842
20	3	$17,0,0^{+*}$	0.0192,0.0870,0.8938	-0.2360,-0.0970,0.3330	0.5449	0.1438	0.1247
20	4	$0,0,0,16^{-/}$	-0.5393,-0.5145,-0.4660,2.5198	-0.3210,-0.3141,-0.2938,0.9289	1.1973	0.3161	0.5373
20	4	$16,0,0,0^{+*}$	0.5287,0.0440,0.1899,0.7132	-0.1961,-0.1059,-0.0290,0.3309	0.3305	0.1097	0.0524

(contd.)

Table 10.6: Coefficients, variances, and covariance of BLUEs for selected schemes for extreme value distribution (contd.)

n	m	Scheme	Coefficients(μ^*)	Coefficients(σ^*)	$\frac{Var(\mu^*)}{\sigma^2}$	$\frac{Var(\sigma^*)}{\sigma^2}$	$\frac{Cov(\mu^*,\sigma^*)}{\sigma^2}$
20	5	$0,0,0,0,15^{-/}$	-0.3374,-0.3234,-0.2943,-0.2553,2.2104	-0.2385,-0.2361,-0.2236,-0.2045,0.9028	0.7031	0.2337	0.3355
20	5	$15,0,0,0,0^{+}$	0.0614,0.0369,0.0982,0.2204,0.5831	-0.1686,-0.1000,-0.0664,0.0105,0.0325	0.2401	0.0918	0.0236
20	5	$0,15,0,0,0^{*}$	-0.0584,0.1689,0.0464,0.1779,0.6652	-0.0936,-0.1699,-0.0676,0.0038,0.3274	0.2757	0.0903	0.0610
20	6	$0,0,0,0,0,14^{-/}$	-0.2264,-0.2173,-0.1984,-0.1723,-0.1408, 1.9551	-0.1888,-0.1886,-0.1807,-0.1674,-0.1500, 0.8755	0.4560	0.1841	0.2248
20	6	$14,0,0,0,0,0^{+}$	0.0618,0.0349,0.0674,0.1263,0.2215, 0.4881	-0.1475,-0.0925,-0.0741,-0.0383,0.0354, 0.3170	0.1905	0.0802	0.0092
20	6	$0,14,0,0,0,0^{*}$	-0.0302,0.1449,0.0349,0.0916,0.2040, 0.5548	-0.0789,-0.1590.-0.0699,-0.0385,0.0330, 0.3133	0.2042	0.0757	0.0328
25	2	$0,23^{-/}$	-2.7197,3.7197	-0.9799,0.9799	8.2493	0.9799	2.7296
25	2	$23,0^{+*}$	-0.1321,1.1321	-0.2982,0.2982	1.5277	0.2330	0.4268
25	3	$0,0,22^{-/}$	-1.1039,-1.0595,3.1633	-0.4881,-0.4747,0.9628	2.9046	0.4847	1.1030
25	3	$22,0,0^{+*}$	0.0169,0.0804,0.9027	-0.2256,-0.0876,0.3132	0.5613	0.1311	0.1258
25	4	$0,0,0,21^{-/}$	-0.6198,-0.5988,-0.5568.2.7754	-0.3235,-0.3181,-0.3019,0.9435	1.4762	0.3196	0.6173
25	4	$21,0,0,0^{+*}$	0.0509,0.0376,0.1868,0.7247	-0.1900,-0.0965,-0.0246,0.3111	0.3377	0.1004	0.0549
25	5	$0,0,0,0,20^{-/}$	-0.3985,-0.3868,-0.3616,-0.3271,2.4740	-0.2410,-0.2390,-0.2291,-0.2139,0.9230	0.8828	0.2371	0.3960
25	5	$20,0,0,0,0^{+}$	0.0604,0.0310,0.0934,0.2196,0.5955	-0.1653,-0.0916,-0.0601,0.0119,0.3051	0.2438	0.0844	0.0265
25	5	$0,20,0,0,0^{*}$	-0.0635,0.1673,0.0376,0.1737,0.6850	-0.0844,-0.1673,-0.0612,0.0053,0.3076	0.2882	0.0819	0.0653
30	2	$0,28^{-/}$	-2.9118,3.9118	-0.9832,0.9832	9.3175	0.9832	2.9201
30	2	$28,0^{+*}$	-0.1307,1.1307	-0.2842,0.2842	1.5748	0.2167	0.4222
30	3	$0,0,27^{-/}$	-1.2007,-1.1617,3.3624	-0.4901,-0.4790,0.9691	3.3470	0.4873	1.1994

(contd.)

Table 10.6: Coefficients, variances, and covariance of BLUEs for selected schemes for extreme value distribution (contd.)

n	m	Scheme	Coefficients(μ^*)	Coefficients(σ^*)	$\frac{Var(\mu^*)}{\sigma^2}$	$\frac{Var(\sigma^*)}{\sigma^2}$	$\frac{Cov(\mu^*,\sigma^*)}{\sigma^2}$
30	3	$27,0,0^{+*}$	0.0153,0.0760,0.9087	-0.2177,-0.0808,0.2985	0.5733	0.1218	0.1258
30	4	$0,0,0,26^{-/}$	-0.6850,-0.6668,-0.6292,2.9810	-0.3252,-0.3207,-0.3072,0.9531	1.7297	0.3220	0.6823
30	4	$26,0,0,0^{+*}$	0.0493,0.0332,0.1849,0.7325	-0.1852,-0.0896,-0.0215,0.2962	0.3432	0.0936	0.0563

Table 10.7: Efficiencies and traces for trace-optimal censoring schemes: Extreme value (II) distribution (Line 1: Efficiency; Line 2: Trace)

$n \setminus m$	2	3	4	5	6	7	8	9	10	11	12	$n-2$	$n-1$	n
15	1.3747	1.2061	1.1060	1.0347	1.0153	1.0018	1.0000	1.0000	1.0000	1.0000	1.0000	1.0000	1.0000	1.0000
	1.0056	0.5117	0.3529	0.2748	0.2282	0.1968	0.1744	0.1585	0.1470	0.1384	0.1319	0.1269	0.1230	0.1201
20	1.5030	1.3221	1.1991	1.1181	1.0652							1.0000	1.0000	1.0000
	1.0130	0.5064	0.3455	0.2664	0.2193							0.0925	0.0906	0.0891
25	1.6155	1.4256	1.2884	1.1925								1.0000	1.0000	1.0000
	1.0141	0.5026	0.3406	0.2612								0.0728	0.0717	0.0707
30	1.7132	1.5172	1.3705										1.0000	1.0000
	1.0135	0.4995	0.3371										0.0593	0.0587

Table 10.8: Efficiencies and determinants for determinant-optimal censoring schemes: Extreme value (II) distribution (Line 1: Efficiency; Line 2: Determinant)

$n \setminus m$	2	3	4	5	6	7	8	9	10	11	12	$n-2$	$n-1$	n
15	1.2443	1.1067	1.0435	1.0042	1.0050	1.0001	1.0000	1.0000	1.0000	1.0000	1.0000	1.0000	1.0000	1.0000
	0.0548	0.0256	0.0162	0.0116	0.0090	0.0073	0.0061	0.0053	0.0047	0.0042	0.0038	0.0036	0.0033	0.0031
20	1.3143	1.1538	1.0740	1.0418	1.0230							1.0000	1.0000	1.0000
	0.0433	0.0201	0.0127	0.0090	0.0069							0.0019	0.0018	0.0017
25	1.3719	1.1941	1.1031	1.0659								1.0000	1.0000	1.0000
	0.0364	0.0168	0.0106	0.0075								0.0011	0.0011	0.0011
30	1.4209	1.2285	1.1286										1.0000	1.0000
	0.0318	0.0146	0.0092										0.0008	0.0007

Table 10.9: Coefficients, variances, and covariance of BLUEs for selected schemes for extreme value (II) distribution

n	m	Scheme	Coefficients (μ^*)	Coefficients (σ^*)	$\frac{Var(\mu^*)}{\sigma^2}$	$\frac{Var(\sigma^*)}{\sigma^2}$	$\frac{Cov(\mu^*,\sigma^*)}{\sigma^2}$
15	2	$0,13^{-/}$	-2.2897,3.2897	-2.9044,2.9044	0.6754	0.7070	0.6397
15	2	$12,1^{+}$	0.0345,0.9655	-0.8524,0.8524	0.5487	0.4569	0.4422
15	2	11,2*	-0.2429,1.2429	-1.0973,1.0973	0.5352	0.4732	0.4455
15	3	$0,0,12^{-}$	-0.8766,-0.1520,2.0286	-1.5323,-0.4373,1.9697	0.2815	0.3356	0.2573
15	3	$9,0,3^{+*}$	-0.0469,0.1601,0.8868	-0.8577,-0.0146,0.8724	0.2649	0.2468	0.1995
15	3	$12,0,0^{/}$	0.3750,0.3218,0.3032	-0.5171,0.2187,0.2985	0.3310	0.2459	0.2279
15	4	0,0,0,11	-0.4284,-0.0444,0.0539,1.4188	-1.0582,-0.3234,-0.1193,1.5010	0.1745	0.2159	0.1441
15	4	$7,0,0,4^{+}$	0.0054,0.1020,0.1645,0.7281	-0.7435,-0.1048,0.0653,0.7830	0.1798	0.1732	0.1220
15	4	$0,11,0,0^{-}$	-0.3687,0.9472,0.2217,0.1998	-0.8797,0.4617,0.1874,0.2306	0.2122	0.1949	0.1531
15	4	6,0,0,5*	-0.0671,0.0768,0.1558,0.8344	-0.7960,-0.1433,0.0444,0.8950	0.1762	0.1788	0.1237
15	4	$11,0,0,0^{/}$	0.3623,0.2186,0.2325,0.1886	-0.4871,0.0847,0.2008,0.2015	0.2338	0.1693	0.1430
15	5	0,0,0,0,10	-0.2096,0.0118,-0.0575,0.3192,0.9361	-0.8104,-0.2599,-0.2455,0.2555,1.0602	0.1280	0.1563	0.0915
15	5	$5,0,0,0,5^{+}$	0.0202,0.0851,0.1154,0.1441,0.6352	-0.6646,-0.1417,-0.0124,0.0787,0.7401	0.1383	0.1365	0.0846
15	5	$10,0,0,0,0^{-/}$	0.3450,0.1755,0.1834,0.1668,0.1293	-0.4597,0.0181,0.1268,0.1624,0.1524	0.1848	0.1307	0.1010
15	5	0,0,0,3,7*	-0.2051,0.0137,-0.0064,0.4301,0.7677	-0.8006,-0.2564,-0.1889,0.3603,0.8857	0.1286	0.1547	0.0911
15	6	0,0,0,0,0,9	-0.0949,0.0434,0.0595,0.1090.0.1175, 0.7655	-0.6652,-0.2199,-0.0973,-0.0106,0.0237, 0.9693	0.1078	0.1238	0.0658
15	6	$3,0,0,0,0,6^{+}$	0.0202,0.0767,0.0983,0.1142,0.1133, 0.5772	-0.6020,-0.1630,-0.0469,0.0221,0.0646, 0.7252	0.1138	0.1143	0.0629

(contd.)

Table 10.9: Coefficients, variances, and covariance of BLUEs for selected schemes for extreme value (II) distribution (contd.)

n	m	Scheme	Coefficients (μ^*)	Coefficients(σ^*)	$\frac{Var(\mu^*)}{\sigma^2}$	$\frac{Var(\sigma^*)}{\sigma^2}$	$\frac{Cov(\mu^*,\sigma^*)}{\sigma^2}$
15	6	$9,0,0,0,0,0^{-/}$	0.3254,0.1545,0.1549,0.1441,0.1250, 0.0960	-0.4348,-0.0203,0.0777,0.1206,0.1337, 0.1231	0.1552	0.1074	0.0761
15	6	$0,3,0,0,0,6^{*}$	-0.0761,0.1797,0.0902,0.1105,0.1132, 0.5825	-0.6254,-0.1337,-0.0576,0.0172,0.0613, 0.7381	0.1121	0.1179	0.0652
20	2	$0,18^{-/}$	-2.9299,3.9299	-3.2127,3.2127	0.8055	0.7171	0.7215
20	2	$16,2^{+*}$	-0.2686,1.2686	-1.0371,1.0371	0.5755	0.4376	0.4566
20	3	$0,0,17^{-}$	-1.2170,-0.3252,2.5422	-1.6926,-0.5636,2.2562	0.3283	0.3412	0.2981
20	3	$14,0,3^{+}$	-0.0662,0.1483,0.9180	-0.8177,-0.0084,0.8261	0.2798	0.2266	0.2080
20	3	$13,0,4^{*}$	-0.1775,0.1076,1.0699	-0.9020,-0.0601,-0.9621	0.2765	0.2342	0.2113
20	3	$17,0,0^{/}$	0.3510,0.3295,0.3194	-0.5027,0.2157,0.2869	0.3496	0.2319	0.2395
20	4	0,0,0,16	-0.6673,-0.1647,-0.0307,1.8627	-1.1681,-0.4105,-0.1988,1.7773	0.1947	0.2196	0.1706
20	4	$12,0,0,4^{+}$	-0.0085,0.0895,0.1557,0.7633	-0.7158,-0.0885,0.0627,0.7415	0.1870	0.1584	0.1298
20	4	$0,0,16,0^{-}$	-0.6285,-0.1381,1.5432,0.2234	-1.0651,-0.3640,1.1910,0.2382	0.2113	0.2082	0.1733
20	4	$9,2,0,5^{*}$	-0.1872,0.1797,0.1309,0.8766	-0.8236,-0.0595,0.0260,0.8571	0.1837	0.1691	0.1354
20	4	$16,0,0,0^{/}$	0.3432,0.2151,0.2434,0.1983	-0.4758,0.0864,0.1968,0.1926	0.2423	0.1594	0.1523
20	5	0,0,0,0,15	-0.4028,-0.0834,0.0003,0.0497,1.4363	-0.8987,-0.3277,-0.1672,-0.0695,1.4631	0.1376	0.1603	0.1124
20	5	$10,0,0,0,5^{+}$	0.0123,0.0682,0.1116,0.1330,0.6749	-0.6457,-0.1258,0.0008,0.0696,0.7012	0.1419	0.1246	0.0919
20	5	$0,15,0,0,0^{-}$	-0.3179,0.8208,0.1658,0.1809,0.1504	-0.7066,0.2764,0.0971,0.1678,0.1653	0.1735	0.1393	0.1180
20	5	$0,10,0,0,5^{*}$	-0.3047,0.3958,0.0885,0.1222,0.6983	-0.7524,-0.0128,-0.0303,0.0559,0.7395	0.1412	0.1376	0.1019

(contd.)

Table 10.9: Coefficients, variances, and covariance of BLUEs for selected schemes for extreme value (II) distribution (contd.)

n	m	Scheme	Coefficients (μ^*)	Coefficients (σ^*)	$\frac{Var(\mu^*)}{\sigma^2}$	$\frac{Var(\sigma^*)}{\sigma^2}$	$\frac{Cov(\mu^*,\sigma^*)}{\sigma^2}$
20	5	15,0,0,0,0$^{/}$	0.3310,0.1658,0.1880,0.1763,0.1389	-0.4513,0.0220,0.1267,0.1577,0.1450	0.1885	0.1229	0.1091
20	6	0,0,0,0,0,14	-0.2500,-0.0350,0.0206,0.0528,0.0739, 1.1378	-0.7333,-0.2752,-0.1453,-0.0662,-0.0118, 1.2317	0.1080	0.1256	0.0803
20	6	8,0,0,0,0,6^{+}	0.0183,0.0591,0.0890,0.1048,0.1124, 0.6164	-0.5901,-0.1462,-0.0363,0.0263,0.0656, 0.6806	0.1152	0.1041	0.0697
20	6	14,0,0,0,0,0$^{-/}$	0.3166,0.1412,0.1544,0.1501,0.1334, 0.1043	-0.4290,-0.0152,0.0796,0.1188,0.1291, 0.1168	0.1561	0.1008	0.0834
20	6	0,7,0,0,0,7*	-0.1955,0.2323,0.0666,0.0898,0.1037, 0.7031	-0.6471,-0.1185,-0.0698,0.0027,0.0505, 0.7821	0.1120	0.1128	0.0754
25	2	0,23$^{-/}$	-3.4458,4.4458	-3.4510,3.4510	0.9135	0.7248	0.7824
25	2	21,2^{+*}	-0.2815,1.2815	-0.9947,0.9947	0.6021	0.4120	0.4601
25	3	0,0,22^{-}	-1.4907,-0.4731,2.9639	-1.8166,-0.6613,2.4778	0.3710	0.3456	0.3288
25	3	19,0,3^{+}	-0.0777,0.1432,0.9345	-0.7904,-0.0019,0.7922	0.2903	0.2123	0.2115
25	3	18,0,4*	-0.1868,0.1013,1.0855	-0.8693,-0.0506,0.9199	0.2870	0.2186	0.2143
25	3	22,0,0$^{/}$	0.3350,0.3354,0.3296	-0.4924,0.2139,0.2785	0.3639	0.2219	0.2459
25	4	0,0,0,21^{-}	-0.8584,-0.2685,-0.1068.2.2337	-1.2527,-0.4788,-0.2602,1.9918	0.2163	0.2226	0.1909
25	4	17,0,0,4^{+}	-0.0170,0.0806,0.1557,0.7806	-0.6965,-0.0788,0.0662,0.7091	0.1926	0.1480	0.1333
25	4	14,1,0,6*	-0.2020,0.0883,0.1166,0.9971	-0.8129,-0.1082,0.0136,0.9075	0.1882	0.1587	0.1388
25	4	21,0,0,0$^{/}$	0.3300,0.2136,0.2510,0.2054	-0.4679,0.0880,0.1939,0.1859	0.2497	0.1523	0.1576
25	5	0,0,0,0,20	-0.5521,-0.1641,-0.0586,0.0029,1.7720	-0.9634,-0.3802,-0.2147,-0.1152,1.6736	0.1490	0.1626	0.1273

(contd.)

Table 10.9: Coefficients, variances, and covariance of BLUEs for selected schemes for extreme value (II) distribution (contd.)

n	m	Scheme	Coefficients (μ^*)	Coefficients (σ^*)	$\frac{Var(\mu^*)}{\sigma^2}$	$\frac{Var(\sigma^*)}{\sigma^2}$	$\frac{Cov(\mu^*,\sigma^*)}{\sigma^2}$
25	5	$15,0,0,0,5^{+}$	0.0070,0.0575,0.1084,0.1338,0.6933	-0.6324,-0.1147,0.0068,0.0714,0.6690	0.1451	0.1161	0.0954
25	5	$0,0,20,0,0^{-}$	-0.4921,-0.1227,1.2696,0.1777,0.1675	-0.8245,-0.3128,0.8093,0.1502,0.1778	0.1702	0.1483	0.1300
25	5	$0,14,0,0,6^{*}$	-0.4041,0.4140,0.0656,0.1097,0.8148	-0.7743,-0.0178,-0.0476,0.0401,0.7996	0.1456	0.1335	0.1093
25	5	$20,0,0,0,0^{/}$	0.3209,0.1602,0.1918,0.1825,0.1446	-0.4455,0.0252,0.1266,0.1542,0.1396	0.1925	0.1173	0.1138
30	2	$0,28^{-/}$	-3.8786,4.8786	-3.6450,3.6450	1.0052	0.7310	0.8305
30	2	$26,2^{+*}$	-0.2886,1.2886	-0.9627,0.9627	0.6210	0.3925	0.4604
30	3	$0,0,27^{-/}$	-1.7192,-0.6045,3.3237	-1.9168,-0.7431,2.6599	0.4089	0.3490	0.3532
30	3	$24,0,3^{+}$	-0.0850,0.1405,0.9445	-0.7698,0.0034,0.7664	0.2980	0.2015	0.2129
30	3	$22,0,5^{*}$	-0.2909,0.0592,1.2317	-0.9138,-0.0850,0.9989	0.2943	0.2130	0.2192
30	4	$0,0,0,26^{-}$	-1.0180,-0.3584,-0.1774,2.5538	-1.3215,-0.5341,-0.3124,2.1680	0.2369	0.2251	0.2072
30	4	$22,0,0,4^{+}$	-0.0228,0.0752,0.1565,0.7912	-0.6820,-0.0712,0.0690,0.6843	0.1969	0.1402	0.1350
30	4	$18,1,0,7^{*}$	-0.2610,0.0561,0.0997,1.1052	-0.8302,-0.1256,-0.0015,0.9572	0.1924	0.1533	0.1425
30	4	$26,0,0,0^{/}$	0.3201,0.2132,0.2566,0.2101	-0.4619,0.0895,0.1916,0.1808	0.2558	0.1468	0.1610

Table 10.10: Efficiencies and traces for trace-optimal censoring schemes: Log-normal distribution (Line 1: Efficiency; Line 2: Trace)

$n \setminus m$	2	3	4	5	6	7	8	9	10	11	12	$n-2$	$n-1$	n
15	1.0018	1.0000	1.0000	1.0000	1.0000	1.0000	1.0000	1.0000	1.0000	1.0000	1.0000	1.0000	1.0000	1.0000
	0.8366	0.4325	0.3019	0.2385	0.2014	0.1774	0.1607	0.1486	0.1396	0.1327	0.1274	0.1234	0.1205	0.1187
20	1.0000	1.0007	1.0000	1.0253	1.0000									
	0.6751	0.3756	0.2697	0.2192	0.1739									

Table 10.11: Efficiencies and determinants for determinant-optimal censoring schemes: Log-normal distribution (Line 1: Efficiency; Line 2: Determinant)

$n \setminus m$	2	3	4	5	6	7	8	9	10	11	12	$n-2$	$n-1$	n
15	1.0000	1.0000	1.0000	1.0000	1.0000	1.0000	1.0000	1.0000	1.0000	1.0000	1.0000	1.0000	1.0000	1.0000
	0.0086	0.0043	0.0030	0.0023	0.0020	0.0017	0.0015	0.0014	0.0013	0.0013	0.0012	0.0012	0.0011	0.0011
20	1.0000	1.0000	1.0000	1.0171	1.0000									
	0.0048	0.0026	0.0018	0.0015	0.0011									

Table 10.12: Coefficients, variances, and covariance of BLUEs for selected schemes for log-normal distribution

n	m	Scheme	Coefficients (μ^*)	Coefficients (σ^*)	$\frac{Var(\mu^*)}{\sigma^2}$	$\frac{Var(\sigma^*)}{\sigma^2}$	$\frac{Cov(\mu^*,\sigma^*)}{\sigma^2}$
15	2	0,13	2.8333,-1.8333	-9.0515,9.0515	0.0502	0.7879	-0.1759
15	2	$2,11^{+*}$	2.6045,-1.6045	-7.9216,7.9216	0.0496	0.7870	-0.1744
15	2	$13,0^{-/}$	1.1307,-0.1307	-0.6454,0.6454	0.0939	2.0131	-0.4081
15	3	$0,0,12^{+*}$	1.9257,0.0547,-0.9804	-4.8240,0.2576,4.5664	0.0328	0.4002	-0.0927
15	3	$12,0,0^{-/}$	1.1959,-0.1559,-0.0400	-0.9784,0.7854,0.1930	0.0467	0.8522	-0.1740
15	4	$0,0,0,11^{+*}$	1.6020,0.0732,-0.0333,-0.6418	-3.3829,0.1754,0.3505,2.8570	0.0260	0.2759	-0.0647
15	4	$11,0,0,0^{-/}$	1.2129,-0.1163,-0.0756,-0.0210	-1.0925,0.6302,0.3630,0.0993	0.0338	0.5332	-0.1097
15	5	$0,0,0,0,10^{+*}$	1.4313,0.0819,-0.0161,-0.0437,-0.4533	-2.6500,0.1378,0.2765,0.2894,1.9463	0.0227	0.2157	-0.0507
15	5	$10,0,0,0,0^{-/}$	1.2143,-0.0807,-0.0741,-0.0459,-0.0136	-1.1438,0.5038,0.3626,0.2142,0.0631	0.0278	0.3869	-0.0803
15	6	$0,0,0,0,0,9^{+*}$	1.3258,0.0860,-0.0053,-0.0331,-0.0442, -0.3292	-2.2088,0.1206,0.2314,0.2450,0.2349, 1.3770	0.0207	0.1807	-0.0423
15	6	$9,0,0,0,0,0^{-/}$	1.2084,-0.0509,-0.0657,-0.0503,-0.0317, -0.0098	-1.1695,0.4091,0.3361,0.2351,0.1445, 0.0448	0.0244	0.3032	-0.0634
20	2	$0,18^{+*}$	2.9753,-1.9753	-11.2580,11.2580	0.0311	0.6440	-0.1236
20	2	$18,0^{-/}$	1.1131,-0.1131	-0.6448,0.6448	0.0691	1.9918	-0.3501
20	3	0,0,17	2.1821,-0.2837,-0.8985	-6.7690,1.6841,5.0849	0.0220	0.3539	-0.0723
20	3	$0,1,16^{+*}$	2.1736,-0.3122,-0.8614	-6.7289,1.8622,4.8667	0.0220	0.3536	-0.0722
20	3	$17,0,0^{-/}$	1.1700,-1.3589,-0.0342	-0.9791,0.7884,0.1907	0.0339	0.8391	-0.1487
20	4	$0,0,0,16^{+*}$	1.7962,-0.0965,-0.0480,-0.6518	-4.7148,0.6877,0.5573,3.4698	0.0184	0.2513	-0.0530

(contd.)

Table 10.12: Coefficients, variances, and covariance of BLUEs for selected schemes for log-normal distribution (contd.)

n	m	Scheme	Coefficients (μ^*)	Coefficients (σ^*)	$\frac{Var(\mu^*)}{\sigma^2}$	$\frac{Var(\sigma^*)}{\sigma^2}$	$\frac{Cov(\mu^*,\sigma^*)}{\sigma^2}$
20	4	16,0,0,0$^{-/}$	1.1859,-0.1033,-0.0648,-0.0178	-1.0958,0.6376,0.3603,0.0978	0.0243	0.5241	-0.0937
20	5	0,0,0,0,15	1.5430,0.0329,-0.0125,-0.0636,-0.4999	-3.3032,-0.0335,0.3592,0.1909,2.7865	0.0170	0.2077	-0.0452
20	5	0,0,0,9,6^{+*}	1.5860,-0.0490,-0.0095,-0.2954,-0.2322	-3.6876,0.4555,0.3694,1.7284,1.1344	0.0164	0.2028	-0.0431
20	5	15,0,0,0,0$^{-/}$	1.1888,-0.0743,-0.0638,-0.0391,-0.0115	-1.1499,0.5144,0.3617,0.2117,0.0620	0.0199	0.3798	-0.0685
20	6	0,0,0,0,0,14^{+*}	1.4997,-0.1558,0.0627,-0.0876,0.2712, -0.5903	-3.1263,0.7367,0.0523,0.2888,-0.3612, 2.4098	0.0141	0.1597	-0.0335
20	6	14,0,0,0,0,0$^{-/}$	1.1857,-0.0511,-0.0562,-0.0434,-0.0268, -0.0083	-1.1786,0.4223,0.3363,0.2337,0.1423, 0.0440	0.0174	0.2975	-0.0541

11

Acceptance Sampling Plans

11.1 Introduction

In this Chapter, we discuss the construction of acceptance sampling plans based on progressively Type-II right censored samples. For this purpose, we may either use exact (ML/BLU) estimators of the parameters and their exact distributional properties or use first-order approximate estimators of the parameters and their distributional properties. We consider the exponential and log-normal distributions for illustration with the exponential being an example of the first kind and the log-normal being an example of the second kind.

Owen (1964, 1969) discussed acceptance sampling plans for normal and non-normal distributions, while Kocherlakota and Balakrishnan (1984) proposed robust acceptance sampling plans. Hosono and Kase (1981), Kocherlakota and Balakrishnan (1986), and Balasooriya (1995) all discussed acceptance sampling plans for the exponential distribution. Fertig and Mann (1980) developed sampling plans for Weibull and extreme value distributions. Schneider (1989) discussed failure-censored sampling plans for Weibull and log-normal distributions. Final mention should be made to the recent work of Balasooriya and Balakrishnan (2000) who have developed life-test sampling plans for the log-normal distribution based on progressively Type-II right censored samples.

11.2 The Exponential Distribution

Let us consider the two-parameter exponential distribution with probability density function

$$f(y;\mu,\sigma) = \frac{1}{\sigma}\, e^{-(y-\mu)/\sigma}, \qquad y \geq \mu,\ \sigma > 0, \tag{11.1}$$

where μ is the location parameter and σ is the scale parameter. In the present context, μ is the guarantee period and σ is the mean residual lifetime.

11.2.1 One-Sided Sampling Plans

For one-sided acceptance sampling plans, based on the progressively Type-II right censored sample $Y_{1:m:n}^{(R_1,\cdots,R_m)}, \cdots, Y_{m:m:n}^{(R_1,\cdots,R_m)}$ observed from a lot (denoted from now on by $Y_{1:m:n}, \cdots, Y_{m:m:n}$, for convenience), the lot is accepted if either $\phi_1(Y_{1:m:n}, \cdots, Y_{m:m:n}) \geq L$ or $\phi_2(Y_{1:m:n}, \cdots, Y_{m:m:n}) \leq U$ where the functions ϕ_1 and ϕ_2 are determined so that in each case certain probability statements are satisfied. Depending on whether a right-sided or left-sided acceptance sampling plan is desired, one of the above two inequalities will be used. Let $K_p = -\log(1-p)$ and $K_{1-p} = -\log p$ denote the lower and upper p-percentage points of the standard exponential distribution, respectively. Then, from the equations

$$\Pr(Y \leq L) = p \quad \text{and} \quad \Pr(Y \geq U) = p,$$

we have

$$K_p = \frac{L-\mu}{\sigma} \quad \text{and} \quad K_{1-p} = \frac{U-\mu}{\sigma}\,.$$

Case 1: μ unknown and σ known

We will accept the lot if either $\hat{\mu} + k_1\sigma \geq L$ or $\hat{\mu} + k_2\sigma \leq U$, where $\hat{\mu} = Y_{1:m:n}$ is the MLE of μ and k_1 and k_2 need to be determined so that

$$\Pr(\hat{\mu} + k_1\sigma \geq L) = 1-\gamma \text{ and } \Pr(\hat{\mu} + k_2\sigma \leq U) = 1-\gamma. \tag{11.2}$$

The first probability statement in (11.2) is equivalent to

$$\Pr\left\{\frac{2n(\hat{\mu}-\mu)}{\sigma} \leq 2n(K_p - k_1)\right\} = \gamma. \tag{11.3}$$

Since $\frac{2n(\hat{\mu}-\mu)}{\sigma} = \frac{2n(X_{1:m:n}-\mu)}{\sigma} \stackrel{d}{=} \chi_2^2$, (11.3) readily yields

$$k_1 = K_p + \frac{1}{n}\,\log(1-\gamma) = -\log(1-p) + \frac{1}{n}\,\log(1-\gamma). \tag{11.4}$$

The second probability statement in (11.2) is equivalent to

$$\Pr\left\{\frac{2n(\hat{\mu}-\mu)}{\sigma} \leq 2n(K_{1-p} - k_2)\right\} = 1-\gamma$$

which readily yields

$$k_2 = K_{1-p} + \frac{1}{n}\,\log\gamma = -\log p + \frac{1}{n}\,\log\gamma. \tag{11.5}$$

Case 2: μ known and σ unknown

We will accept the lot if either $\mu + k_1^*\hat{\sigma} \geq L$ or $\mu + k_2^*\hat{\sigma} \leq U$, where $\hat{\sigma} = \frac{1}{m}\sum_{i=1}^{m}(R_i+1)(Y_{i:m:n}-\mu)$ is the MLE of σ and k_1^* and k_2^* need to be determined so that

$$\Pr(\mu + k_1^*\hat{\sigma} \geq L) = 1-\gamma \text{ and } \Pr(\mu + k_2^*\hat{\sigma} \leq U) = 1-\gamma. \tag{11.6}$$

The first probability statement in (11.6) is equivalent to

$$\Pr\left(\frac{2m\hat{\sigma}}{\sigma} \leq \frac{2mK_p}{k_1^*}\right) = \gamma. \tag{11.7}$$

Since $\frac{2m\hat{\sigma}}{\sigma} = 2\sum_{i=1}^{m}(R_i+1)\left(\frac{Y_{i:m:n}-\mu}{\sigma}\right) \stackrel{d}{=} \chi^2_{2m}$, (11.7) readily yields

$$k_1^* = \frac{2mK_p}{\chi^2_{2m,\gamma}} = -\frac{2m\log(1-p)}{\chi^2_{2m,\gamma}}, \tag{11.8}$$

where $\chi^2_{2m,\gamma}$ is the γ-percentage point of the chi-square distribution with $2m$ degrees of freedom, i.e.,

$$\int_0^{\chi^2_{2m,\gamma}} f_{\chi^2_{2m}}(x)dx = \gamma.$$

Similarly, the second probability statement in (11.6) is equivalent to

$$\Pr\left(\frac{2m\hat{\sigma}}{\sigma} \leq \frac{2mK_{1-p}}{k_2^*}\right) = 1-\gamma$$

which readily yields

$$k_2^* = \frac{2mK_{1-p}}{\chi^2_{2m,1-\gamma}} = -\frac{2m\log p}{\chi^2_{2m,1-\gamma}}. \tag{11.9}$$

Case 3: μ and σ both unknown

In this case, we will accept the lot if either $\hat{\mu}+\ell_1\hat{\sigma} \geq L$ or $\hat{\mu}+\ell_2\hat{\sigma} \leq U$, where $\hat{\mu} = Y_{1:m:n}$ and $\hat{\sigma} = \frac{1}{m}\sum_{i=2}^{m}(R_i+1)(Y_{i:m:n}-Y_{1:m:n})$ are the MLEs of μ and σ, respectively, and ℓ_1 and ℓ_2 need to be determined so that

$$\Pr(\hat{\mu}+\ell_1\hat{\sigma} \geq L) = 1-\gamma \text{ and } \Pr(\hat{\mu}+\ell_2\hat{\sigma} \leq U) = 1-\gamma. \tag{11.10}$$

The first probability statement in (11.10) is equivalent to

$$\Pr(Z_1 + \frac{n}{m}\,\ell_1 Z_2 \geq 2nK_p) = 1-\gamma, \tag{11.11}$$

where $Z_1 \stackrel{d}{=} \chi^2_2$ and $Z_2 \stackrel{d}{=} \chi^2_{2m-2}$ independently. Eq. (11.11) can be rewritten as

$$\gamma = \int_0^{2mK_p/\ell_1} f_{\chi^2_{2m-2}}(z)\left(1-e^{-(2nK_p-\frac{n}{m}\,\ell_1 z)/2}\right)dz,$$

where $f_{\chi^2_{2m-2}}(z)$ denotes the probability density function of the chi-square distribution with $2m-2$ degrees of freedom. Upon substituting for $f_{\chi^2_{2m-2}}(z)$ and integrating the resulting expression, we obtain the equation

$$\gamma = F_{\chi^2_{2m-2}}\left(\frac{2mK_p}{\ell_1}\right) - \frac{e^{-nK_p}}{(1-\frac{n}{m}\,\ell_1)^{m-1}}F_{\chi^2_{2m-2}}\left(\frac{2mK_p}{\ell_1}\left(1-\frac{n}{m}\,\ell_1\right)\right), \tag{11.12}$$

where $K_p = -\log(1-p)$ and $F_{\chi^2_{2m-2}}(\cdot)$ denotes the distribution function of χ^2_{2m-2}. Eq. (11.12) needs to be solved iteratively for ℓ_1.

Similarly, the second probability statement in (11.10) is equivalent to

$$\Pr(Z_1 + \frac{n}{m}\,\ell_2 Z_2 \le 2nK_{1-p}) = 1-\gamma$$

which can be rewritten as

$$\begin{aligned} 1-\gamma &= \int_0^{2mK_{1-p}/\ell_2} f_{\chi^2_{2m-2}}(z)\left(1-e^{-(2nK_{1-p}-\frac{n}{m}\,\ell_2 z)/2}\right)dz \\ &= F_{\chi^2_{2m-2}}\left(\frac{2mK_{1-p}}{\ell_2}\right) - \frac{e^{-nK_{1-p}}}{(1-\frac{n}{m}\,\ell_2)^{m-1}} F_{\chi^2_{2m-2}}\left(\frac{2mK_{1-p}}{\ell_2}(1-\frac{n}{m}\,\ell_2)\right), \end{aligned} \tag{11.13}$$

where $K_{1-p} = -\log p$. Eq. (11.13) needs to be solved iteratively for ℓ_2.

11.2.2 Two-Sided Sampling Plans

Two-sided acceptance sampling plans call for acceptance of the lot if $L \le \phi(Y_{1:m:n}, \cdots, Y_{m:m:n}) \le U$, where the function ϕ is determined so that a certain probability statement is satisfied.

Case 1: μ unknown and σ known

We will accept the lot if $L \le \hat{\mu} + k\sigma \le U$, where $\hat{\mu} = Y_{1:m:n}$ is the MLE of μ and k needs to be determined so that

$$\Pr(L \le \hat{\mu} + k\sigma \le U) = 1-\gamma. \tag{11.14}$$

Since $\frac{2n(\hat{\mu}-\mu)}{\sigma} = \frac{2n(Y_{1:m:n}-\mu)}{\sigma} \stackrel{d}{=} \chi^2_2$, we can rewrite (11.14) as

$$\Pr\left\{2n(K_{p_1}-k) \le \chi^2_2 \le 2n(K_{1-p_2}-k)\right\} = 1-\gamma$$

which readily yields

$$e^{-n(K_{p_1}-k)} - e^{-n(K_{1-p_2}-k)} = 1-\gamma$$

or

$$k = \frac{1}{n}\,\log\left(\frac{1-\gamma}{e^{-nK_{p_1}} - e^{-nK_{1-p_2}}}\right), \tag{11.15}$$

where $K_{p_1} = \frac{L-\mu}{\sigma} = -\log(1-p_1)$ and $K_{1-p_2} = \frac{U-\mu}{\sigma} = -\log p_2$.

Case 2: μ known and σ unknown

We will accept the lot if $L \leq \mu + k^*\hat{\sigma} \leq U$, where $\hat{\sigma} = \frac{1}{m}\sum_{i=1}^{m}(R_i+1)(Y_{i:m:n} - \mu)$ is the MLE of σ and K^* needs to be determined so that

$$\Pr(L \leq \mu + k^*\hat{\sigma} \leq U) = 1 - \gamma. \tag{11.16}$$

Since $\frac{2m\hat{\sigma}}{\sigma} = 2\sum_{i=1}^{m}(R_i+1)\left(\frac{Y_{i:m:n}-\mu}{\sigma}\right) \stackrel{d}{=} \chi^2_{2m}$, we can rewrite (11.16) as

$$\Pr\left(\frac{2mK_{p_1}}{k^*} \leq \chi^2_{2m} \leq \frac{2mK_{1-p_2}}{k^*}\right) = 1 - \gamma$$

or

$$F_{\chi^2_{2m}}\left(\frac{2mK_{1-p_2}}{k^*}\right) - F_{\chi^2_{2m}}\left(\frac{2mK_{p_1}}{k^*}\right) = 1 - \gamma. \tag{11.17}$$

Eq. (11.17) needs to be solved iteratively for k^*.

Case 3: μ and σ both unknown

We will accept the lot if $L \leq \hat{\mu} + \ell\hat{\sigma} \leq U$, where $\hat{\mu} = Y_{1:m:n}$ and $\hat{\sigma} = \frac{1}{m}\sum_{i=2}^{m}(R_i + 1)\left(Y_{i:m:n} - Y_{1:m:n}\right)$ are the MLEs of μ and σ, respectively, and ℓ needs to be determined so that

$$\Pr(L \leq \hat{\mu} + \ell\hat{\sigma} \leq U) = 1 - \gamma. \tag{11.18}$$

Since $Z_1 = \frac{2n(\hat{\mu}-\mu)}{\sigma} \stackrel{d}{=} \chi^2_2$ and $Z_2 = \frac{2m\hat{\sigma}}{\sigma} \stackrel{d}{=} \chi^2_{2m-2}$ independently, we can rewrite (11.18) as

$$\Pr\left(2nK_{p_1} \leq Z_1 + \frac{n}{m}\,\ell Z_2 \leq 2nK_{1-p_2}\right) = 1 - \gamma. \tag{11.19}$$

Eq. (11.19) yields

$$\begin{aligned} 1-\gamma \;=\; & \Pr\left(\frac{2mK_{p_1}}{\ell} \leq \chi^2_{2m-2} \leq \frac{2mK_{1-p_2}}{\ell}\right) \\ & + e^{-nK_{p_1}} \int_0^{2mK_{p_1}/\ell} \frac{1}{2^{m-1}\Gamma(m-1)}\, e^{-z(1-\frac{n}{m}\ell)/2} z^{m-2}\, dz \\ & - e^{-nK_{1-p_2}} \int_0^{2mK_{1-p_2}/\ell} \frac{1}{2^{m-1}\Gamma(m-1)}\, e^{-z(1-\frac{n}{m}\ell)/2} z^{m-2}\, dz. \end{aligned} \tag{11.20}$$

Once we determine the constant ℓ for specified values of p_1, p_2 and $1-\gamma$, we will readily obtain the required two-sided acceptance sampling plan that accepts the lots with the specified probability of $1-\gamma$ while controlling the lower p_1 and upper p_2 percentages.

Remark 11.1 The factors in the one-sided acceptance sampling plans coincide with the one-sided tolerance limits which have been studied earlier by Bain and Antle (1970), Guenther, Patil and Uppuluri (1976), and Engelhardt and Bain (1978) based on complete samples from the exponential distribution. This problem is, of course, related to the determination of confidence limits on reliability discussed earlier by Grubbs (1971) and Pierce (1973).

Remark 11.2 In the case of the exponential distribution, the exact determination of the factors required in the acceptance sampling plans is made possible due to the explicit form of the estimators of the parameters based on progressively Type-II right censored samples and the nice distributional properties that these estimators possess. For most other distributions, such an exact determination of the factors becomes impossible in which case approximate methods may be used. The following section discusses such a scenario.

11.3 The Log-Normal Distribution

Let us consider the two-parameter log-normal distribution with probability density function

$$f(y;\mu,\sigma) = \frac{1}{\sqrt{2\pi}\,\sigma y}\, e^{-\frac{1}{2\sigma^2}(\log y-\mu)^2}, \quad y \geq 0. \tag{11.21}$$

If Y is lognormally distributed as in (11.21), then $Z = \log Y$ is distributed as $Normal(\mu,\sigma^2)$. For an elaborate discussion on various properties and applications of the log-normal distribution, one may refer to Crow and Shimizu (1988) and Johnson, Kotz and Balakrishnan (1994).

Let $Y_{1:m:n}^{(R_1,\cdots,R_m)},\cdots,Y_{m:m:n}^{(R_1,\cdots,R_m)}$ be the progressively Type-II right censored sample observed from the log-normal distribution in (11.21). Let $Z_{1:m:n}^{(R_1,\cdots,R_m)},\cdots,$ $Z_{m:m:n}^{(R_1,\cdots,R_m)}$ be the natural logarithms of the progressively Type-II right censored sample $Y_{1:m:n}^{(R_1,\cdots,R_m)},\cdots,Y_{m:m:n}^{(R_1,\cdots,R_m)}$; then, $Z_{1:m:n}^{(R_1,\cdots,R_m)},\cdots,Z_{m:m:n}^{(R_1,\cdots,R_m)}$ is simply a progressively Type-II right censored sample from the $Normal(\mu,\sigma^2)$ distribution.

In this case, since neither the MLEs nor the BLUPs are in explicit form, we will use the first-order approximate BLUEs μ^* and σ^* discussed earlier in Chapter 6. These estimators are

$$\mu^* = \sum_{i=1}^{m} a_i Z_{i:m:n}^{(R_1,\cdots,R_m)} \quad \text{and} \quad \sigma^* = \sum_{i=1}^{m} b_i Z_{i:m:n}^{(R_1,\cdots,R_m)} \tag{11.22}$$

which are, as seen earlier in Chapter 6, approximations to the exact BLUEs given by

$$\mu^* = -\,\boldsymbol{\mu}^T\boldsymbol{\Gamma}\boldsymbol{Z} \quad \text{and} \quad \sigma^* = \mathbf{1}^T\boldsymbol{\Gamma}\boldsymbol{Z}. \tag{11.23}$$

The variances and covariance of the approximate BLUEs in (11.22) are

$$\mathrm{Var}(\mu^*) = \frac{\sigma^2}{n}\,\gamma_{11},\ \mathrm{Var}(\sigma^*) = \frac{\sigma^2}{n}\,\gamma_{22} \quad \text{and} \quad \mathrm{Cov}(\mu^*, \sigma^*) = \frac{\sigma^2}{n} = \gamma_{12}, \tag{11.24}$$

where γ_{11}, γ_{22} and γ_{12} are approximations of

$$\gamma_{11} = \frac{n}{\Delta}\,(\boldsymbol{\mu}^T \boldsymbol{\Sigma}^{-1} \boldsymbol{\mu}),\ \gamma_{22} = \frac{n}{\Delta}\,(\mathbf{1}^T \boldsymbol{\Sigma}^{-1} \mathbf{1}) \quad \text{and} \quad \gamma_{12} = -\frac{n}{\Delta}\,(\boldsymbol{\mu}^T \boldsymbol{\Sigma}^{-1} \mathbf{1}). \tag{11.25}$$

Now, let L denote the actual one-sided lower specification limit of a product with log-normal lifetime distribution given in (11.21). Next, according to an agreement between the consumer and the producer, lots with fraction of defectives $p \le p_\alpha$ are presumed good and the consumer accepts these lots with probability at least $1 - \alpha$; furthermore, lots with fraction of defectives $p \ge p_\beta$ are presumed bad and the consumer rejects these lots with probability at least $1 - \beta$. In quality control terminology, α and β are known as *producer's risk* and *consumer's risk*, respectively.

Upon employing the well-known Lieberman and Resnikoff (1955) procedure, for a specified lower limit L, a lot under consideration is accepted if

$$\mu^* - \ell\sigma^* \ge L', \tag{11.26}$$

where $L' = \log L$ and ℓ is known as the *acceptability constant.* We may approximate the distribution of

$$\frac{(\mu^* - \ell\sigma^*) - (\mu - \ell\sigma)}{\sqrt{\frac{\sigma^2}{n}(\gamma_{11} + \ell^2\gamma_{22} - 2\ell\gamma_{12})}} \tag{11.27}$$

by a standard normal. Then, for two given points $(p_\alpha, 1-\alpha)$ and (p_β, β) on the asymptotic OC (operating characteristic) curve together with a pre-specified *degree* of censoring at the time of the i-th failure, $q_i = \frac{R_i}{n}$, $i = 1, 2, \cdots, m$, we obtain the required acceptance sampling plan as one with

$$\ell = \frac{z_{p_\alpha} z_{1-\beta} - z_{p_\beta} z_\alpha}{z_\alpha - z_{1-\beta}} \quad \text{and} \quad n = \left(\frac{z_\alpha - z_{1-\beta}}{z_{p_\alpha} - z_{p_\beta}}\right)^2 (\gamma_{11} + \ell^2\gamma_{22} - 2\ell\gamma_{12}), \tag{11.28}$$

where $z_\delta = \Phi^{-1}(\delta)$ and $\Phi(\cdot)$ is the standard normal cumulative distribution function. Thus, in general, for given $(p_\alpha, 1-\alpha)$, (p_β, β) and censoring fractions $q_i = \frac{R_i}{n}$ $(i = 1, 2, \cdots, m)$, one can determine (n, ℓ) from (11.28) for the required acceptance sampling plan. Even though the number of progressive censoring patterns is enormously large even for moderately large n, Balasooriya and Balakrishnan (2000) have tabulated these plans for some specific censoring schemes and with (p_α, p_β) matching with MIL-STD-105D; see, for example, Schneider (1986).

Remark 11.3 Note that the specification of two points on the operating characteristic curve allows the determination of the acceptability constant ℓ as well as the sample size n. A similar approach can be adopted for other location-scale distributions.

Remark 11.4 In practical applications, it is often useful to have an idea of the total test time. In the case of progressively Type-II censored sampling plans, one can estimate the total test time by approximating $E(Y_{m:m:n}^{(R_1,\cdots,R_m)})$ through $\exp\left\{\mu + \sigma E\left(Z_{m:m:n}^{(R_1,\cdots,R_m)}\right)\right\}$. Often, in life-testing experiments, one will have some information on μ and σ from past data based on an acceptable quality level of p_α, and this will provide a conservative estimate for the expected test length.

Example 11.1 In order to illustrate this reliability sampling plan, let us consider the specifications given by Schneider (1989) in his illustration of conventional Type-II right censored sampling plan. Given two points ($p_\alpha = 0.001$, $1 - \alpha = 0.99$) and ($p_\beta = 0.10$, $\beta = 0.01$) on the OC curve and the lower specification limit of $L = 457$ hours with 90% censoring, Schneider (1989) assumed the lifetime distribution to be log-normal.

To keep the total degree of censoring at 90%, suppose that censoring fractions $q_1 = 45\%$, $q_2 = \cdots = q_{m-1} = 0$ and $q_m = 45\%$ are chosen. Under the given specifications, the values of z_{p_α}, z_{p_β}, z_α and $z_{1-\beta}$ are -3.0902, -1.2816, -2.3263 and 2.3263, respectively. Then, from (11.28), we find the acceptability constant ℓ to be $\ell = 2.1859$. Formulae in (11.25), with the use of first-order approximations to the means, variances and covariances of progressively Type-II right censored order statistics from the normal distribution, yield $\gamma_{11} = 12.9530$, $\gamma_{22} = 5.1699$ and $\gamma_{12} = 7.3378$. Eq. (11.28) then gives $n = 37$. This means that a total of $n = 37$ test specimens are to be put on test simultaneously and at the time of the first failure $R_1 = 16$ surviving units are to be withdrawn, while at the time of the fourth failure ($m = 4$) the test is to be terminated by censoring $R_4 = 17$ remaining surviving units to make the total degree of censoring 90%.

Instead of using the above progressive censoring scheme, had we used conventional Type-II right censoring with $q_m = 90\%$ keeping all other specifications the same as before, we would have $\gamma_{11} = 18.9218$, $\gamma_{22} = 7.2449$ and $\gamma_{12} = 10.8916$ in which case the total sample size turns out to be $n = 40$. This means that a total of $n = 40$ test specimens are to be put on test simultaneously and at the time of the fourth failure ($m = 4$) the test is to be terminated by censoring $R_m = 36$ remaining surviving units. We note with interest that upon using the MLEs of μ and σ under the same specifications, Schneider (1989) obtained $m = 5$. As pointed out by Balasooriya and Balakrishnan (2000), the difference of 1 (in m) may be attributed to the two types of estimation methods and also the approximations used in the computation of the factors γ_{11}, γ_{22} and γ_{12} in (11.25).

Bibliography

Abramowitz, M. and Stegun, I. A. (Eds.) (1965). *Handbook of Mathematical Functions with Formulas, Graphs, and Mathematical Tables*, Dover Publications, New York.

Aggarwala, R. (1996). *Advances in life-testing: Progressive censoring and generalized distributions*, Ph.D. Thesis, McMaster University, Hamilton, Ontario, Canada.

Aggarwala, R. and Balakrishnan, N. (1996). Recurrence relations for single and product moments of progressive Type-II right censored order statistics from exponential and truncated exponential distributions, *Annals of the Institute of Statistical Mathematics*, **48**, 757–771.

Aggarwala, R. and Balakrishnan, N. (1998a). Some properties of progressive censored order statistics from arbitrary and uniform distributions with applications to inference and simulation, *Journal of Statistical Planning and Inference*, **70**, 35–49.

Aggarwala, R. and Balakrishnan, N. (1998b). Relationships among moments and product moments of progressive Type-II censored order statistics in samples from two related populations with applications to inference, *Submitted for publication.*

Aggarwala, R. and Balakrishnan, N. (1998c). Some properties of progressive censored order statistics from the Pareto distribution with applications to inference, *Submitted for publication.*

Aggarwala, R. and Balakrishnan, N. (1998d). Recurrence relations for single and product moments of progressively Type-II right censored order statistics from Pareto and power function distributions, *Submitted for publication.*

Aggarwala, R. and Balakrishnan, N. (1999). Maximum likelihood estimation of the Laplace parameters based on progressive Type-II censored samples, In *Advances in Methods and Applications of Probability and Statistics* (N.

Balakrishnan, Ed.), Gordon and Breach Publishers, Newark, New Jersey (to appear).

Aggarwala, R. and Childs, A. (1999). Conditional inference for the parameters of Pareto distributions when observed samples are progressively censored, In *Advances in Stochastic Simulation Methods* (N. Balakrishnan, V. B. Melas and S. M. Ermakov, Eds.), Birkhäuser, Boston.

Aitchison, J. and Brown, J. A. C. (1957). *The Lognormal Distribution*, Cambridge University Press, Cambridge, England.

Arnold, B. C. (1983). *Pareto Distributions*, International Co-operative Publishing House, Fairland, Maryland.

Arnold, B. C. and Balakrishnan, N. (1989). *Relations, Bounds and Approximations for Order Statistics*, Lecture Notes in Statistics – **53**, Springer-Verlag, New York.

Arnold, B. C., Balakrishnan, N. and Nagaraja, H. N. (1992). *A First Course in Order Statistics*, John Wiley & Sons, New York.

Bain, L. J. (1978). *Statistical Analysis of Reliability and Life-Testing Models — Theory and Practice*, Marcel Dekker, New York.

Bain, L. J. and Antle, C. E. (1970). Inferential procedures for the Weibull and generalized gamma distributions, *ARL Report No. 70-0266*, Wright Patterson Air Force Base, Dayton, Ohio.

Bain, L. J. and Engelhardt, M. (1991). *Statistical Analysis of Reliability and Life-Testing Models*, Second edition, Marcel Dekker, New York.

Balakrishnan, N. (1989). Recurrence relations among moments of order statistics from two related sets of independent and non-identically distributed random variables, *Annals of the Institute of Statistical Mathematics*, **41**, 323–329.

Balakrishnan, N. (1990). Improving the Hartley-David-Gumbel bound for the mean of extreme order statistics, *Statistics & Probability Letters*, **9**, 291-294.

Balakrishnan, N. (1993). A simple application of binomial–negative binomial relationship in the derivation of sharp bounds for moments of order statistics based on greatest convex minorants, *Statistics & Probability Letters*, **18**, 301-305.

Balakrishnan, N. and Basu, A. P. (Eds.) (1995). *The Exponential Distribution: Theory, Methods and Applications*, Gordon and Breach Publishers, Amsterdam, The Netherlands.

Balakrishnan, N. and Chan, P. S. (1992). Order statistics from extreme value distributions, I: Tables of means, variances and covariances, *Communications in Statistics – Simulation and Computation*, **21**, 1199–1217.

Balakrishnan, N. and Chen, W. W. S. (1997). *CRC Handbook of Tables for Order Statistics from Inverse Gaussian Distributions with Applications*, CRC Press, Boca Raton, Florida.

Balakrishnan, N. and Chen, W. W. S. (1999). *Handbook of Tables for Order Statistics from Lognormal Distributions with Applications*, Kluwer Academic Publishers, Amsterdam, The Netherlands.

Balakrishnan, N. and Cohen, A. C. (1991). *Order Statistics and Inference: Estimation Methods*, Academic Press, San Diego.

Balakrishnan, N., Cramer, E. and Kamps, U. (1999). Inequalities for means and variances of progressive Type II censored order statistics, *Submitted for publication.*

Balakrishnan, N., Cramer, E., Kamps, U. and Schenk, N. (1999). Progressive Type II censored order statistics from exponential distributions, *Submitted for publication.*

Balakrishnan, N. and Cutler, C. D. (1995). Maximum likelihood estimation of Laplace parameters based on Type-II censored samples, In *Statistical Theory and Applications: Papers in Honor of Herbert A. David* (H. N. Nagaraja, P. K. Sen and D. F. Morrison, Eds.), pp. 141-151, Springer-Verlag, New York.

Balakrishnan, N., Govindarajulu, Z. and Balasubramanian, K. (1993). Relationships between moments of two related sets of order statistics and some extensions, *Annals of the Institute of Statistical Mathematics*, **45**, 243–247.

Balakrishnan, N. and Joshi, P. C. (1981). Moments of order statistics from doubly truncated power function distribution, *Aligarh Journal of Statistics*, **1**, 98–105.

Balakrishnan, N. and Joshi, P. C. (1982). Moments of order statistics from doubly truncated Pareto distribution, *Journal of the Indian Statistical Association*, **20**, 109–117.

Balakrishnan, N. and Joshi, P. C. (1984). Product moments of order statistics from the doubly truncated exponential distribution, *Naval Research Logistics Quarterly*, **31**, 27–31.

Balakrishnan, N., Malik, H. J. and Ahmed, S. E. (1988). Recurrence relations and identities for moments of order statistics, II: Specific continuous distributions, *Communications in Statistics – Theory and Methods*, **17**, 2657-2694.

Balakrishnan, N. and Maswadah, M. S. (1999a). Structural inference based on Type-II progressively censored samples from Fréchet-type extreme value distribution, *Submitted for publication.*

Balakrishnan, N. and Maswadah, M. S. (1999b). Structural inference based on Type-II progressively censored samples from Rayleigh distribution, *Submitted for publication.*

Balakrishnan, N. and Mi, J. (1999). Stochastic order and MLE of the mean of exponential distribution, *Submitted for publication.*

Balakrishnan, N. and Rao, C. R. (1997a). Large-sample approximations to the best linear unbiased estimation and best linear unbiased prediction based on progressively censored samples and some applications, In *Advances in Statistical Decision Theory and Applications* (S. Panchapakesan and N. Balakrishnan, Eds.), pp. 431-444, Birkhäuser, Boston.

Balakrishnan, N. and Rao, C. R. (1997b). A note on the best linear unbiased estimation based on order statistics, *The American Statistician*, **51**, 181–185.

Balakrishnan, N. and Rao, C. R. (Eds.) (1998a). *Order Statistics: Theory and Methods*, Handbook of Statistics – **16**, North-Holland, Amsterdam, The Netherlands.

Balakrishnan, N. and Rao, C. R. (Eds.) (Eds.) (1998b). *Order Statistics: Applications*, Handbook of Statistics - **17**, North-Holland, Amsterdam, The Netherlands.

Balakrishnan, N. and Rao, C. R. (1999). On the efficiency properties of BLUEs, *Journal of Statistical Planning and Inference* (to appear).

Balakrishnan, N. and Sandhu, R. A. (1995). A simple simulational algorithm for generating progressive Type-II censored samples, *The American Statistician*, **49**, 229-230.

Balakrishnan, N. and Sandhu, R. A. (1996). Best linear unbiased and maximum likelihood estimation for exponential distributions under general progressive Type-II censored samples, *Sankhyā, Series B*, **58**, 1–9.

Balakrishnan, N. and Sultan, K. S. (1998). Recurrence relations and identities for moments of order statistics, In *Handbook of Statistics, Vol. 16: Order*

Statistics: Theory and Methods (N. Balakrishnan and C. R. Rao, Eds.), pp. 149–228, North-Holland, Amsterdam, The Netherlands.

Balasooriya, U. (1995). Failure-censored reliability sampling plans for the exponential distribution, *Journal of Statistical Computation and Simulation*, **52**, 337–349.

Balasooriya, U. and Balakrishnan, N. (2000). Life-test sampling plans for log-normal distribution based on progressively censored samples, *IEEE Transactions on Reliability* (to appear).

Cacciari, M. and Montanari, G. C. (1987). A method to estimate the Weibull parameters for progressively censored tests, *IEEE Transactions on Reliability*, **R-36**, 87–93.

Cacoullos, T. and Papathanasiou, V. (1985). On upper bounds for the variance of functions of random variables, *Statistics & Probability Letters*, **3**, 175-184.

Chandrasekar, B. and Balakrishnan, N. (1999). On a multiparameter version of Tukey's linear sensitivity measure and its properties, *Submitted for publication.*

Chandrasekar, B., Leo Alexander, T. and Balakrishnan, N. (1999). Equivariant estimation for the parameters of an exponential model based on Type II progressive censored sampling, *Submitted for publication.*

Chatterjee, S. K. and Sen, P. K. (1973). Nonparametric testing under progressive censoring, *Calcutta Statistical Association Bulletin*, **22**, 13-50.

Childs, A. and Balakrishnan, N. (1996). Conditional inference procedures for the Laplace distribution based on Type-II right censored samples, *Statistics & Probability Letters*, **31**, 31-39.

Childs, A. and Balakrishnan, N. (1997). Maximum likelihood estimation of Laplace parameters based on general type-II censored samples, *Statistical Papers*, **38**, 343-349.

Cohen, A. C. (1963). Progressively censored samples in life testing, *Technometrics*, **5**, 327–329.

Cohen, A. C. (1966). Life testing and early failure, *Technometrics*, **8**, 539–549.

Cohen, A. C. (1975). Multi-censored sampling in the three parameter Weibull distribution, *Technometrics*, **17**, 347–351.

Cohen, A. C. (1976). Progressively censored sampling in the three parameter log-normal distribution, *Technometrics*, **18**, 99–103.

Cohen, A. C. (1991). *Truncated and Censored Samples — Theory and Applications*, Marcel Dekker, New York.

Cohen, A. C. and Norgaard, N. J. (1977). Progressively censored sampling in the three parameter gamma distribution, *Technometrics*, **19**, 333–340.

Cohen, A. C. and Whitten, B. J. (1988). *Parameter Estimation in Reliability and Life Span Models*, Marcel Dekker, New York.

Crow, E. L. and Shimizu, K. (Eds.) (1988). *Lognormal Distributions*, Marcel Dekker, New York.

David, F. N. and Johnson, N. L. (1954). Statistical treatment of censored data. I. Fundamental formulae, *Biometrika*, **41**, 228–240.

David, H. A. (1981). *Order Statistics*, Second edition, John Wiley & Sons, New York.

DiCiccio, T. (1987). Approximate inference for the generalized gamma distribution, *Technometrics*, **29**, 33–40.

Doganaksoy, N. and Balakrishnan, N. (1997). A useful property of best linear unbiased predictors with applications to life testing, *The American Statistician*, **51**, 22–28.

Engelhardt, M. and Bain, L. J. (1978). Tolerance limtis and confidence limits on reliability for the two-parameter exponential distribution, *Technometrics*, **20**, 37–39.

Epstein, B. and Sobel, M. (1953). Life testing, *Journal of the American Statistical Association*, **48**, 486–502.

Epstein, B. and Sobel, M. (1954). Some theorems relevant to life testing from an exponential distribution, *Annals of Mathematical Statistics*, **25**, 373–381.

Fertig, K. W. and Mann, N. R. (1980). Life-tests sampling plans for two-parameter Weibull populations, *Technometrics*, **22**, 165–177.

Fisher, R. A. (1934). Two new properties of mathematical likelihood, *Proceedings of the Royal Society, Series A*, **144**, 285–307.

Fraser, D. A. S. (1968). *The Structure of Inference*, John Wiley & Sons, New York.

Gajjar, A. V. and Khatri, C. G. (1969). Progressively censored samples from log-normal and logistic distributions, *Technometrics*, **11**, 793–803.

Gibbons, D. I. and Vance, L. C. (1983). Estimators for the 2-parameter Weibull distribution with progressively censored samples, *IEEE Transactions on Reliability*, **R-32**, 95–99.

Goldberger, A. S. (1962). Best linear unbiased prediction in the generalized linear regression model, *Journal of the American Statistical Association*, **57**, 369-375.

Govindarajulu, Z. (1963). Relationships among moments of order statistics in samples from two related populations, *Technometrics*, **5**, 514–518.

Graybill, F. A. (1983). *Matrices with Applications in Statistics*, Second edition, Wadsworth, Belmont, California.

Grubbs, F. E. (1971). Approximate fiducial bounds on reliability for the two parameter negative exponential distribution, *Technometrics*, **13**, 873–876.

Guenther, W. C., Patil, S. A. and Uppuluri, V. R. R. (1976). One-sided β-content tolerance factors for the two parameter exponential distribution, *Technometrics*, **18**, 333–340.

Gupta, S. S., McDonald, G. C. and Galarneau, D. I. (1974). Moments, product moments and percentage points of the order statistics from the lognormal distribution for samples of size twenty and less, *Sankhyā, Series B*, **36**, 230-260.

Hahn, G. J. and Meeker, W. Q., Jr. (1991). *Statistical Intervals: A Guide for Practitioners*, John Wiley & Sons, New York.

Harter, H. L. (1970). *Order Statistics and Their Uses in Testing and Estimation, Vol. 2: Estimates Based on Order Statistics of Samples From Various Populations*, U.S. Government Printing Office, Washington, D.C.

Harter, H. L. and Balakrishnan, N. (1996). *CRC Handbook of Tables for the Use of Order Statistics in Estimation*, Second edition, CRC Press, Boca Raton.

Herd, R. G. (1956). *Estimation of the parameters of a population from a multi-censored sample*, Ph. D. Thesis, Iowa State College, Ames, Iowa.

Horn, P. S. and Schlipf, J. S. (1986). Generating subsets of order statistics with applications to trimmed means and means of trimmings, *Journal of Statistical Computation and Simulation*, **24**, 83–97.

Hosono, H. O. Y. and Kase, S. (1981). Design of single sampling plans for doubly exponential characteristics, In *Frontiers in Quality Control* (H. J. Lenz, G. B. Wetherill and P. T. Wilrich, Eds.), pp. 94–112, Physica-Verlag, Berlin, Germany.

Johnson, N. L., Kotz, S. and Balakrishnan, N. (1994). *Continuous Univariate Distributions – Vol. 1*, Second edition, John Wiley & Sons, New York.

Johnson, N. L., Kotz, S. and Balakrishnan, N. (1995). *Continuous Univariate Distributions – Vol. 2*, Second edition, John Wiley & Sons, New York.

Joshi, P. C. (1978). Recurrence relations between moments of order statistics from exponential and truncated exponential distributions, *Sankhyā, Series B*, **39**, 362–371.

Joshi, P. C. (1979). A note on the moments of order statistics from doubly truncated exponential distribution, *Annals of the Institute of Statistical Mathematics*, **31**, 321–324.

Joshi, P. C. (1982). A note on the mixed moments of order statistics from exponential and truncated exponential distributions, *Journal of Statistical Planning and Inference*, **6**, 13–16.

Kambo, N. S. (1978). Maximum likelihood estimators of the location and scale parameters of the exponential distribution from a censored sample, *Communications in Statistics – Theory and Methods*, **7**, 1129-1132.

Kamps, U. (1995a). *A Concept of Generalized Order Statistics*, Teubner, Stuttgart, Germany.

Kamps, U. (1995b). A concept of genrealized order statistics, *Journal of Statistical Planning and Inference*, **48**, 1–23.

Kamps, U. and Cramer, E. (1999). On distributions of generalized order statistics, *Submitted for publication.*

Kaplan, E. L. and Meier, P. (1958). Nonparametric estimation from incomplete observations, *Journal of the American Statistical Association*, **53**, 457-481.

Kocherlakota, S. and Balakrishnan, N. (1984). Two sided acceptance sampling plans based on MML estimators, *Communications in Statistics — Theory and Methods*, **13**, 3123–3131.

Kocherlakota, S. and Balakrishnan, N. (1986). One- and two-sided sampling plans based on the exponential distribution, *Naval Research Logistics Quarterly*, **33**, 513–522.

Kotz, S. and Johnson, N. L. (Eds.) (1991). *Breakthroughs in Statistics*, Vols. 1 and 2, Springer-Verlag, New York.

Lawless, J. F. (1973). On the estimation of safe life when the underlying life distribution is Weibull, *Technometrics*, **15**, 857–865.

Lawless, J. F. (1978). Confidence interval estimation for the Weibull and extreme value distributions, *Technometrics*, **20**, 355–368.

Lawless, J. F. (1982). *Statistical Models & Methods for Lifetime Data*, John Wiley & Sons, New York.

Lieberman, G. J. and Resnikoff, G. J. (1955). Sampling plans for inspection by variables, *Journal of the American Statistical Association*, **50**, 457–516.

Lloyd, E. H. (1952). Least-squares estimation of location and scale parameters using order statistics, *Biometrika*, **39**, 88-95.

London, D. (1988). *Survival Models*, Actex Publications, Connecticut.

Lurie, D. and Hartley, H. O. (1972). Machine-generation of order statistics for Monte Carlo computations, *The American Statistician*, **26**, 26–27.

Lurie, D. and Mason, R. L. (1973). Empirical investigation of general techniques for computer generation of order statistics, *Communications in Statistics*, **2**, 363-371.

Majumdar, H. and Sen, P. K. (1978). Nonparametric tests for multiple regression under progressive censoring, *Journal of Multivariate Analysis*, **8**, 73-95.

Malmquist, S. (1950). On a property of order statistics from a rectangular distribution, *Skandinavisk Aktuarietidskrift*, **33**, 214-222.

Mann, N. R. (1969a). Exact three-order-statistic confidence bounds on reliable life for a Weibull model with progressive censoring, *Journal of the American Statistical Association*, **64**, 306–315.

Mann, N. R. (1969b). Optimum estimators for linear functions of location and scale parameters, *Annals of Mathematical Statistics*, **40**, 2149-2155.

Mann, N. R. (1971). Best linear invariant estimation for Weibull parameters under progressive censoring, *Technometrics*, **13**, 521–533.

Mann, N. R., Schafer, R. E. and Singpurwalla, N. D. (1974). *Methods for Statistical Analysis of Reliability and Life Data*, John Wiley & Sons, New York.

Montanari, G. C. and Cacciari, M. (1988). Progressively-censored aging tests on XLPE-insulated cable models, *IEEE Transactions on Electrical Insulation*, **23**, 365–372.

Moriguti, S. (1953). A modification of Schwarz's inequality with applications to distributions, *Annals of Mathematical Statistics*, **24**, 107-113.

Nagaraja, H. N. (1994). Tukey's linear sensitivity and order statistics, *Annals of the Institute of Statistical Mathematics*, **46**, 757-768.

Nelson, W. (1982). *Applied Life Data Analysis*, John Wiley & Sons, New York.

Owen, D. B. (1964). Control of percentages in both tails of the normal distribution, *Technometrics*, **6**, 377–387.

Owen, D. B. (1969). Variable acceptance sampling with non-normality, *Technometrics*, **11**, 631–637.

Papathanasiou, V. (1990). Some characterizations of distributions based on order statistics, *Statistics & Probability Letters*, **9**, 145-147.

Pierce, D. A. (1973). Fiducial, frequency, and Bayesian inference for the two parameter negative exponential distribution, *Technometrics*, **15**, 249–253.

Ramberg, J. S. and Tadikamalla, P. R. (1978). On the generation of subsets of order statistics, *Journal of Statistical Computation and Simulation*, **7**, 239-241.

Rao, C. R. (1973). *Linear Statistical Inference and Its Applications*, Second edition, John Wiley & Sons, New York.

Rényi, A. (1953). On the theory of order statistics, *Acta Math. Acad. Sci. Hung.*, **4**, 191–231.

Ringer, L. J. and Sprinkle, E. E. (1972). Estimation of the parameters of the Weibull distribution from multicensored samples, *IEEE Transactions on Reliability*, **R-21**, 46–51.

Robinson, J. A. (1983). Bootstrap confidence intervals in location-scale models with progressive censoring, *Technometrics*, **25**, 179–187.

Sampford, M. R. (1952). The estimation of response-time distributions, Part II, *Biometrics*, **9**, 307–369.

Scheffé, H. and Tukey, J. W. (1945). Non-parametric estimation—I. Validation of order statistics, *Annals of Mathematical Statistics*, **16**, 187–192.

Schneider, H. (1986). *Truncated and Censored Samples from Normal Populations*, Marcel Dekker, New York.

Schneider, H. (1989). Failure-censored variables-sampling plans for log-normal and Weibull distributions, *Technometrics*, **31**, 199–206.

Schucany, W. R. (1972). Order statistics in simulation, *Journal of Statistical Computation and Simulation*, **1**, 281-286.

Sen, P. K. (1986). Progressively censored data analysis, In *Encyclopedia of Statistical Sciences* – **7** (Eds., S. Kotz and N. L. Johnson), pp. 299-303, John Wiley & Sons, New York.

Sherif, A. and Tan, P. (1978). On structural predictive distribution with Type-II progressively censored Weibull data, *Statistische Hefte*, **19**, 247–255.

Sinha, A. N. and Sen, P. K. (1982). Tests based on empirical processes for progressive censoring schemes with staggering entry and random withdrawal, *Sankhyā, Series B*, **44**, 1-18.

Sinha, A. N. and Sen, P. K. (1983). Staggering entry, random withdrawal and progressive censoring schemes: Some nonparametric procedures, In *Statistical Applications and New Directions: Proceedings of the Indian Statistical Institute Golden Jubilee Conference* (J. K. Ghosh and J. Roy, Eds.), pp. 531-547, Indian Statistical Institute, Calcutta, India.

Srikantan, K. S. (1962). Recurrence relations between the PDF's of order statistics and some applications, *Annals of Mathematical Statistics*, **33**, 169–177.

Sukhatme, P. V. (1937). Tests of significance for samples of the χ^2 population with two degrees of freedom, *Annals of Eugenics*, **8**, 52–56.

Tadikamalla, P. R. and Balakrishnan, N. (1998). Computer simulation of order statistics, In *Handbook of Statistics, Vol. 16: Order Statistics: Theory and Methods* (N. Balakrishnan and C. R. Rao, Eds.), pp. 65–72, North-Holland, Amsterdam, The Netherlands.

Thomas, D. R. and Wilson, W. M. (1972). Linear order statistic estimation for the two-parameter Weibull and extreme value distributions from Type-II progressively censored samples, *Technometrics*, **14**, 679–691.

Tietjen, G. L., Kahaner, D. K. and Beckman, R. J. (1977). Variances and covariances of the normal order statistics for sample sizes 2 to 50, In *Selected Tables in Mathematical Statistics* – **5**, pp. 1–73, American Mathematical Society, Providence, Rhode Island.

Tukey, J. W. (1965). Which part of the sample contains the information? *Proceedings of the National Academy of Sciences, USA*, **53**, 127-134.

Viveros, R. and Balakrishnan, N. (1994). Interval estimation of life characteristics from progressively censored data, *Technometrics*, **36**, 84–91.

Viveros, R. and Sprott, D. A. (1987). Allowance for skewness in maximum likelihood estimation with application to the location-scale model, *Canadian Journal of Statistics*, **15**, 349–361.

Wingo, D. R. (1973). Solution of the three-parameter Weibull equations by constrained modified quasilinearization (progressively censored samples), *IEEE Transactions on Reliability*, **R-22**, 96–102.

Wingo, D. R. (1993). Maximum likelihood methods for fitting the Burr Type XII distribution to multiply (progressively) censored life test data, *Metrika*, **40**, 203–210.

Wong, J. Y. (1993). Simultaneously estimating the three Weibull parameters from progressively censored samples, *Microelectronics and Reliability*, **33**, 2217–2224.

Author Index

Subject Index